The Science of Astrobiology

Cellular Origin, Life in Extreme Habitats and Astrobiology

Volume 20 (Second Edition)

Julian Chela-Flores

The Science of Astrobiology

A Personal View on Learning to Read the Book of Life

Springer

Julian Chela-Flores
The Abdus Salam International
Centre for Theoretical Physics
P.O. Box 586
34014 Trieste
Italy
chelaf@ictp.trieste.it

ISSN 1566-0400
ISBN 978-94-007-1626-1 e-ISBN 978-94-007-1627-8
DOI 10.1007/978-94-007-1627-8
Springer Dordrecht Heidelberg London New York

Library of Congress Control Number: 2011934255

Printed on acid-free paper

Springer is part of Springer Science+Business Media (www.springer.com)

The cupola in the West Atrium of St. Mark's Basilica in Venice, Italy representing the biblical interpretation of Genesis
(Cf., also pp. 215-216 at the beginning of Part 4: The destiny of life in the universe. With kind permission of the Procuratoria of St. Mark's Basilica.)

For Sarah Catherine Mary

Table of contents

Table of contents vii

Preface xvii

Acknowledgements xxi

Recommendations to the readers xxiii

INTRODUCTION
The cultural and scientific context of astrobiology

I.1 Early attempts to read the Book of Life 3

ARISTARCHUS OF SAMOS AND HIPPARCHUS 4

NICHOLAS OF CUSA (CUSANUS) 4

NICHOLAS COPERNICUS 4

GIORDANO BRUNO 5

CHARLES DARWIN 6

I.2 Some pioneers of the science of astrobiology 8

ALEXANDER OPARIN 8

STANLEY MILLER 10

SIDNEY W. FOX 11

CYRIL PONNAMPERUMA 12

JOHN ORO, 1ST MARQUESS OF ORÓ 13

I.3 Three strategies for astrobiological exploration 14

I.4 Inserting philosophy into astrobiology 17

I.5 Emphasizing the role of biology in astrobiology 17

I.6 Inserting physics in biology especially in astrobiology 18

I.7 Inserting the earth sciences in astrobiology 19

I.8 New fingerprints of early life 23

I.9 Inserting paleontology into astrobiology 24

Supplementary Reading 26

References 27

THE BOOK OF LIFE

PART 1: THE ORIGIN OF LIFE IN THE UNIVERSE

CHAPTER 1.
From cosmic to chemical evolution

1.1 Beginnings of a rational cosmology based on physical constraints 31

1.2 Cosmological models 32

1.3 The cosmic microwave background 34

1.4 The growth of studies on the origin of life 35

1.5 Origin of the elements: from the big bang to the interior of stars 37

1.6 Supernovae: the source of biogenic dust 39

1.7 Molecular clouds and circumstellar disks 42

1.8 Stellar and biological evolution 43

Supplementary Reading 45

References 46

CHAPTER 2.
From chemical to prebiotic evolution

2.1 Organic Cosmochemistry 47

2.2 Precursor biomolecules in interplanetary dust 48

2.3 Cosmic dust and comets: their role in astrobiology 49

2.4 Origin of the Solar System: the Rosetta mission 51

2.5 Origin of the terrestrial planets 52

2.6 Origin of the Jovian planets 53

2.7 Origin of the moons in the outer Solar System 55

Supplementary Reading 55

References 56

CHAPTER 3.
Sources of life's origin: A search for biomarkers

3.1 The Moon and its water content as a mirror of the early Earth 57

3.2 Biogenic elements on asteroids 63

3.3 Biogenic elements on comets 64

3.4 Biogenic elements on meteorites 66

3.5 The Murchison meteorite 68

3.6 Biogenic elements in extreme terrestrial environments 70

3.7 A terrestrial analog to Europa in Antarctica: The dry valley lakes 72

3.8 A second terrestrial analog to Europa near the South Pole: Lake Vostok 75

3.9 A third terrestrial analog to Europa in the Canadian Arctic 76

Supplementary Reading 76

References 77

CHAPTER 4.
From prebiotic evolution to the emergence of single cells

4.1 Which are the macromolecules of life? 79

4.2 The primitive Earth 80

4.3 The origin of the first cell 81

4.4 A biochemical relic of the earliest stages of life 82

4.5 Symmetry in the cosmos 83

4.6 Symmetry in the life sciences 85

ASYMMETRY IN THE ANIMAL WORLD 85

ASYMMETRY IN THE MICROSCOPIC WORLD: CHIRALITY 85

HOMOCHIRALITY AS A BIOMARKER OF LIFE 86

ABDUS SALAM'S VISION OF SYMMETRY IN NATURE 87

THE ORIGIN OF CHIRALITY AND THE MURCHISON METEORITE 88

4.7 An analogy with languages: the genetic code 89

4.8 The origin of cellular organelles 90

4.9 From chemical to cellular evolution 91

4.10 The influence of the ancient Sun on the early evolution of life 91

4.11 Eventual deeper insights into the origin of life from solar missions 93

Supplementary Reading 98

References 98

THE BOOK OF LIFE

PART 2: EVOLUTION OF LIFE IN THE UNIVERSE

CHAPTER 5.
From the age of prokaryotes to the emergence of eukaryotes

5.1 Could prokaryotes have emerged on the early Earth 4,4 Gyr ago? 103

5.2 The dawn of unicellular organisms 104

5.3 An early emergence of life makes the Moon a major target for astrobiology 105

5.4 A unique event lost somewhere deep in the Archean 106

5.5 The evolution of the Earth's atmosphere: The Great Oxidation Event 106

5.6 The role of oxygen and iron in eukaryogenesis 108

5.7 First appearance of metazoans in the Paleoproterozoic 109

5.8 The origin of chromosomes 110

5.9 The Molecular Clock hypothesis 111

5.10 A transition in the pathway towards intelligence 112

5.11 In which environments can extremophiles survive in our Solar System? 113

5.12 Mechanisms of evolution beyond natural selection 113

5.13 The adaptation of life to extreme environments 115

Supplementary Reading 116

References 116

CHAPTER 6.
The evolution of intelligent behavior

6.1 Modern taxonomy emphasizes single-celled organisms 119

6.2 The phenomenon of the eukaryotic cell 120

6.3 The phenomenon of multicellularity 121

6.4 Evolution of the hominoids 122

6.5 Evolution of intelligent behavior in the hominids 124

6.6 Did external events contribute to the evolution of intelligence on Earth? 126

6.7 The pathway towards communication 127

6.8 The emergence of language and music in modern humans 128

Supplementary Reading 129

References 130

THE BOOK OF LIFE

PART 3: THE DISTRIBUTION OF LIFE IN THE UNIVERSE

CHAPTER 7.
On the possibility of biological evolution on Mars

7.1 What have we learnt from previous missions to Mars? 135

7.2 An Eden-like early Martian environment 136

7.3 The early Viking Missions 137

7.4 Mars Pathfinder and the Sojourner Rover 139

7.5 The Mars Global Surveyor 139

7.6 Mars Express 141

7.7 The Phoenix Mission 142

7.8 The question of the origin, evolution and distribution of Martian life 144

7.9 Mars Odyssey and the Exploration Rover Opportunity 146

7.10 Is methane a possible biomarker in the Martian atmosphere? 147

7.11 Testing for life on Mars on terrestrial analogs 148

7.12 Mars in the second and third decades of the 21st century 149

Supplementary Reading 149

References 150

CHAPTER 8.
On the possibility of biological evolution on the moons of Jupiter

8.1 The discovery of Europa 151

8.2 The Galileo mission 154

8.3 Tentative inventory of organic elements in the Europan ocean 157

8.4 The habitability of Europa 159

8.5 What are the constraints on the putative Europan biotope? 159

8.6 Technological challenges for reading further into the Book of Life 160
THE ORIGINAL HYDROBOT-CRYOBOT 160
ADVANCED DESIGN FOR HYDROBOTS 162
PENETRATORS; SURFICIAL LANDERS FOR EUROPA 162
8.7 Returning to the Jovian System 164
8.8 Are there biosignatures revealing a second biology on Europa? 164
8.9 Sulfur patches and space weather in the neighborhood of Europa 166
Supplementary Reading 168
References 168

CHAPTER 9.
On the possibility of biological evolution on the moons of Saturn

9.1 The discovery of Titan 171
9.2 The Voyagers: The first missions to Saturn and Titan 172
9.3 The Cassini-Huygens Mission 173
9.4 Titan and Enceladus 175
9.5 The atmosphere and hydrosphere of Titan 177
9.6 Titan's surface 179
9.7 Is there a liquid ocean on Titan? 181
9.8 The plumes and habitability of Enceladus 182
9.9 An eventual return to Titan and Enceladus 183
9.10 The habitability of Titan 184
Supplementary Reading 185
References 185

CHAPTER 10.
How different would life be elsewhere?

10.1 Possible degrees of evolution of life 187
10.2 The search for other solar systems 188
10.3 How are extrasolar planets found? 189
10.4 Which are likely habitable zones? 190
10.5 The multiplicity of exoplanets in our galactic neighborhood 191

10.6 What is a planet? 191

10.7 The CoRoT mission 193

10.8 The Kepler mission 193

10.9 The discovery of the first super-Earths in habitable zones 194

10.10 Exomoonology: the habitability of satellites around other stars 194

10.11 Potentially habitable worlds: planets like Earth 197

Supplementary Reading 199

References 199

CHAPTER 11.
The search for the evolution of exointelligence in worlds around other stars

11.1 What is our place in the universe? 201

11.2 Intelligent behavior is intimately related with the need to communicate 202

11.3 The Drake Equation 202

11.4 Progress in instrumentation 206

11.5 SETI on the Moon 208

11.6 From the first neuron to brains 209

11.7 Beyond geocentricism 210

11.8 The Fermi Paradox 211

11.9 SETI: towards the future 212

Supplementary Reading 213

References 214

THE BOOK OF LIFE

PART 4: THE DESTINY OF LIFE IN THE UNIVERSE

CHAPTER 12.
Is the destiny of life inexorably linked with intelligence?

12.1 The origin of the neuron: a first step in the evolution of intelligence 217

12.2 The origin of metazoans: a second step in the evolution of intelligence 218

12.3 Convergence and contingency in evolutionary biology 218

12.4 Biological evolution on other worlds 220

12.5 Evolution of intelligent behavior in aquatic media 221

12.6 Testing the evolution of microorganisms in the Solar System 222

12.7 Precursors of the evolution of exointelligence—where should we look? 224

12.8 Can the origin of metazoans be detected elsewhere in the Solar System? 225

12.9 Back to Europa for further constraining the Drake Equation 228

12.10 Convergence between SETI and the exploration of our Solar System 228

Supplementary Reading 232

References 232

CHAPTER 13.
Cultural frontiers of astrobiology

13.1 The frontiers of science, philosophy and theology 235

13.2 Positivism and the Vienna Circle 236

13.3 Position of humans in the totality of all earthly species 237

13.4 Is evolution more than a hypothesis? 238

13.5 What is specific to a human being? 238

13.6 Are there trends in evolution? 240

13.7 Cultural implications of discovering extraterrestrial life 241

13.8 Relevance and implications of discovering life elsewhere 242

13.9 Consequences of other intelligent behavior 243

Supplementary Reading 244

References 245

CHAPTER 14.
When astrobiology meets philosophy

14.1 Pasteur, Darwin and Wallace 247

14.2 A philosophical issue: design in biology 248

14.3 Towards a common interest in astrobiology and the humanities 249

14.4 Is there life elsewhere in the universe? 250

14.5 Can our intelligence be repeated elsewhere? 251

14.6 Some of the larger issues when philosophy meets astrobiology 252

Supplementary Reading 254

References 255

CHAPTER 15.
Why we may be unable to read the complete Book of Life

15.1 Evolution of the universe and its contents including the living process 257

15.2 Constraints on the universality of biology 259

15.3 Extending Newton's mechanics into the microscopic domain 260

15.4 From quantum electrodynamics to quantum chromodynamics 262

15.5 The unification of elementary forces 263

15.6 Hopes for clarification from new instrumentation 264

15.7 Influential steps other than quantum field theory 265

15.8 String theory as a basis for reading the Book of Life 266

15.9 Questions that lie in the frontier where astrobiology meets philosophy 269

Supplementary Reading 269

References 269

CHAPTER 16.

An intelligible universe with the science of astrobiology

16.1 The meaning of reality 271

16.2 The Philosophy of Kant and Hegel 272

16.3 Is the universe a well-determined ordered system? 273

16.4 Logical positivism and consilience 274

16.5 Are we approaching the end of biogeocentrism? 275

16.6 Is reductionism inevitable in the physical and in the life sciences? 275

16.7 Difficulties reading Part 3 of the Book of Life 277

16.8 Perhaps the Third Part of the Book of Life can be intelligible 278

Supplementary Reading 279

References 279

EPILOGUE

Learning to read the Book of Life: An interdisciplinary process 281

Acronyms and abbreviations 287

Illustrated glossary 289

General index

Index of illustrations 319

Index of Tables 325

Alphabetical index 327

About the author 335

Books by author 336

Preface

Since the publication of *The New Science of Astrobiology* in the year 2001, two significant events have taken place raising the subject from the beginning of the century to its present maturity a decade later (Chela-Flores, 2001). Firstly, at that time the Galileo Mission still had two years to complete its task, which turned out to be an outstanding survey of the Jovian system, especially of its intriguing satellite Europa. Secondly, the successful outcome of the Cassini Huygens Mission on its way to Saturn went beyond all expectations of the European Space Agency (ESA) and the United States of America National Aeronautics and Space Administration (NASA). Cassini had been launched four years earlier and Huygens was to land on Titan three years after the publication of our first edition.

Besides, we had no idea that another satellite of Saturn, Enceladus, was going to lure the scientific community with the most surprising jets of water exuding an air of mystery, hinting at a submerged inhabitable ocean of salty water. Ahead of the date of publication of our book was the awareness of the Earth-like surface morphology and hydrosphere of Titan with its prominent lake system. It dawned upon us that Titan was the fourth body of the Solar System that possibly contained a water ocean, thus joining our planet and Jupiter's three Galilean satellites: Ganymede, Callisto and Europa. Titan appears now to be the only planetary body besides the Earth to have persistent, almost permanent liquid bodies on its surface. These surprising aspects of Titan were unknown to us in 2001. In our first edition, at the end of Chapter 9, we anticipated the possibility of the emergence of an autochthonous biology on Titan, but left our readers with a word of caution to wait until the present post Cassini-Huygens era before advancing further hypotheses on this most basic issue of astrobiology. We now discuss this issue in Chapter 9. The new scientific landmarks of the first decade of the present century warrant a new look at the same subject matter. The urgency of this undertaking is emphasized by the much deeper insights that we have gained into the geophysics of Mars. Some remarkable events include the analysis of its surface by the Mars Reconnaissance Orbiter and the Martian water ice exposed by the Phoenix Mars Lander in 2008, not forgetting the much clearer views that are now emerging on Martian paleolimnology. Some of this progress due largely to the stunning images retrieved by ESA's Mars Express and subsequent missions to the Red Planet.

We have sadly witnessed during the three decades preceding the publication of the first edition of this book, *The New Science of Astrobiology*, a most unfortunate missed opportunity regarding the acquisition of further insights into our own satellite. In the present Second Edition this particular topic is being reviewed, namely the ongoing revolution of interest in the Moon. In October 2007, Japan sent up the Kaguya spacecraft. A month later, China's Chang'e-1 entered lunar orbit. This was followed by India's Chandrayaan-1, whose objective was to map not only the surface of the moon, but also what lies underneath. We are convinced that promising new instrumentation, such as the penetrator technology, especially the work of the British Penetrator Consortium, will be able to demonstrate in the coming decade communication and navigation technologies that will support the eventual return to the Moon and the exploration of other planets and satellites of the inner and outer Solar System.

Beyond the terrestrial planets this renewed interest in our own satellite will be useful as a platform to improve on the achievements of the Galileo Mission. The benefits will concern all the satellites of Jupiter, especially Europa's intriguing non-water elements on its icy surface that will be elucidated by the forthcoming Europa-Jupiter System Mission (EJSM). With combined efforts focused on our own satellite by India, China, Japan, Russia, the United Kingdom, ESA and NASA, *The Science of Astrobiology* has new exciting results and discoveries to review. Consequently, at the present time it is appropriate to attempt to reflect on current scientific progress with a broad canvas ranging from cosmic evolution to the implications in the humanities of the inevitable discovery of universality of biology.

There is still the challenge of taking into account the search for exo-intelligence—intelligent behavior elsewhere in the universe—with the help of the gigantic leaps in radio astronomy that are expected to come early in the 2020s from the Square Kilometer Array (SKA discussed in Chapter 11). Since the year 2001, only six years after the seminal discovery of a planet of the star 51 Pegasi that was reported in our first edition, a whole spectrum of further "exoplanets" have arisen amongst our galactic neighbors: Jovian-like gaseous giants, super Neptunes and super-Earths are constantly being added to the astronomer's catalogs. With the Kepler Space Telescope, a NASA Mission launched in 2009, Earth-like planets are now within reach of observation. Not only a large number of exoplanets are now known, but also solar systems with several planets each have been identified. The possibility of searching for exomoons will also be raised in Chapter 10. This new broad vision of the cosmos and its possible habitability has given additional strong support to the search for intelligent behavior with the tools of the bioastronomers. One of the factors of the Drake Equation—the number of possible inhabitable planets—is slowly, but steadily coming to our attention, especially with Kepler.

The many books that are now available represent another significant progress in astrobiology. In 2001 we found it difficult to identify a single-author book especially written on astrobiology. They were not generally known (not even to the present author). Now the situation is much better and our bibliography at the end of this volume is consequently much richer, providing our readers with a most enjoyable, instructive and comprehensive view of our subject (cf., the bibliographic references listed at the end of the Preface offer the reader a small sample of a growing list of books).

The writing this second edition of the book coincided in its first stages with the double anniversary of Charles Darwin, the 200th anniversary of his birth and the 150th anniversary of the publication of The Origin of Species, possibly the most influential book ever published in any branch of science. We have profited from this remarkable coincidence, in order to underline "the biology" of astrobiology (cf., Introduction). Being a multidisciplinary subject, astrobiology sometimes regretfully neglects the life sciences, as there are so many other aspects to keep in mind, such as chemical evolution, the earth sciences, the physical sciences and cutting-edge technology.

Finally, the emphasis we attempted to imprint on our previous book *The New Science of Astrobiology* made a very modest effort in setting the scientific subject appropriately amongst other sectors of culture that are the natural frontiers of astrobiology. These boundaries are philosophy and theology, branches of the humanities asking similar questions to the basic issues of astrobiology (origin, evolution, distribution and destiny of life in the universe). In the meantime I have dedicated a full volume to this aspect of astrobiology in *A Second Genesis Stepping-Stones Towards the Intelligibility of Nature* (Chela-Flores, 2009). We have tried in this new edition of *The Science of Astrobiology* to benefit from the experience gained during these last decade while enjoying the multiple fascinating aspects of astrobiology and its cultural frontiers.

Julian Chela-Flores,
Salita di Contovello, Trieste, Italy

References

Basiuk, V. A. (2010) *Astrobiology Emergence, Search and Detection of Life*. American Scientific Publishers, Stevenson Ranch, California, USA.

Chela-Flores, J. (2001) *The New Science of Astrobiology From Genesis of the Living Cell to Evolution of Intelligent Behavior in the Universe*. Series: Cellular Origin, Life in Extreme Habitats and Astrobiology , Band 3 Kluwer Academic Publishers: Dordrecht, The Netherlands, 251 pp.

Chela-Flores, J. (2009) *A second Genesis: Stepping-stones towards the intelligibility of nature,* World Scientific Publishers, Singapore, 248 pp.

Gilmour, I. and Sephton, M. A. (editors) (2004) *An Introduction to Astrobiology*, Cambridge University Press.

Horneck, G. and Rettberg, P. (eds.) (2007) *Complete Course in Astrobiology,* Wiley-VCH, Berlin, 414 pp.

Impey, C. (2007) *Our Search for Life in the Universe*, Random House 416 pp.

Impey, C. ed. (2010) Talking about life: conversations on astrobiology. Cambridge University Press, Cambridge, 408 pp.

Jastrow, R. and Rampino, M. (2008), *Origins of life in the universe,* Cambridge University Press, 395 pp.

Lunine, J. (2005) *Astrobiology: A Multi-Disciplinary Approach,* Addison-Wesley 450 pp.

Acknowledgements

The author is indebted to many scientists for their influence and their insights in astrobiology and its frontiers: firstly, the Nobel Laureate Abdus Salam's shared interest and discussions in the 1990s on the origin of chirality in the life sciences (Chapter 4). Secondly, he acknowledges the influence of his co-directors in the Trieste Series on Chemical Evolution and the Origin of Life (astrobiology's earlier name towards the end of the 20th century): Cyril Ponnamperuma, François Raulin and Tobias Owen, as well as its many participants.

He is particularly privileged not only for having had the opportunity to devote his life to science, but his studies and research have benefitted from a fortunate opportunity that has been given to him for participating in activities in the humanities: In 1998 His Eminence Cardinal Carlo Maria Martini invited him to take part in the *Chair of Non-Believers*, in order to discuss the frontiers of astrobiology and the humanities. The same year the Venezuelan philosopher Ernesto Mayz Vallenilla invited him to hold the UNESCO Chair of Philosophy at the Instituto de Estudios Avanzados in Caracas. In addition, the author has benefitted especially from the eminent American astronomer George Coyne's participation in the Trieste events, where he discussed some of the most difficult issues at the frontier of astrobiology and the humanities, including closely-related theological issues (Chapter 13).

He would also like to thank colleagues from many nations that have helped him to appreciate the many fields that make up astrobiology. Due to either their areas of expertise, or their areas of influence, his vision of the subject has improved. He would like to mention especially the Spanish philosopher, Roberto Aretxaga-Burgos, for his collaboration over the last 7 years and particularly for our most recent one that has led to a much-improved version of chapters 13, 14 and 16. Thanks are due to the Venezuelan geneticist and astrobiologist Harold P. De Vladar, the Mexican biophysicist Moisés Santillán and the Indian physicist Santosh Chidangil for their advice in the subject matter of Chapter 12.

The Science of Astrobiology has also improved due to the help of the following scientists in many relevant issues of which the author was not familiar: Aranya Bhattacherjee (Physics, India), Suman Dudeja (Chemistry, India), Narendra Kumar (Condensed Matter Physics, India), Mauro Messerotti (Astronomy, Italy), Nevio Pugliese (Paleontology, Italy), Joseph Seckbach (Microbiology, Israel), Vinod Tewari (Micropaleontology, India) and Claudio Tuniz (Paleoanthropology, Italy).

For their kind and generous advice on parts of the manuscript acknowledgement is also due to the following scientists: Drs. Roberto Aretxaga-Burgos, Cristiano Cosmovici, Marco Fulle, Robert Gowen, Robert Greenberg, Mauro Messerotti, François Raulin and Giovanni Vladilo. But for anyone who is familiar with the process of preparing a manuscript of over 350 pages it should be clear that no matter how much kind collaboration his collegues and friends might offer, the work will not be free of errors, all of which are entirely the author's responsibility.

The administrative personnel of ICTP contributed significantly to this work: Johannes Grassberger and Marco Marcor (The Information, Communication and Technology Section), Ms Anna Triolo (Public Relations Office), Lucio Visentin and Dora Tirana (The Marie Curie Library).

And last, but not least, thanks are due to his wife, Sarah for her friendship, patience, perseverance, critical spirit, permanent support, guidance and encouragement.

Recommendations to the readers

The science of astrobiology is one of the most appealing aspects of culture that we believe should be accessible to everyone. However, the personal point of view expressed in this work makes a humble attempt to insert astrobiology inside its frontiers with the other branches of our cultural heritage. While writing this second edition of our work that was published a decade ago (when astrobiology could still be qualified with the adjective "new"), we firmly believe that any reader can benefit from the level we have chosen.

However, some guidance to my readers is still necessary. We have not shied away from placing astrobiology within a real scientific context of physics down to the subnuclear level, not neglecting the implications of the still-to-gain consensus theoretical framework (Chapter 15). Right from the Introductory Chapter, we have not been timid and glossed over the wonderful clarity and simplicity provided by the science of biogeochemistry. Neither have we hidden from our readers a Chapter 4 that includes the wonderful, yet complex concept of symmetry at the frontier of molecular biology and quantum mechanics that fascinated our distinguished colleague, who shared many radical views with us at the Trieste Center in the time frame of 1986-1994: the Nobel Laureate Professor Abdus Salam. Finally, due to certain fortunate circumstances mentioned in the *Acknowledgements* he has also felt that it was possible to include three chapters (13, 14, and 16) on the critical issues that arise when astrobiology meets philosophy.

This ambition of ours that science communication is still possible at this most difficult level that we have set ourselves has, we believe, a pedagogical solution. Each chapter has been preceded with an introductory paragraph that is followed in each case by an advice to the readers to use extensively the Illustrated glossary that has been provided: It contains some 300 entries, including some images and biographies of significant scientists that have influenced his way of understanding the science of astrobiology. Most of these scientists have participated in meetings at the Trieste Center, in which the author has attended during his career, since attending Graduate School at the University of London, when he was offered the unique opportunity of participating in the 1968 Trieste *Symposium on Contemporary Physics* (Salam, 1969). Some of the scientists that appear as protagonists of the physical sciences, as well as in the birth and growth of astrobiology, were participants in the events organized at the Center. The astrobiology meetings were organized with his colleagues Tobias Owen, Cyril Ponnamperuma and Francois Raulin (cf., Epilogue).

In addition, we have provided a rich General Index that contains a detailed list of over one hundred illustrations and almost 50 tables, besides the traditional name index. There is also an extensive bibliography supporting the text and also including suggestions of supplementary readings in each chapter to give the opportunity to the readers to go deeper into the wide range of topics that are relevant. Taking into account all of these additions to the 16 chapters, Introduction and Epilogue, we believe that the readers will be able to make sense of the fascinating topics in the physical, space, earth and life sciences that together make up the science of astrobiology.

Reference

Abdus Salam (ed.) (1969). *Contemporary Physics: Trieste Symposium 1968.* (2 volumes). International Atomic Energy Agency, Vienna. STI/PUB/214.

THE SCIENCE OF ASTROBIOLOGY

Introduction

The cultural and scientific context of astrobiology

Our main objective in writing this book has been to present the science of astrobiology in a manner accessible to non-specialists, but attempting at all stages to insert its scientific achievements in an adequate cultural context. We return to the beginnings of the subject in a brief historical account. In fact, a difficult task confronts the student of astrobiology in trying to comprehend the gradual emergence of our subject. The basic questions that astrobiology has to face have been with us since the beginning of culture itself. At the same time the reader would have cause to complain if we had not dwelled on the origins of astrobiology, since it is one of the most fascinating aspects of the history of science. We complete the introduction with an account of astrobiology in a spectrum ranging from the humanities to the space, earth and life sciences.

The reader is advised to refer especially to the following entries in the Illustrated glossary:
Apollo Program, biogeochemistry, Book of Life, chert, Crick (Francis), Galileo Mission, isotopic fractionation, Keynes (Sir Richard Darwin), mass-independent fractionation, Schopf (J. William).

I.1 Early attempts to read the Book of Life

Attempts to read the "Book of Life", the ever-present wish to comprehend the position of humans in the universe, have been a deep concern of humanity since the beginning of civilization. Going back to the Greek Golden Age we could have already separated the Book of Life in several chapters that covered the origin, evolution, distribution and destiny of life in the universe. Only in recent times the Book of Life has also been an objective of a scientific discipline. It coincides with the subject matter of the present book *The Science of Astrobiology*. In ancient times its first pages were beginning to be comprehensible even to our earliest ancestors, when agriculture arose soon after the end of the last glacial period. We began to understand the plant kingdom. In the Milesian School (6th century BC) we find attempts to read several other pages. Later on in the 4th century BC both Plato and Aristotle raised relevant questions still unanswered to the present day. At the end of the Middle Ages, Giordano Bruno expressed views that are still at the center of astrobiologist's main focus of research. We will come back to this great thinker below. Before the recent emergence of astrobiology perhaps the largest steps ever taken in our search for the position of humans in the universe were provided by Nicholas Copernicus in the 16th century and by Charles Darwin in the 19th century. In addition, about one hundred years later Francis Crick and James Watson with their contemporaries interpreted for us key issues that deal with the molecular mysteries of life (cf., Chapter 4, Sec. 4.7). The historical development of the basic concepts that astrobiology has made its own, can best be appreciated against the historical background of the whole range of sciences of the universe. We begin our account by returning, once again to Classical Greece.

ARISTARCHUS OF SAMOS AND HIPPARCHUS

The first important contribution to our present understanding of the science of astrobiology can be traced back to ancient Greece. Aristarchus of Samos lived approximately form 310 to 230 BC. He was one of the last of the Ionian scientists that founded the studies of philosophy. He lived over 23 centuries before our time, and 18 centuries before Copernicus. In spite of such vast time gaps, Aristarchus already formulated a complete Copernican hypothesis, according to which the Earth and other planets revolve round the Sun; but in so doing, Aristarchus asserted, the Earth rotates on its axis once every 24 hours. A lunar crater is named for him.

The heliocentric theory did not prosper in antiquity. Instead, the influential astronomer Hipparchus, who flourished from 161 to 126 BC, adopted and developed a non-heliocentric theory ("epicycles"), which was going to dominate the ancient world right into the Middle Ages. Indeed, Hipparchus went beyond Aristarchus geocentric cosmology with a spherical Earth at the centre of the universe. From his point of view, all of the solar system bodies including the Sun, rotated around the Earth daily. Ptolemy defended the final form of the ancient model of the Solar System in the second century AD.

NICHOLAS OF CUSA (CUSANUS)

The Italian cardinal Cusanus (1401-1464) flourished in the middle of the 15th century AD. He had been ordained a priest about 1440. Pope Nicholas V had made Cusa a cardinal in Brixen, Italy. At the time of his priesthood he published a significant work: *De Docta Ignorantia* ("On learned ignorance"). In this book he described the learned man as one who recognizes his own ignorance. But perhaps from the point of view of the evolution of our concepts of the cosmos we should underline that Cusanus denied the one infinite universe centered on Earth, thereby anticipating the heliocentric revolution of Copernicus Digges and others the following century.

He also claimed that the Sun was made of the same elements as Earth. He spoke of "a universe without circumference or center". The work of Cusanus touched on a theological question. In his system, all celestial bodies are suns representing to the same extent the manifestation of God's creative power. Such dialogue between a scientific question (cosmology) and theology (divine action) would lead in the subsequent century, to a conjecture that anticipated the formulation of the central question of astrobiology: the existence of a plurality of inhabited worlds.

NICHOLAS COPERNICUS

The Polish cleric Nicholas Copernicus (1473-1543) published his heliocentric theory in 1543. During his stay at the University of Padua from 1501 to 1503, Copernicus had been influenced by the sense of dissatisfaction of the Paduan instructors with the Ptolemaic and Aristotelian systems (Bertola, 1992). The new century was a time in which scientists were requiring a sense of simplicity that classical philosophy could no longer provide. The main work of Copernicus, *De revolutionibus orbium coelestium* ("On the revolution of the heavenly spheres") was published a few months before his own death in 1543. In this influential work Copernicus placed the Sun at the center of the Solar System and inserted both the Sun and planets inside a sphere of fixed stars (cf., Fig. I.1).

Copernicus deliberately appeared to accept the tenets of Aristotle's universe, as proposed in his *De Caelo,* a cosmos inserted within a system of revolving planets, surrounded by a sphere of fixed stars. There is ample evidence that the Polish scientist knew of the heliocentric hypothesis of Aristarchus. Indeed, Bertrand Russell argues that the almost forgotten hypothesis of the Ionian philosopher did encourage Copernicus, by finding ancient authority for his innovation (Russell, 1991). It would, of course, be Johannes Kepler, who was born in Germany in 1571, the scientist that would initiate a breakthrough with the discovery of the elliptical orbit. Subsequently Isaac Newton would provide the theory of gravitation that put the heliocentric theory of Copernicus on solid mathematical bases.

Fig. I.1 The heliocentric view of the universe according to Copernicus, in which he included a sphere of fixed stars (Bertola, 1995). This last difficulty with the Copernican universe was discussed in the writings of Giordano Bruno.

GIORDANO BRUNO

Giordano Bruno (1548-1600) was a philosopher, astronomer, and mathematician. He is best remembered for intuitively going beyond the heliocentric theory of Copernicus, which still maintained a finite universe with a sphere of fixed stars. As we have already seen in Chapter 10, from the point of view of astrobiology his anticipation of the multiplicity of worlds has been amply confirmed since 1995. In that year the first detection of extrasolar planets was announced. But what is more significant regarding Bruno's intuition is that he also conjectured that such worlds would be inhabited by living beings. Astrobiology, the science of life in the universe is just concerned with this key question, still without a convincing answer. Bruno was influenced by the philosophy of Cusanus (cf., previous Section). According to Hilary Gatti (Gatti, 1999) his attention on the philosopher that preceded him was probably due to the clarity of the concepts stated by the 15th century cardinal. (An articulate expression of Bruno's reading of Cusanus can be consulted in his third Oxford Dialogue: *De l'infinito, universo e mondi ("On the infinite universe and worlds").*

Besides, it can be said that Bruno's major innovation was his refusal to accept that the Solar System is contained in a cosmos bounded by a finite sphere of fixed stars. To sum up, Bruno proposes an infinite cosmos, populated by an infinite number of worlds. This proposal was first outlined in his first Oxford Dialogue: *The Ash Wednesday Supper.* Altogether Bruno wrote three Italian dialogues, which are relevant to our discussion. (The subsequent set of three additional dialogues refers to different matters.)

His work took place during a visit to England in 1584. These writings were in fact stimulated by controversial debates at the University of Oxford. We should emphasize his third dialogue in which Bruno developed valuable concepts first introduced in *The Ash Wednesday Supper.* His cosmological vision matured in Bruno's writings long before the science of astrometry allowed this concept to be brought within the scientific domain. Bruno's remarkable intuition is underlined by the large number of extra solar planets known to be circling around other stars (cf., Chapter 10). Galileo Galilei shared the position of Bruno in his own writings, such as the "*Starry Messenger*". It was in The Ash Wednesday Supper that he went beyond the heliocentric theory, in so far as he assumed that the universe is infinite, with innumerable worlds similar to those we know in the Solar System. He went on to anticipate Galileo Galilei in the consideration of the Bible as an archive of ethics rather than a scientific book that would help us to comprehend the universe. The right pathway for a rational cosmology was way ahead, as we shall begin to discuss in Chapter 1.

CHARLES DARWIN

The second objective of astrobiology is the evolution of life in the universe. Since life on Earth is included in its cosmic scope, any account of the history of astrobiology should take a closer look at the origin of the concept of evolution itself. Erasmus Darwin (1731-1802), a freethinker who lived through the Enlightenment, is best remembered for its two volumes, essentially a medical treatise. Darwin preferred to express some of his scientific ideas in verse. In "Zoonomia or the Laws of Organic Life" (1794–96) Erasmus Darwin anticipated some original evolutionary thinking, somewhat along the lines of J. -B. Lamarck. Charles Darwin (1809-1882) was the grandson of Erasmus and went much further.

Fortunately, for understanding the possible influence of his ancestor on Charles Darwin we have an important document from the Trieste series of conferences. In fact, during the Fifth Trieste Conference, the neurophysiologist Sir Richard Darwin Keynes (Charles Darwin's great-grandson, cf., Fig. I.2 and the Illustrated Glossary) presented a significant lecture. Darwin Keynes pointed out the possible influence of Erasmus on his grandson Charles (Keynes, 1998): *"I had previously read the Zoonomia of my grandfather, in which similar views are maintained, but without producing any effect on me. Nevertheless it is probable that the hearing rather early in life such views maintained and praised may have favored my upholding them under a different form in my "Origin of Species".*

Charles Darwin went on to complain on the degree of speculation in proportion to the facts given. In any case the final form of Darwinism is entirely founded on the overwhelming set of supporting observations included in The Origin of Species.

Fig. I.2 Sir Richard Darwin Keynes, the distinguished electrophysiologist, during his visit to the 1997 Trieste Astrobiology Series of conferences, where he discussed in detail the implications of his ancestor's natural history researches during the Beagle trip

In fact, in 1858 Alfred Wallace includes an essay on species that is extremely close to Darwin's theory of natural selection. Charles Lyell and his life-long friend the botanist Sir Joseph Dalton Hooker (1817-1911) present the work of Darwin and Wallace at the London Linnean Society in the absence of both authors. This leads to related publications in the Proceedings of the Linnean Society. In 1859 At long last the great book of Darwin is published on 24 November: "*On the Origin of Species by means of Natural Selection, or the Preservation of Favoured Races in the Struggle for Life*" (Darwin, 1859). The publisher is John Murray. But, at the same time it is really interesting to remark that Darwin himself avoided the problem of the origin of life itself. In his book *The Descent of Man,* which he published in 1871, he stated that:

"In what manner the mental powers were first developed in the lowest organisms, is as hopeless an enquiry as how life itself first emerged. These are problems for the distant future, if they are ever to be solved by man."

In his letters he insisted on his position with respect to the study of the origin of life. In the same year he wrote a letter to Hooker, which is frequently quoted:

"It is often said that all the conditions for the first production of a living organism are now present, which could ever have been present. But if (and oh what a big if) we could conceive in some warm little pond, with all sorts of ammonia and phosphoric salts, light, heat, electricity, etc., present, that a protein compound was chemically formed, ready to undergo still more complex changes, at the present day such matter would be instantly devoured or absorbed, which would not have been the case before living creatures were formed."

About this time he went on to the publication of the controversial book, we mentioned above: "The Descent of Man". In a letter written in 1871 to Hooker, Darwin stated his seminal ideas on the origin of life itself, but acknowledged that it was still too early to face this challenging problem. (In fact, his ideas only were inserted into biology over a century later in the present new science of astrobiology.) This was a very reasonable attitude for the late 19th century, before experimental science began to address the question of the origin of life with Oparin and others (cf., next section).

However, unconscious of the fact, Darwin influenced a young Russian scientist who went to meet him at Down House. The name of the young scientist was K. A. Timiryazev (1843-1920). Timiryazev went on to become a leading botanist and Alexander Oparin's teacher. His influence on young Alexander was crucial for his pioneering work in establishing the experimental approach to the study of the origin of life on Earth (Schopf, 1999).

I.2 Some pioneers of the science of astrobiology

ALEXANDER OPARIN

Alexander Ivaanovich Oparin was a pioneer of the studies of chemical evolution at a very early stage that eventually laid the basis of astrobiology (cf., I.3). Cyril Ponnamperuma asked a question (Ponnamperuma, 1995):

"Why was it that Oparin made such a difference?"

The answer is to be searched in the fact that Oparin was an unusual scholar, close to what sometimes is called a "renaissance man". He faced a problem that was basically philosophical in nature, but brought to bear upon it history as well as science. Oparin referred both to the philosophy of Aristotle, as well as to later writings of Saint Augustine. In spite of his biochemical training, Oparin was familiar with various fields that we have encountered in "The Science of Astrobiology": from astronomy to chemistry, from geology to biology, from philosophy to theology. Oparin obtained his first degree from the Moscow State University in 1917. Soon afterwards became a professor of biochemistry. In 1924 he postulated theory of life on Earth developing through gradual chemical evolution of carbon-based molecules in primordial soup.

In 1935, he collaborated with Aleksei Bakh in the foundation of the Biochemistry Institute by the USSR Academy of Sciences. Already in the 1920s Oparin influenced the studies of the origin of life in different directions, outstanding amongst which we should underline:

Fig. I.3 Alexander Oparin (left) and Cyril Ponnamperuma (right) at the Cortina D'Ampezzo meeting. This event was held in 1963, after the Wakulla Springs Conference, organized by the International Radiation Research Conference (Chadha, 2001). There was a special symposium on the Origin of Life

• There is no basic difference between a microbial life and condensed matter, a point that at present makes biogeochemistry an essential discipline in differentiating true fossils from pseudofossils (cf., Chapter 1).
The apparent complexity of microbial and multicellular life has arisen in the process of evolution through natural selection.
• Taking into account the recent discovery of methane in the atmospheres of Jupiter and the other giant planets (cf., Chapters 8 and 9), Oparin postulated that the infant Earth had possessed a strongly reducing atmosphere, containing methane, ammonia, hydrogen, and water vapor. In his opinion, these were the raw materials for the evolution of life.
• In the Archean there were solutions of organic substances. Gradually, as the result of growth and increased complexity of the molecules, new properties have come into being and a new colloidal-chemical order was imposed on the more simple organic chemical relations. He succeeded in making preliminary models of the primitive cells.

Besides such basic contributions to astrobiology that we have referred to briefly in Chapter 1, Oparin was also a good organizer of international meetings. In 1957 Oparin organized the first international conference on the origin of life. That meeting established the pattern of study of our subject. In 1963 a second conference was held in Florida (Wakulla Springs Conference). On that occasion the International Society for the Study of the Origin of Life (ISSOL) was founded. Oparin was its first President, while Sidney Fox was its Vice-president and Cyril Ponnamperuma its Secretary.

STANLEY MILLER

Stanley Miller was only a second year graduate student at the University of Chicago, when he published a remarkable paper in 1953 on the generation of amino acids (cf., Fig. I.4). It was a simple experiment that attempted to reproduce conditions similar to those in the early Earth, when life first originated.

Fig. I.4 Stanley Miller lecturing at the Sixth Trieste Conference in 2000

He was under the guidance of Harold Urey (cf., Chapter 4, "The Primitive Earth"), who had done fundamental research in nuclear physics. Urey was responsible for the discovery of an isotope of hydrogen: deuterium. He received the Nobel Prize for this work. Urey had subsequently suggested that the early Earth had conditions favorable for the formation of organic compounds. As a subject of his doctoral thesis Miller demonstrated experimentally that amino acids, the building blocks of the proteins could be formed without the intervention of man in environmental conditions, which we have called prebiotic—similar to those of the earliest stages in the evolution of the Earth.

The corresponding geologic period will be referred to in chapters 2-4 as the Archean. Miller's work was an important step in the growth of the subject of chemical evolution. Throughout his long and productive career Miller has inspired younger generations with his research on a wide spectrum of topics on chemical evolution.

His Trieste lectures were memorable for the many participants to those events. Sadly, Miller passed away soon after the celebration in Trieste of the 50th anniversary of the first synthesis of amino acids in his 1953 experiments. His lecture was the Abdus Salam Memorial Lecture on: "The Beginning of Chemical Evolution Experiments".

SIDNEY W. FOX

Sidney Fox (1912-1998) brought vitality and excitement to his work on the origin of life. He participated in three Trieste Conferences (1994, 1995 and 1997, cf., Fig. I.5).

Fig. I.5 Sidney Fox during his first visit to the Fourth Trieste Conference in 1995

He was particularly well appreciated for many reasons: In spite of the presence amongst many researchers at the conference site in Miramare, Fox stood out as a uniquely persistent scientist. He had been making significant contributions valuable to the field of the origin, evolution and distribution of life in the universe for over half a century. The publication of his sequence analysis of amino acid residues in 1945 was to be followed by an enormous output of influential papers, which were written in collaboration with over sixty associates. The study of the origin of life has gained many insights due to his work. He was the first to synthesize a protein by heating amino acids under conditions found here on Earth. He also showed that these new thermal proteins, when placed in water, would self-organize into a primitive cell. Fox was the first vice president of ISSOL. He had urged Oparin, its first president, that the time was ripe for a society that would bring together specialists from many countries and disciplines. He organized numerous conferences including, as we have mentioned above, the Wakulla Springs Conference in 1963. This meeting brought together both pioneers of the studies of the origin of life: Oparin and the English biologist John Burdon Sanderson Haldane (1892-1964). Sidney Fox also produced many books related to the origin of life. By persevering in the field of chemical evolution for so long, he will be remembered as one of the chief pioneers of the emerging field of astrobiology. His contribution to chemical evolution is mentioned in Sec. 4.3.

CYRIL PONNAMPERUMA

Cyril Ponnamperuma (1923-1994) was born in Ceylon, now Sri Lanka (cf., Fig. I.6).

Fig. I.6 Cyril Ponnamperuma lecturing during the First Trieste Conference in October 1992

He went to Birkbeck College, in the University of London, where he came under the influence of the Irish physicist John Desmond Bernal (1901-1971), a well-known crystallographer who had published in 1949 an early, but influential paper. It concerned the idea that the organic compounds that serve as a basis of life were formed when the Earth had an atmosphere of methane, ammonia, water and hydrogen.

Afterwards he continued his studies in the United States, where in 1962 he received a doctorate in chemistry from the University of California, Berkeley under the direction of the Nobel Laureate Melvin Calvin (1911-1997). In 1962, he was honored with a National Academy of Science resident associateship with NASA at Ames Research Center. In 1963 he joined NASA's Exobiology Division and take over the helm of the Chemical Evolution Division. He was selected as the principal investigator for analysis of lunar soil brought to earth by Project Apollo that we have already mentioned in Chapter 3. Ponnamperuma's work with Calvin led to a series of papers on the synthesis of DNA components, extending in a significant way the pioneering work initiated by Stanley Miller. In 1971 he joined the Maryland faculty as head of the Laboratory of Chemical Evolution, which he directed until his death. He became principal organic analysis investigator for the Apollo project and also worked on the Viking and Voyager programs. He left NASA to join the Maryland faculty. Ponnamperuma played a major role in NASA's early experiments for detecting life on Mars at the time of the Viking missions (cf., Chapter 7). His analysis of meteorites showed that the basic chemicals of life were not confined to the Earth. In the analysis of the meteorite that fell in Murchison, Australia in 1969 (cf., Chapter 3), he and his co-workers provided evidence for extraterrestrial amino acids and hydrocarbons.

Cyril Ponnamperuma played a leading role in ISSOL from 1977 till 1986. He was the recipient of awards from several countries. These included the first Alexander Ivanovich Oparin Medal awarded by ISSOL at its triennial meetings for the best "sustained scientific research program" in the origin of life field (in 1980).

Since 1990 he was closely associated with the ICTP. He interacted with Abdus Salam in research on the chirality of amino acids, and tested some of these new ideas in his own Laboratory of Chemical Evolution. This collaboration was the initiation of a longer association with the ICTP in the form of the series of conferences on various aspects of astrobiology that continued from 1992 till the end of his life in 1994 (Navarro-Gonzalez, 1998).

JOHN ORO, 1ST MARQUESS OF ORÓ

He was one of the pioneers in astrobiology and was known to the academic community as John Oro (1923-2004, cf., Fig. I.7). He did his main work in the USA at the University of Houston However, he was born in Spain and his name was Joan Oró i Florensa, named 1st marquess of Oró by Royal Decree. He was a biochemist from Catalonia (Spain), whose research has been of importance in understanding the first chapter of astrobiology: origin of life. From the 1960s he worked with NASA on the Viking missions which explored the planet Mars. His work was essential in the analysis of samples of Martian soil, and suggested that there was in fact no life on Mars. One of his most important contributions was the prebiotic synthesis of the nucleobase adenine (a key component of nucleic acids) from hydrogen cyanide. This was achieved during the period 1959–1962 and stands, together with the Miller-Urey experiment, as one of the fundamental results of prebiotic chemistry. It opened up a research area eventually leading to the complete synthesis of other components of nucleic acids. He was also the first scientist pointing towards comets as the carriers of organic molecules to our early biosphere. This conjecture (Oro, 1961) is largely accepted today). He was awarded, among other honors, the Alexander Ivanovich Oparin Medal Award from the ISSOL (awarded at the 1986 Berkeley meeting). Oro played also an important role in the Viking Missions to Mars (cf., Chapter 7, The Early Viking Missions).

Fig. I.7 John Oro during the 4th Trieste Conference in 1995

I.3 Three strategies for astrobiological exploration

Three strategies have been devised for the search for extraterrestrial life: the study of the cellular makeup of exotic organisms on Earth; the search for organic matter and living micro-organisms beyond Earth; and the use of radio telescopes to detect signals of intelligent behavior in the universe. The following three strategies have provided the road map for a great journey of discovery that represents some of the most fascinating scientific research currently taking place for finding out the real place that planet Earth, and all life that ever evolved on our small planet has in the universe (cf., Fig. I.8).

The first strategy has focused on understanding how life began on Earth. Research has shown exotic organisms living in inhospitable environments as ocean-floor bottoms,

Antarctic glacial sheets, and volcanic lava streams—all of which display temperatures and pressures that may have been present during the Earth's 'fire and ice' formation billions of years ago. Perhaps one of the most unexpected recent discoveries has been that there are whole underground ecosystems, which to a large extent are independent of sunlight, extending our old concept of what was a habitable zone in a given solar system. Research into our own origins not only broadens our appreciation of the enormous diversity of life here on Earth, but it may help us understand the environmental extremes that simple organisms can tolerate. Such extremes may be found on other celestial bodies making it more probable that life can exist elsewhere.

Fig. I.8 The Earth as seen by the astronauts of Apollo 17

The second strategy for deciding if we are not alone in the universe is a search for the simplest forms of organic matter—amino acids or proteins—that may be embedded in the ancient rock of planets, comets or meteorites, or suspended interstellar clouds. The search has focused on three celestial bodies besides our own: Mars, Europa (a moon of Jupiter), and Titan (a moon of Saturn). The discovery of meteorites from Mars suggests that all the terrestrial planets (Mercury, Venus, the Earth, the Moon and Mars) at one stage in the past may have been in biological intercourse. There is compelling evidence that liquid water has flowed in the geologically recent past on Mars, (or may even be flowing now).

The Galileo Space Mission has provided ample evidence for an ocean on Europa underneath its frozen surface. In a few years from now we expect substantial information to unveil the secrets of the hidden surface of Titan, a world with an atmosphere that resembles our own planet when life first emerged on it. The main interest of research on Earth has shifted toward discovering fossilized remnants of life that existed billions of years ago when our world was an inhospitable environment where only thermophilic organisms (those that could withstand extreme temperatures) could survive. One of the primary goals of astrobiology is to determine whether life ever existed in places other than the Earth and, if so, what were the environmental conditions that made it possible.

Research, continues for clues that could reveal the existence of life forms beyond Earth. The discovery of an independent life form, a separate tree of life from our own on Mars or Europa, would not only be fascinating in its own right, but shed revealing light on the microorganisms that inhabited the Earth more than four billion years ago.

The third strategy used in the hunt for life beyond Earth relies on radio telescopes such as the huge one at the National Astronomy and Ionosphere Center in Arecibo, USA. These huge 'dishes' actually have two roles to play: First and foremost, they help to examine wavelengths that cannot be seen by the human eye; for example, radio waves and microwaves. Such information has proven essential for understanding the movement and behavior of planets and stars. Second, radio telescopes also seek anomalies in microwaves and radio waves wafting across the universe. Such anomalies may represent the imprint of intelligent life in the heavens beyond.

Thus far astronomers have been scanning the radio and microwave spectrum for four decades with no reliable signal from an extraterrestrial civilization. But that doesn't mean the initiative is likely to be abandoned. The public's fascination with the search for extraterrestrial life, combined with the vast reaches of outer space, where earth-bound scientific research continues to find new extrasolar planets and even solar systems, keeps the hope alive that 'somewhere out there' there is intelligent life that will some day send us a signal. If a signal is ever received, this much is certain: It will be one of the most remarkable and influential discoveries ever made, whose implications will range from science, philosophy and even theology.

Our position in the Solar System began to be widely realized by the layman after the extraordinary tour of he Voyagers in the 1970s to 1980s (cf., Fig. I. 9).

Fig. I.9 The Earth as part of the Solar System in a combined photographic work retrieved by the Voyager missions in the period 1976-1989

I.4 Inserting philosophy into astrobiology

Bernard–Henri Levy, a member of the group of the "New Philosophers", introduced one of the simplest definitions of philosophy: a philosopher is one who develops, produces and organizes concepts (Levy, 2009). From this point of view, astrobiology, as presented in the present book can benefit from philosophy (cf., chapters 13 to 16). Astrobiology, being possibly one of the most interdisciplinary of sciences, once reviewed as part of our cultural heritage, raises some questions in which theoretical philosophy can contribute to organize the different approaches to the basic questions of the origin, evolution, distribution and destiny of life in the universe.

Rather than facing the complex problem of the origin of life on Earth-the main scientific approach in the last 60 years-by means of the time-honored reductionist methods of chemical evolution introduced by Oparin in the 1920s, but with the significant work of Stanley Miller in the early 1950s, the main thesis of recent work is instead to focus on the origin of habitable ecosystems on Earth and elsewhere (Chela-Flores, 2010). We attempt to organize apparently different approaches to the emergence of life on Earth: chemical evolution versus the space exploration of the Solar System by the space agencies of USA, Russia, Europa, India and China. By identifying such ecosystems, either on Mars, Europa, Titan or Enceladus, we would be in a better position to plan rationally the future exploration of the Solar System and beyond and eventually learn about our own origins by having an example of a second Genesis. Early during the evolution of the Solar System, terrestrial planets and similar environments may have presented an analogous geophysical context that eventually favored the emergence of life on Earth in the Early Archean, but when in the same period other evolutionary lines may have emerged elsewhere. Around other stars life may have started earlier than on our own solar system, or later, but we shall develop in some detail in Chapter 8 the case for searching a second Genesis on the outer Solar System, especially on Europa.

I.5 Emphasizing the role of biology in astrobiology

The present work presents some reflections on astrobiology, in particular on the transition from inert matter to intelligence, as we know it on Earth. It also discusses the possibility of it occurring elsewhere as well, particularly in other environments of solar systems. This possibility may be probed with the forthcoming series of space missions, which at the time of writing are being planned in Europe well into the next decade with the Cosmic Vision program. This is a way of "looking ahead, building on a solid past, and working today to overcome the scientific, intellectual and technological challenges of tomorrow". Research in astrobiology is a major component Cosmic Vision that is strongly based on the basic sciences. We would like, before beginning the main body of the work, to consider the relation between the basic sciences that serve to support the sciences of the origin, evolution, distribution and destiny of life in the universe.

For simplicity, and in agreement with a growing tendency in the specialized community of researchers, we shall refer to this subject as 'astrobiology', reserving the expressions 'exobiology' and 'bioastronomy' for the specific topics discussed in the third part of the Book of Life (cf., Table of Contents).

In his very influential book Ernst Mayr argues, correctly in our view, against biology being an imperfect stepchild of the physical sciences (Mayr, 1982). The history of the classification of living or extinct organisms into groups, a discipline known as taxonomy is a good point in question, for its traditional functions have been the identification of particular specimens and the classification of living things into a satisfactory conceptual order.

In Aristotle's time the basis of such order was not understood. For deep insights we had to wait for the advent of the birth of modern biology in the middle of the 19th century, with the work of Charles Darwin. In *The Origin of Species* a true revolution was brought about by incorporating a very wide range of biological phenomena within the scope of natural explanation, including an appropriate approach to the study of 'taxonomy', namely, the theory, practice and rules of classification of living or extinct organisms into groups, according to a given set of relationships. Darwin's theorizing in biology may have begun in 1838 (in his Notebook D), although for reasons that have been well documented elsewhere, the publication of his seminal book was delayed until 1859.

The intellectual revolution initiated by Darwin has produced a far-reaching change in our thinking, even outside the strict boundaries of the natural sciences, but the radical departure from the biological science of his time consisted in the formulation simultaneously with Alfred Wallace of a mechanism, natural selection, for the evolution of life. At a time when the prevalent ideas in the scientific community were mathematical principles and physical laws, biology was enriched with the concepts of chance and probability. In spite of the fact that it is widely appreciated that biology is not the imperfect stepchild of the physical sciences, physics itself nevertheless has had a valuable influence on the life sciences, particularly when a century after the publication of *The Origin of Species,* the molecular bases of genetics were searched for with precise experiments.

I.6 Inserting physics in biology especially in astrobiology

During the 1940s physicists and physical chemists made vital contributions to the transition of biology into an experimental phase, in which traditional physical sub disciplines were to make modern molecular biology possible. The theoretical physicist, Max Delbrück, had contacts with radiation biologists (Fischer and Lipson, 1988). Learning about mutations, he foresaw the possibility of linking them to the lesions produced by a physical agent acting on the atoms of the still-to-be-discovered genetic material.

The theoretical physicist, Erwin Schrödinger, working during that decade at the Dublin Institute for Advanced Studies, popularized this important step in the development of modern science. At Trinity College, Dublin, he delivered, a set of lectures known to us through his book "*What is Life?*" (Schrodinger, 1967). In those early years, when science was still ignorant of the chemical nature of the carrier of genetic information, a generation of physicists was responsible for the birth of the new biology, together with colleagues trained in the more traditional branches of the biological sciences. This group included, Sir Francis Crick, Nobel Laureate and co-discoverer of the DNA structure, who graduated in physics at University College,

London (cf., Fig. I.10). Lewis Wilbert has referred to Crick as "the genius theoretical biologist of our age" (Wolpert, 1990).

The origin of life is currently one of the most attractive interdisciplinary fields. In spite of dealing with the origin of biology, major components come from the physical, as well as the earth sciences. It is not difficult to understand the reasons for the ever-growing popularity of the field. They are implicit in the facts that are gradually emerging regarding the first steps of life on Earth.

Fig. I.10 Francis Crick (right) and Abdus Salam (left) during the Symposium on Contemporary Physics, Trieste, June 1968

I.7 Inserting the earth sciences in astrobiology

J. William Schopf discovered colonies of bacterial microfossils in Western Australia in the early 1990s. A date of some 3500 million years before the present (Myr BP) may be assigned to these fossils. The structures are called stromatolites and today they are formed by a group of ancient microorganisms called cyanobacteria (Schopf, 1993).

However, according to geochemical work of Manfred Schidlowski (Schidlowski, 1995), and others since then, including Gustaf Arrhenius and co-workers (Mojzsis *et al.*, 1996), it is not to be excluded that photosynthetic bacteria might be as old as some of the oldest extant rocks available from the Isua peninsula in Greenland. These are amongst the oldest rocks on Earth; the first identification of these rocks was due to Stephen Moorbath and colleagues. Their work has contributed to a long, still ongoing process of retrieving rocks dating from 3800 Myr BP from a very ancient geologic era, the early Archean. This date should be compared with the age of the Earth itself, of some 4600 Myr BP (Moorbath, 1995).

According to Schidlowski the 'smoking gun' is the carbon content of the fossils available, since they are biased towards an isotope of carbon, which is normally left behind as the result of the metabolism of cyanobacteria. (Such isotopic enrichment is characteristic of those bacteria that are able to sustain themselves through the process of photosynthesis.) What is clear from the work of the above specialists in the earth sciences is the following: micropaleontologists, biogeochemists and geochronologists have argued that the early Earth biota was dominated by cyanobacteria. As we mentioned above, these are prokaryotic microorganisms, which were at one time called blue-green algae. A transcendental aspect of their physiology is that they were able to extract hydrogen from water in photosynthesis liberating oxygen.

Cyanobacteria were so successful that, as time passed, free oxygen accumulated in the atmosphere, thereby transforming radically the primitive Earth atmosphere, which according to most scholars originally resembled the one we observe today on Titan. We shall return to this interesting satellite in Chapter 9. Suffice it to say at this stage that during the early geologic history of the Earth, the reactants freely available in the environment, such as iron, had been used up. This showed up in the geologic record as a transition from "banded iron" formation (this will be discussed in more detail in Chapter 5) to rocks largely colored in red. This second group of rocks is called "red beds". To give a compelling name to the colossal transformation of the Earth, this increment in atmospheric oxygen has appropriately been referred to as the 'oxygen holocaust' (Margulis and Sagan, 1987), in order to convey the idea that micro-organisms, other than cyanobacteria, had no defenses against the new atmosphere that was developing. The only passive remedy for life on Earth was evolution into new forms that tolerated, and even thrived on oxygen. This led to the onset of the highly evolved cellular blueprint (eukaryotic), which will be the main theme of this book.

There are many constraints that the development of modern science and technology impose on the eventual understanding of the emergence of life in the universe. Paramount amongst these limitations is instrumentation. Specifically, the direct search for life beyond the Earth is feasible in the near future mainly within the Solar System, on Mars, Europa, Titan and Enceladus. In these locations we have a number of missions with many instruments that are capable of retrieving accurate information to elucidate the question of life on Mars. The Galileo mission from 1995 till 2003 has given us suggestive insights for Europa, the Jovian satellite (Fig. 1.11 and Chapter 8). The Cassini-Huygens mission after 2004 brought the two satellites of Saturn to the closer scrutiny of astrobiology (Fig. 1.12 and Chapter 9).

Outside the Solar System there are a large number of stars that are known to bear planets. This is an early stage in the search for extra-solar planets and exomoons. This activity has a history of less than two decades, since the first exoplanet was found around Pegasi-51. But we have to wait for the results of missions such as COROT and Kepler to begin providing Earth-like planets in order to proceed with the detailed spectroscopic analysis of the incoming beams of starlight to decide which are the most reliable biomarkers (cf., Chapter 10).

Fig. I.11 Galileo's encounter with Jupiter's satellite Io in an artistic interpretation, showing volcanic activity on its surface

Fig. I.12 An artist's early conception shows the moment the Huygens probe detached itself from Cassini late in 2004 to begin its descent in Titan's atmosphere.

These are a few of the reasons why the earth sciences take a prominent position in astrobiology. A very promising branch is this respect is geochemistry. Indeed, isotope variations that can be retrieved form our own solar system consist of samples from the Moon, Mars and the small bodies (comets and asteroids). They provide us with a key to unlock the mysteries of the evolution of our solar system. A powerful tool that is available is that of the stable isotope geochemistry. Through this window we can even retrieve data from events that occurred prior to the formation of the Sun itself. This is the case of grains that have been found inside meteorites, such as grains of carbide and diamond; but our greatest ally in our search for biosignatures is the activity of microbial life that we have learnt to recognize as the producers of phenomena that are not possible to be imitated by abiotic events. Of all the biogenic elements, sulfur has the most relevant isotopic fractionation for the detection of traces of biogenic activity (Chela-Flores 2006): Once a primordial satellite internal silicate nucleus (for example, on Europa) had entered their corresponding geochemical cycles, their initial isotope mixtures began to be redistributed. The Earth upper mantle and crust are believed to reflect broadly the isotopic distribution patterns of chondritic meteorites. A similar mechanism would equally apply to the silicate core of Europa. The new point we make here requires the definitions of the sulfur and carbon fractionation parameters. For the isotopic fractionation of sulfur we restrict our attention to ^{32}S and ^{34}S. We define:

$$\delta\,^{34}S = [(^{34}S/^{32}S)_{sa} / (^{34}S/^{32}S)_{st} - 1] \times 10^3\ [^0/_{00}, \mathrm{CDT}]$$

For simplicity this function will be referred to as the "delta ^{34}S parameter", or simply as the "delta 34-S parameter". Its value is close to zero when the sample coincides with the corresponding value of the Canyon Diablo meteorite that is a triolite (FeS), abbreviated as CDT. This parameter allows a comparison of a sample (sa) with the standard (st) CDT. The relevant terms are the dominant sulfur isotope (^{32}S) and the next in abundance (^{34}S). In fact, $(^{34}S/^{32}S)_{st}$ coincides with the average terrestrial fraction of the two most abundant isotopes of sulfur. We obtain positive values of the delta ^{34}S parameter when by comparison we have a larger quantity of the less abundant isotope ^{34}S. For the fractionation of the carbon stable isotopes we require the delta ^{13}C $[^0/_{00}$, PDB] parameter that is defined as follows:

$$\delta\,^{13}C = [(^{13}C/^{12}C)_{sa} / (^{13}C/^{12}C)_{st} - 1] \times 10^3\ [^0/_{00}, \mathrm{PDB}]$$

The value of delta 13-C parameter is close to zero when the sample coincides with the PeeDee belemnite standard (PDB), in which $(^{13}C/^{12}C) = 88.99$. (The delta 13-C parameter is defined as equaling 0.00 $^0/_{00}$.)

This parameter can be used as a good biomarker. On the Earth biota, for instance, there is ample evidence that photosynthetic bacteria, algae and plants have typical significant deviations that yield values of up to - 30 and beyond, due to biological processes (Schidlowski *et al.,* 1983). These results are analogous to the deviations shown by fractionation due to bacterial sulfate reduction. Yet, negative values of the delta 13-C parameter do not arise exclusively from biogenic sources: in lunar fines we know that life is absent, significant negative deviations in the delta 13-C parameter do in fact occur, but are absent in the corresponding sulfur parameter (Kaplan, 1975). Thus, without prior knowledge whether we are in the presence of life in a given environment, negative values of the delta 13- C parameter do not arise exclusively from biogenic sources. For this reason we have mentioned above that sulfur is more relevant for studying possible biomarkers.

A closer and more detailed understanding of the origin of life on Earth has forced upon us a more significant view on the Late Archean and Early Proterozoic evolution of trophic relations in the microworld. Presently we are more aware that hydrothermal vents in the Earth oceans may have played a role in the origin and evolution of the three domains of life (cf., Chapter 6, fig. 6.1). Indeed, it is possible that throughout evolution entire ecosystems depend on geothermal, rather than solar energy. This is not only evident on the Earth, but this is also likely to be the case on the other oceans of the solar system, as for example on the moons of Jupiter: Europa and Ganymede. On these satellites this particular hypothesis for the origin of life may be tested in the foreseeable future. After the initial proposal of the LAPLACE mission (Blanc *et al.,* 2009), the Europa-Jupiter System Mission is now being seriously considered by the main space agencies of Europe, the United States, Japan and Russia (cf., Fig. 1.13 and Chapter 8)

Fig. I.13 Left: ESA's Cosmic Vision Programme: The preliminary Laplace Mission. Right: The Europa Jupiter System Mission (EJSM), a natural extension of the Laplace Mission for the exploration of the Jovian System involving all the major Space Agencies

The primary sources of organic matter for the microbial autotrophs are photosynthesis, methanogenesis and sulfate reduction. In the limited space available it is most important to single out and highlight that the group of sulfate reducers may contain some of the earliest forms of life on Earth. Thus, this special form of metabolism may hold the key to understanding the primordial state of life, since sulfate-reducers are in deeply rooted branches of the phylogenetic tree of life (Shen *et al.,* 2001). The morphological simplicity of the primitive sulfate reducers is one drawback in probing the fossilized remains of these microbes. Instead we must rely on the science of biogeochemistry when our objective is to enquire on the antiquity of life and its trophic relations. From the early papers of Manfred Schildowski and co-workers the stable isotope geochemistry of sulfur and the other biogenic elements (H, C, O, N) has been reviewed extensively (Strauss and Beukes, 1996).

I.8 New fingerprints of early life

For a proper understanding of the Archean S-isotopic record we should first realize that the abiotic fractionation of carbon isotopes could produce effects comparable to geomicrobiological effects. However, the situation is more favorable for sulfur. Indeed, microorganisms mediate the reduction of sulfate to sulfide. The resulting fractionations can be reliably taken as good markers for the geological record, especially for the Archean S-isotopic record, where we have hinted that the oldest signatures for life are to be retrieved. The biology that underlies this significant aspect of our quest for the evolution of trophic relations is as follows:

The preferential use of ^{32}S over ^{34}S by microorganisms depletes the sulfide in the environment of ^{34}S with respect to the original sulfate. Several species of bacteria and Archaea can make this happen via the metabolic pathway known as dissimilatory sulfate reduction. Sulfur itself is not incorporated into cell, but it ends up in the oxidation of organic matter. In normal marine sediments of sulfate, the fractionation can range from 10 to 49 ‰, but this effect can be as large as 70 ‰, which leaves an ample margin for distinguishing the microbial activity in rocks at a hydrothermal vent and the abiotic fractionations.

A few natural processes are not subject of fractionation entirely due to the mass difference of the isotopes that enter a chemical reaction. In meteorites the first phenomena were observed where factors other than isotopic mass differences showed up in its chemistry (Clayton *et al.*, 1973).

These early observations in planetary science exposed the fallacy that we should always expect the mass-dependent relationships: in the case of the oxygen isotopes $\delta^{17}O = 0.5\ \delta^{18}O$, or in the case of sulfur isotopes $\delta^{33}S = 0.5\ \delta^{34}S$ as it would happen if the only factor involved were the mass difference of the isotopes. Some of the causes of the mass-independent contributions to the fractionation (MIF) are known. A striking presence of MIF in rocks of Archean and Paleoproterozoic age, but absent from the rock record subsequently. This phenomenon is related with the lack of oxygenation in the early atmosphere. Deviations are most conveniently introduces in terms of the capital delta parameter:

$$D^{33} = \delta^{33}S - \lambda\, \delta^{34}S$$

where λ is the main parameter that characterizes the MDF, namely, 0.515 for thermochemical equilibrium. Terrestrial sulfates have been shown to exhibit MIF. These differences are measurable and are documented in multiple isotope systems of oxygen (Angert *et al.,* 2004) and sulfur (Farquhar *et al.,* 2003). As a result, natural samples can show measurable and systematic variations in $\Delta^{33}S$ ($=\delta^{33}S–0.515 \times \delta^{34}S$) as well as $\Delta^{36}S$ ($=\delta^{36}S–1.90 \times \delta^{34}S$), even when $\delta^{34}S$ values are identical (Ono *et al.,* 2006).

I.9 Inserting paleontology into astrobiology

The S-isotopic record of sulfide and sulfate in Archean sedimentary rocks ranges from Isua of ~3.8 Ga BP (pyrite in banded-iron formations) and ~3.47 Ga BP (barite deposits). In these early times the sulfate reducers were beginning to leave measurable traces, but some difficulties have still to be fully understood, as to their sources and the role of atmospheric contributions. In conclusion the stable isotope geochemistry of the ~3.47 Ga barium sulfate (barite) deposit suggests that reactions mediated by microorganisms were already fractionating sulfur much in the same way as present day sulfate-reducing microorganisms.

South Africa is in a very favorable location for understanding the whole range of life on Earth, from the first appearance of bacteria to the evolution of Man. Ancient rock formations are retrieved from part of a continent that has succeeded in retaining crustal stability. These continental regions are called "cratons". Some mountainous formations of igneous rocks derive their name from the fact that much of the mineral content is chlorite. Hence, green color is predominant and is due to hydrated magnesium, iron and aluminosilicates. Such mountain ranges are called greenstone belts. The presence of chlorite in greenstone belts is produced as a result of physical and chemical changes that have taken place on the rocks due to their residence time at different substrata during their long history. These changes are simply referred to as "metamorphism". The Barbeton Greenstone Belt is located at the eastern edge of the Kaapvaal Craton in South Africa. It lies around the town of Barbeton, some 360 Km east of Johannesburg. With sediments dating back to 3.6 Gyr BP, these are some of the oldest sites where reliable fingerprints of life stand out from their geological setting. In other similar, and even older, cratons there is controversy over the identification of early signatures of life. They shall be considered in subsequent sections, including the Pilbara Craton in Western Australia and, as we have seen in the previous section, the Isua Greenstone Belt near the Gothabsfiord region in West Greenland. Returning to South Africa, the isotopic analysis of the fossils of ancient microorganisms has centered on the $\delta^{13}C$ parameter in metamorphosed areas that are geologically consistent and discontinuous with their neighboring areas. (These regions are called "terranes".) These carbon-related geochemical markers are difficult to interpret, so we have searched for confirmation of biogenicity from various alternatives, such as additional morphologically preserved cellular materials, chemical, and other isotopic traces of early life.

A first step in this search for alternative hints of true biogenicity is to follow up the evolution of the so-called "heteroatoms" (namely, those atoms present in the rock minerals that differ from carbon). The objective of this line of research is to follow up the gradual reduction of the number of non-carbon atoms due to the gradual and efficient process of metamorphism. Graphite is a naturally occurring form of crystalline carbon. The ultimate effect of metamorphism is its tendency to induce carbonates substances to produce graphite.

The Barbeton Greenstone Belt has produced some significant archives of fossils that are contained in a type of sedimentary rock that is known as "chert". This type of sediment is rich in crystalline silica (SiO_2). Indeed, some chert has been retrieved from South African rock units that are used as basis for rock mapping. Such stratigraphical units are referred to as "formations" that may happen to be gathered into similar units. We speak about such conglomerations as "groups"; in turn, such groups may be part of a larger geological classification known as "supergroups".

South Africa has a well-known example of a significant group known as the Overwacht Group, which includes the Hooggenoeg Formation. Later on we shall meet in Western Australia the Pilbara Supergroup, which includes two important geologic landmarks: the Warrawoona Group and the Dresser Formation. Both of them will be discussed in some detail later on, especially since the deepest nodes in the Tree of Life have both been identified in the Dresser Formation: these microorganisms are a mesophilic sulfate reducing microbe producing H_2S, and a methane producing microbe (commonly known as "methanogen"). Returning to the Hooggenoeg Formation, we recognize it as a rock structure about 3000 m thick, including lavas dating back to 3,470 Ma. In these cherts there are layers of carbonaceous material that can be traced back to bacterial mat precursors (Walsh and Westall, 2009). Measurements of $\delta^{13}C$ and nitrogen to carbon atom ratios (N/C) have been studied by a combination of two remarkable experimental techniques:

• Laser-Raman Spectroscopy (LRS), which is a spectroscopic technique used to study vibrational, rotational, and other low-frequency modes in a system by means of inelastic scattering of monochromatic laser in the visible, near infrared, or near ultraviolet range. The laser light interacts excitations in the system under scrutiny, including phonons. As a consequence of the scattering the energy of the laser photons are shifted. Such shift in energy gives information about the phonon modes in the system.

• Secondary Ion Mass Spectrometry (SIMS). This is a technique that is used to analyze the composition of thin films by sputtering the surface of the specimen with a focused primary ion beam. Then the ejected *secondary ions* (and hence the name) are collected for measurements with a mass spectrometer in order to determine various properties. They may include isotopic and molecular composition of the surface. SIMS is able to detect elements present in the parts per billion ranges, thus becoming an invaluable tool for the analysis of microfossils preserved in Hooggenoeg and elsewhere, when primary textures are preserved down to submicron scales.

Two trends have clearly emerged from the experimental approaches: Firstly, extreme metamorphism would be expected to yield very low values of N/C (the graphite limit). Secondly, biogenicity would be revealed by values of $\delta^{13}C$ that are very far removed from the expected isotopic fractionation that would necessarily take place in thermal equilibrium. The chert samples retrieved form Barbeton fulfill both criteria and betray their biological precursors (Van Zuilen, 2006).

Supplementary Reading

Barrow, J. D., Conway Morris, S, Freeland, S. J. and Harper, C. L. (eds.) (2008) *Fitness of the cosmos for life: Biochemistry and fine-tuning.* Cambridge University Press.

Bertola, F. (1995) *Imago Mundi, la rappresentazione del cosmo attraverso i secoli,* Biblos, Cittadella PD, Italy.

Chadha, M. S. (2001) *Reminiscences-Pont-a-Mousson-1970 to Trieste-2000,* in J. Chela-Flores, T. Owen, and F. Raulin (eds.) *The First Steps of Life in the Universe.* Proceedings of the Sixth Trieste Conference on Chemical Evolution. Trieste, Italy, 18-22 September. Kluwer Academic Publishers: Dordrecht, The Netherlands.

Clarke, A.C. (1993) *By space possessed,* Victor Gollancz, London.

Conway-Morris, S. (1998) *The crucible of creation,* Oxford University Press, London.

Conway-Morris, S. (2003) *Life's Solution Inevitable Humans in a Lonely Universe.* Cambridge University Press.

Crick, F. (1981) *Life Itself,* MacDonald & Co, London.

Davies, P.C.W (1998) *The Fifth Miracle The search for the origin of life,* Allan Lane The

De Duve, C. (1995) *Vital Dust. Life as a cosmic imperative.* Basic Books: New York

De Duve, C. (2005) *Singularities Landmarks on the Pathway of Life.* Cambridge University Press.

De Grasse, T. and Goldsmith, D. (2004) Origins Fourteen Billion Years of Cosmic Evolution. W. W. Norton, New York. 345 pp.

Dennett, D. C. (1995) *Darwin's dangerous idea Evolution and the Meanings of Life,* Penguin, London.

Desmond, A. and Moore, J. (1991) *Darwin,* M. Joseph: London .

Fischer, E. P. and Lipson. W.W. (1988) *Thinking about Science (The life of Max Delbruck),* Norton, New York.

Gale, J. (2009) *Astrobiology of Earth: the emergence, evolution, and future of life on a planet in turmoil*. Oxford: Oxford University Press, 245 p.
Jastrow, R. and Rampino, M. R. (2008) *Origins of life in the universe,* Cambridge University Press, 395 pp.
Konhauser, K. (2007) *Introduction to geomicrobiology*. Blackwell Publishing, Oxford, 425 pp.
Mayr, E. (1991) *One Long Argument Charles Darwin and the Genesis of Modern Evolutionary Thought,* Penguin Books, London.
Moorhead, A. (1971) *Darwin and the Beagle,* Penguin: London.
Ricci, S. (2000) *Giordano Bruno nell'Europa del Cinquecento,* Salerno Editrice, Roma.
Rizzotti, M. (1996) *Defining Life The central problem of theoretical biology,* The University of Padova, Padua.
Rollinson, H. R. (2007) *Early Earth systems: A geochemical approach*. Blackwell, Oxford. 285 p.
Sagan, C. (1980) *Cosmos,* Random House: New York.
Schopf, J. William (1999) *Cradle of Life*. Princeton University Press: Princeton.
Schrodinger, E. (1967) *What is life?* Cambridge University Press, London.

References

Angert, A., Cappa, C. D. and DePaolo, D. J. (2004) Kinetic O-17 effects in the hydrologic cycle: indirect evidence and implications. *Geochim. Cosmochim. Acta* **68**, 3487–3495.
Bertola, F. (1992) Seven centuries of astronomy in Padua, in *From Galileo to the Stars,* Biblos Edizione, Cittadella (PD), Italy, pp.71-74.
Bertola, F. (1995) *Imago Mundi*, Biblos, Cittadella PD, Italy, p. 141.
Blanc, M., and the LAPLACE consortium (2008) LAPLACE: a mission to Europa and the Jupiter System, Astrophysical Instruments and Methods, in press. Berlin A digital version can be consulted in http://www.ictp.it/~chelaf/ss186.html; A full list of team members is available at http://www.ictp.it/~chelaf/ss164.html.
Chadha, M. (2001) Private communication.
Chela-Flores, J. (2006) The sulfur dilemma: are there biosignatures on Europa's icy and patchy surface? *International Journal of Astrobiology* **5**, 17-22.
Chela-Flores, J. (2010) Instrumentation for the search of habitable ecosystems in the future exploration of Europa and Ganymede. *International Journal of Astrobiology* **9**, 1-8.
Clayton, R. N., Grossman, L. and Mayeda, T. K. (1973) *Science* **182**, 485-488.
Darwin, C. (1859) *The origin of species by means of natural selection or the preservation of favored races in the struggle for life.* London, John Murray. Reprinted by Oxford World's Classics (1998), G. Beer (ed.), Oxford University Press.
Farquhar, J., Johnston, D.T., Wing, B.A., Habicht, K.S., Canfield, D.E., Airieau, S. and Thiemens, M.H. (2003) Multiple sulphur isotopic interpretations of biosynthetic pathways; implications for biological signatures in the sulphur isotope record. *Geobiology* **1**, 27–36.
Fischer, E. P. and Lipson, C. (1988) *Thinking about Science (The life of Max Delbruck).* W.W. Norton, New York.
Gatti, H. (1999) *Giordano Bruno and Renaissance Science,* Cornell University Press, Ithaca.
Kaplan, I.R. (1975) Stable isotopes as a guide to biogeochemical processes. *Proc. R. Soc. Lond.* B **189**, 183-211.
Keynes, R.D. (1998) The Theory of Common Descent, in Chela-Flores, J. and Raulin, F. (eds.), *Exobiology: Matter, Energy, and Information in the Origin and Evolution of Life in the Universe,* Kluwer Academic Publishers, Dordrecht, The Netherlands, pp. 35-49.
Levy, B. -H. (2009) Albert Camus: Perchè I sartriani devono dar ragione al filosofo artista. Corriere della Sera 30 December, pp. 38-39.

Margulis, L. and Sagan D. (1987) *Microcosmos Four Billion Years of Evolution form Our Microbial Ancestors,* Allen & Unwin, London, pp. 99-114.

Mayr, E. (1982). *The Growth of Biological Thought. Diversity, Evolution and Inheritance.* Belknap Press/Harvard University Press, Cambridge, Mass.

Mojzsis, S. J., Arrhenius, G., McKeegan, K.D., Harrison, T.M., Nutman, A.P. and Friend, C.R. (1996) Evidence for life on Earth before 3,800 million years ago, *Nature* **384,** 55-59;

Moorbath, S. (1995) Age of the oldest rocks with biogenic components, in Ponnamperuma, C. and Chela-Flores, J. (eds.). (1995) *Chemical Evolution: The Structure and Model of the First Cell,* Kluwer Academic Publishers, Dordrecht, The Netherlands. pp. 85-94.

Navarro-González, R (April 1998) In memoriam Cyril Andrew Ponnamperuma 1923-1994. *Origins of life and evolution of the biosphere* **28** (2), 105–8.

Ono, S., Wing, B. A., Johnston, D., Farquhar, J. and Rumble, D. (2006) Mass-dependent fractionation of quadruple sulfur isotope system as a new tracer of sulfur biogeochemical cycles. *Geochim. Cosmochim. Acta* **70**, 2238–2252.

Oro, J. (1961) Comets and the formation of biochemical compounds on the primitive earth. *Nature* **190,** 389-390.

Ponnamperuma, C. (1995) The origin of the cell from Oparin to the present day, Ponnamperuma, in C. and Chela-Flores, J. (eds.), *Chemical Evolution: The Structure and Model of the First Cell.* Kluwer Academic Publishers, Dordrecht, The Netherlands, pp.3-9.

Russell, B. (1991) *History of Western Philosophy,* G. Allen & Unwin, London, England.

Schidlowski, M. (1995) Early Terrestrial Life: Problems of the oldest record, in Chela-Flores, J., M. Chadha, A. Negron-Mendoza, and T. Oshima (eds.). *Chemical Evolution: Self-Organization of the Macromolecules of Life,* A. Deepak Publishing, Hampton, Virginia, USA, pp. 65-80

Schidlowski, M., Hayes, J.M. and Kaplan, I.R. (1983) Isotope inferences of ancient biochemistries: carbon, sulfur, hydrogen, and nitrogen. In: J.W. Schopf, Editor, Earth's Earliest Biosphere: Its Origin and Evolution, Princeton Univ. Press, Princeton, pp. 149–186.

Schopf, J.W. (1993) Microfossils of the Early Archean Apex Chert: New Evidence of the Antiquity of Life. *Science* **260,** 640-646.

Schopf, J. W. (1999) *Cradle of Life the discovery of Earth's earliest fossils,* Princeton University Press, Princeton, p. 112.

Schrodinger, E. (1967) *What is life*? Cambridge University Press.

Shen, Y., Buick, R., Canfield, D. E. (2001) Isotopic evidence for microbial sulphate reduction in the Early Archaean era. Nature 410, 77 – 81.

Strauss, H. and Beukes, N. (1996) Carbon and sulfur isotopic compositions of organic carbon and pyrite in sediments from the Transvaal Supergroup, South Africa. *Precambrian Res.* **79**, 57–71.

Van Zuilen, M. (2006) Traces of Early Life–A Geochemist's View. Société Française d'Astobiologie. http://www.exobiologie.fr/index.php/vulgarisation/geologie-vulgarisation/traces-of-early-life-a-geochemist's-view/

Walsh, M. M. and Westall, F. (2009) Disentangling the microbial fossil record in the Barbeton Greenstone Belt: A cautionary tale. "From Fossils to Astrobiology", Ed. J. Seckbach, Cellular Origins, Life in Extreme Habitats and Astrobiology, Springer, Dordrecht, The Netherlands, pp. 27-37.

Wolpert, L. (1990) *New Scientist* **125** (No. 1707) 64.

THE BOOK OF LIFE

PART 1: THE ORIGIN OF LIFE IN THE UNIVERSE

The Kepler mission is a NASA space observatory in heliocentric orbit trailing the Earth. This is a major instrument in the search for possible habitats where life could prosper in terrestrial-like planets. These exoplanets can be discovered by looking for a periodic dimming in stellar brightness, caused by a planet passing in front of a star.

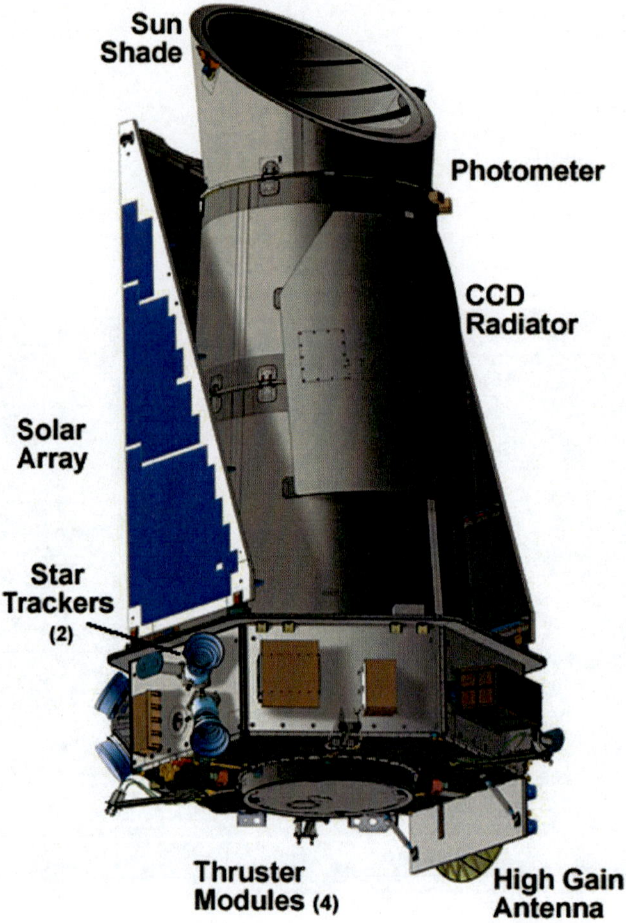

The components of the Kepler telescope. The problem of the origin of life in the universe, already hinted at by Charles Darwin in a famous letter is going past a new landmark. Beyond the pioneering footsteps in the chemistry laboratories of Oparin, Calvin, Miller, Fox, Ponnamperuma and Oro we are entering a new phase of discoveries that to a large extent will be modified by Kepler, as the question of the origin of life in the universe has shifted to developing the right instrumentation in order to reach deeper insights into our own origins.

1

From cosmic to chemical evolution

After the first large step taken by the pioneers of astrobiology described in the Introduction, a second large step is still ongoing in the development of origin of life studies. It consists of taking the subject from the laboratories of the 'card-carrying' organic chemists, where Oparin had introduced it, into the domain of the life, earth and space scientists. We discuss the progress in chemical evolution in the context of the space sciences.

The reader is advised to refer especially to the following entries in the Illustrated glossary: *Age of the universe, angular momentum, Chandrasekhar Limit, chemosynthesis, Einstein, evolution (cosmic), habitable zone, Hertzprung-Russell diagram, Hoyle (for the introduction of the big bang model of cosmic evolution), Hubble (Edwin), hydrocarbons, Kuiper Belt, Lagrangian points, Orion Nebula, Kepler mission, photosphere, Planck Space Observatory, pulsar (neutron star), RNA, spectral type of a star, supernova, Wilson (Robert), WMAP.*

1.1 Beginnings of a rational cosmology based on physical constraints

The heliocentric theory of Nicholas Copernicus was published posthumously in 1543. During his stay at the University of Padua from 1501 to 1503, Copernicus had been influenced by the sense of dissatisfaction of the Paduan instructors with the systems of both Aristotle and Ptolemy. The new century was a time in which scientists were requiring a sense of simplicity that classical philosophy could no longer provide. The main work of Copernicus, "*On the revolution of the heavenly spheres*", was published a few months before his own death in 1543. In this influential work Copernicus placed the Sun at the center of the Solar System and inserted both the Sun and planets inside a sphere of fixed stars. Copernicus deliberately appeared to accept the tenets of Aristotle's universe, as proposed in his book, a cosmos inserted within a system of revolving planets, surrounded by a sphere of fixed stars. There is ample evidence that the Polish scientist knew of the heliocentric hypothesis of Aristarchus. Indeed, Bertrand Russell argues that the almost forgotten hypothesis of the Ionian philosopher did encourage Copernicus, by finding ancient authority for his innovation.

With Thomas Digges (1546-1595), the English mathematician and astronomer we are in the presence of a major step forward in the ascent of man towards a true understanding of his position in the universe. His main contribution to the question of

J. Chela-Flores, *The Science of Astrobiology: A Personal View on Learning to Read the Book of Life*, Cellular Origin, Life in Extreme Habitats and Astrobiology 20, DOI 10.1007/978-94-007-1627-8_1,

the plurality of worlds was to depart from the Aristotelian model of fixed spheres of stars. He translated part of "*On the revolution of the heavenly spheres*" and introduced the idea of an infinite universe with the stars at varying distances in an infinite space. From Digges work we can anticipate the clearer statement later made by Bruno of a universe in which a multitude of worlds were rotating around stars not unlike our own Sun. Copernicus did not anticipate this concept.

We have seen in the Introduction that with Giordano Bruno we reach the closing of a cycle in the gradual maturing of the science of cosmology that had begun with Anaximander 2000 years earlier. Bruno is best remembered for intuitively going beyond the heliocentric theory of Copernicus, which still maintained a finite universe with a sphere of fixed stars. From the point of view of astrobiology his anticipation of the multiplicity of worlds has been amply confirmed since 1995. In that year the first detection of an extrasolar planet was announced, as we shall discuss more fully in Chapter 10. But what is more significant regarding Bruno's intuition is that he also conjectured that such worlds would be inhabited by living beings, suggesting the possibility of a second Genesis in a modern cosmological context. His ideas went further than those of Digges. Astrobiology, the science of life in the universe is just concerned with this key question, still without a convincing answer. It can be said that one of Bruno's major innovation, anticipated by Digges, was his refusal to accept that the Solar System is contained in a cosmos that is bounded by a finite sphere of fixed stars.

To sum up, Bruno proposes an infinite cosmos, populated by an infinite number of worlds. In his third Oxford dialogue (cf., Introduction) Bruno developed valuable concepts that matured in Bruno's writings long before the science of astrometry allowed the concept of a multitude of solar systems to be brought within the scientific domain at the very closing of the 20th century.

1.2 Cosmological models

For our first topic we must return to the first instants of cosmic expansion (Coyne, 1996). Edwin Hubble discovered in 1929 that large groups of stars (galaxies), which in addition contain interstellar matter and nebulae, were moving away from us and that the velocity V of recession of such galaxies is proportional to their distance d from us. (This is known as Hubble's Law.) This implies if we assume isotropy that the universe as a whole is expanding. The distances that measure this phenomenon are normally expressed in megaparsecs (1 pc = 3.26 ly; 1 ly, in turn, denotes the distance that light travels in a year; 1 megaparsec, Mpc = 10^6 pc). In other words, $V = H_0 d$. The constant of proportionality, which is known as the Hubble constant H_0 is, consequently, given by the ratio of the speed of recession of the galaxy to its distance; H_0 represents quantitatively the current rate of expansion of the universe. Determination of the value of H_0 (expressed in units of kilometers per second per million *parsec,* megaparsec, Mpc) has been difficult and the true value is still the subject of controversy (Kennicutt Jr., 1996). A growing consensus, including data from a large sample of relatively near supernovae leads to the value 74.2 ± 3.6 kms^{-1} Mpc^{-1} (Freedman and Madore, 2010).

The current estimates of the Hubble constant in the standard cosmological model (cf., the Friedmann model below) of an expanding universe implies an age of the universe in a range of over 10 thousand million years (Gyr), the precise values depending on the particular assumption we may adopt for the matter density present in the universe. An alternative approach measures the abundance of radioactive elements found in stars. Astronomers search for very old stars and estimate the abundance of a given radioactive element. From nuclear physics the time of decay of the element allows an estimate of the age of the galaxy itself. This method uses in particular the elements uranium and thorium. The current estimate puts the age of the universe $t_0 = 13.7 \pm 0.5$ Gyr (Rago, 2000; Sneden, 2001; Freedman and Madore, 2010).

Unfortunately, this state of affairs raises some difficulties, as there are globular clusters in our galaxy whose age is estimated to be 13-17 Gyr. Ours is a typical spiral galaxy, not unlike others that are well known form the images provided by the Hubble Space Telescope (cf., Fig. 1.1).

Fig. 1.1 The large spiral galaxy NGC 4414 image completed in 1999

On the other hand, life on Earth extends back only to some 4 Gyr before the present (BP), while the age of the solar system is 4.6 Gyr BP. From the point of view of the origin of life on Earth we therefore do not need to follow the cosmological arguments in detail. The chemical evolution scenario faces no particular difficulties with the above values of H_0. The theory of gravitation formulated by Albert Einstein is known as General Relativity (GR). Cosmological models may be discussed within the context of this theory in terms of a single function R of time t. This function may be referred to, quite appropriately, as a 'scale factor', a measure of the size of the universe. Sometimes, when referring to the particular solution the expression 'radius of the universe' may be preferred for the function *R*. As the universal expansion sets in, *R* is found to increase in a model that assumes homogeneity in the distribution of matter (the 'substratum'), as well as isotropy of space. The functional dependence of *R,* as a function of time *t,* is a smooth increasing function for a specific choice of two free parameters, which have a deep meaning in the GR theory of gravitation, namely, the curvature of space and the cosmological constant.

The Russian mathematician Alexander Friedmann found the functional behavior of the scale factor R in 1922. This solution is also attributed to Howard Percy Robertson and A.G. Walker for their work done in the 1930s. Such a (standard) model is referred to as the Friedmann model. In fact, *R* is inversely proportional to the substratum temperature T.

Hence, in the context of the expansion of the universe we have been discussing, since *R* is also found to increase with time *t* (cf., the previous paragraph), *T* decreases; this model implies, therefore, that as *t* tends to zero (the 'zero' of time) the value of the temperature *T* is large. (The temperature goes to infinity, as t tends to zero.) In other words, the Friedmann solution suggests that there was a 'hot' initial condition. As the function *R* represents a scale of the universe (in the sense we have just explained), the expression "big bang", due to Sir Fred Hoyle, has been adopted for the beginning of the universe in the Friedmann model (cf., Glossary "Hoyle"). The almost universal acceptance of big bang cosmology is due to its observational support.

1.3 The cosmic microwave background

The big bang model tells us that as time *t* increases, the universe cools down to a certain temperature, which at present is close to 3° K. This discovery took place in 1964 by Arno Penzias and Robert Wilson; they provided solid evidence that the part of the universe that surrounds us is presently illuminated by "$T = 3°$ K" radiation, the cosmic background, but since it has a typical wavelength of about 2 mm, it is referred to as the cosmic microwave background, CMB.

The CMB may be confidently considered to be a cooled remnant from the hot early phases of the universe. It has an "isotropic" distribution, in other words, its temperature does not vary appreciably independent of the direction in which we are observing the celestial sphere (the accuracy of this statement is 10 parts per million, ppm). The isotropy is a consequence firstly of the uniformity of cosmic expansion, secondly, of its homogeneity when its age was 300,000 years and temperature of 3,000K. On the other hand, in 1992 more precise measurements of the $T = 3°$ K radiation, began to be made by means of the satellite called the Cosmic Background Explorer (COBE): When the accuracy of the isotropy was tested with more refined measurements, it was found that there was some degree of anisotropy after all - the temperature did vary according to the direction of observation (one part in 100,000). This fact is interpreted as evidence of variations in the primordial plasma, a first step in the evolution of galaxies. Further accuracy in understanding the deviations from isotropy of the CMB (and hence a better understanding of the early universe) can be expected in the next few years. The Microwave Anisotropy Probe (MAP)—an initiative of the National Aeronautics and Space Administration (NASA)— extended the precise observations of the CMB to the entire sky. MAP was in a solar orbit of some 1,5 million kilometers. This was extended by the European Space Agency (ESA) firstly to WMAP (cf., Glossary) and subsequently to the Planck mission, which we have already encountered in the introductory page of their chapter. This mission has made accurate measurements of the CMB over a broad range of far-infrared wavelengths (cf., Fig. 1.2).

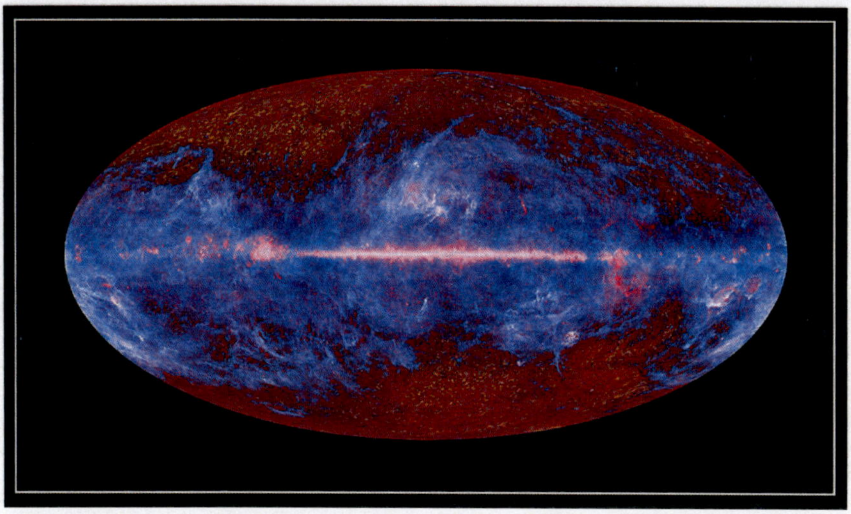

Fig. 1.2 This is the first all-sky image from the Planck telescope. It was received in 2009 to survey the "oldest light" in the cosmos. It shows the CMB as well as other light sources, such as gas and galactic dust from the Milky Way. Further processing by Planck will allow elucidating further the contribution from the CMB

This will enable the extraction essentially all of the information that can be obtained from the CMB. Planck is expected to provide answers to many fundamental questions about the early history and evolution of our universe Planck was designed to image the anisotropies of the Cosmic Background Radiation Field over the whole sky, with unprecedented sensitivity and angular resolution. Planck is the source of information that is relevant to several cosmological and astrophysical questions. For example, testing theories of the early universe and the origin of cosmic structure. Planck was launched in May 2009. After 50 days, as foreseen, Planck entered its final orbit around the Second Lagrangian point of the Sun–Earth system (L2), at a distance of 1.5 million kilometers from Earth. In January 2010, ESA's advisory bodies approved an extension of Planck operations to continuously acquire high-quality science data until at least the end of 2011.

1.4 The growth of studies on the origin of life

The subject of astrobiology has come of age due to the space missions of the last 40 years: Mariner, Apollo, Viking, Voyagers and particularly the most recent ones, Galileo, which is currently sending valuable information on the Jovian system and Cassini-Huygens that will soon yield its first results (cf., Chapters 8 and 9). Further missions are being planned. New areas of research have come into existence as a consequence of this change of emphasis, such as *planetary protection,* whose aim is to prepare in advance for the sample-return missions from Mars and Europa; in the case of

Mars, this event can only take place in the second decade of the present century. But by restricting the effort to the satellites of Mars the sample-return becomes feasible, for example the Phobos-Grunt mission (cf., Table 1.1).

Table 1.1 Some of the space missions. In Chapter 7, Tables 7.1 and 7.7 we mention additional missions that have explored Mars

Mission	*Launch*	*Mission end*
Mars Global Surveyor	7/11/1996	31/1/2000
Mars Pathfinder	4/12/1996	10/3/1998
Cassini-Huygens Mission Orbiter of Saturn and probe into Titan atmosphere	15/10/1997	-
Mars Global Surveyor-Climate Orbiter	10/12/1998	Mission lost
Galileo Orbiter of Jupiter and probe into the Jovian atmosphere		
Mars Global Surveyor-Polar Lander	8/1/1999	3/12/1999 Mission lost
Stardust (Comet dust sample return)	6/2/1998	15/1/2006
Rosetta (ESA comet lander)	1/7/2003	1/7/2013

(Based on Meyer, 1996 with more recent data)

It should be stressed that there are two kinds of planetary protection:

• Protecting the other planet from us. This has been a requirement in previous missions such as the Viking mission to Mars in 1976.

• Protecting us from possible exogenous contamination. This is needed only in sample-return missions, as discussed here.

The experiments and sample retrieval from Europa are in the process of discussion. Indeed, such matters were discussed at an early NASA and National Science Foundation (NSF) meeting in 1996 (*Europa Ocean Conference,* San Juan Capistrano, California). Donald DeVincenzi at NASA-Ames, John Rummel at the Marine

Biological Laboratory and Margaret Race at the SETI Institute have pioneered planetary Protection.

A second discipline that has arisen from the exploration of the Solar System is *comparative planetology*, a new subject that allows us to learn more about phenomena in each specific planet or satellite by having several examples to examine. Since the advent of the discovery of planets around other stars (cf., Chapter 10), comparative planetology has gone beyond the frontiers of our single Solar System. A good local example is the phenomenon of volcanism, which is remarkably different in the Jovian satellite Io as compared to what is observed on Earth. These studies are needed in the formulation and development of devices, which in principle may allow the search for Solar System relics of the earliest stages of the evolution of life. A very brief summary of some of the relevant missions of exploration of the Solar System that has allowed us to reach a mature outlook on comparative planetology are listed in Table 1.1, but Chapters 7 – 9 will put some of these efforts in their proper context.

1.5 Origin of the elements: from the big bang to the interior of stars

As we have seen above, according to the big bang model, initially the temperature of the primordial matter was so elevated that atomic, nuclear or subnuclear stability was impossible. In less than one million years after the beginning of the general expansion, the temperature T was already sufficiently low for electrons and protons to be able to form hydrogen atoms. At an earlier epoch there would have been too much thermal energy for the electromagnetic force to be able to bound up nuclei and electrons.

Up to that moment these elementary particles were too energetic to allow atoms to be formed. Once 'recombination' of electrons and protons was possible, due to falling temperatures, thermal motion was no longer able to prevent the electromagnetic interaction from forming hydrogen atoms. This is the 'moment of decoupling' of matter and radiation. At this stage of universal expansion the force of gravity was able to induce the hydrogen gas to coalesce into stars and galaxies.

Hans Bethe proposed a series of nuclear reactions in the interior of stars. From 1935-1938, he studied nuclear reactions and reaction cross-sections, carbon-oxygen-nitrogen cycle, leading to his important contribution to stellar nucleosynthesis. His aim was to understand the nuclear reactions that are responsible for the luminosity of the Sun and other stars (cf., Fig. 1.3).

The subsequent development of this theory explains how all the elements we know can be built up through nuclear processes inside stars, and by giant stellar explosions. The underlying phenomenon consists of high-energy collisions between atomic nuclei and elementary particles that have been stripped off their corresponding atoms, or even nuclei, due to the presence of the enormous thermal energy in the core of the star.

At such high temperatures nuclear fusion may occur. In other words, fusion of two light atomic nuclei into a single heavier nucleus can take place at high temperatures by a collision of the two interacting particles, with the consequent release of a relatively large amount of energy. In the interior of stars such reactions — thermonuclear — involve nuclear fusion, in which the reacting bodies have sufficient energy to initiate and sustain the process.

Fig. 1.3 Hans Albrecht Bethe (1906 – 2005) the German-American physicist, and Nobel laureate in physics for his work on the theory of stellar nucleosynthesis during the Symposium on Contemporary Physics, Trieste, June 1968

One example is provided by a series of nuclear reactions that induce hydrogen nuclei (essentially single 'elementary' particles called protons) to fuse into helium nuclei. A helium nucleus is heavier than the proton; in fact, it consists of two protons and two slightly more massive particles called neutrons. But the total mass of the helium nucleus is actually a tiny bit less than the sum of the masses of the four particles. In the process of forging the nucleus from its components, the missing mass is converted directly into energy, following Einstein's famous equation, $E = mc^2$, where E denotes energy, c is the velocity of light and m is the mass. This is the energy that produces sunlight. This process, in addition, releases other particles and some energy. After a long phase (measured in millions of years) dominated by such conversion, or 'burning' of hydrogen into helium, the star evolves: its structure becomes gradually that of a small core, where helium and the heavier elements accumulate. Since both temperature and density of the core increase, the pressure balance with the gravitational force is maintained. The star itself increases in size. It becomes what is normally called a "red giant", because at that stage of their evolution they are changed into a state of high luminosity and red color. During this long process, from a young star to a red giant, the elements carbon and oxygen, and several others, are formed by fusing helium atoms (cf., Table 1.2).

1.6 Supernovae: the source of biogenic dust

As we shall see later, solar systems originate out of interstellar dust, which can be considered mostly as 'biogenic dust', namely dust constituted mainly out of the fundamental elements of life such as carbon (C), hydrogen (H), oxygen (O) and nitrogen (N), and a few others. It is for this reason that the late stage of stellar evolution is so relevant. In fact, just before a star explodes into its supernova stage, all the elements that have originated in its interior out of thermonuclear reactions are expelled, thus contributing to the interstellar dust. At a later stage in the stellar evolution the expanding dust and gas forms typical 'planetary nebulae (cf., Fig. 1.4).

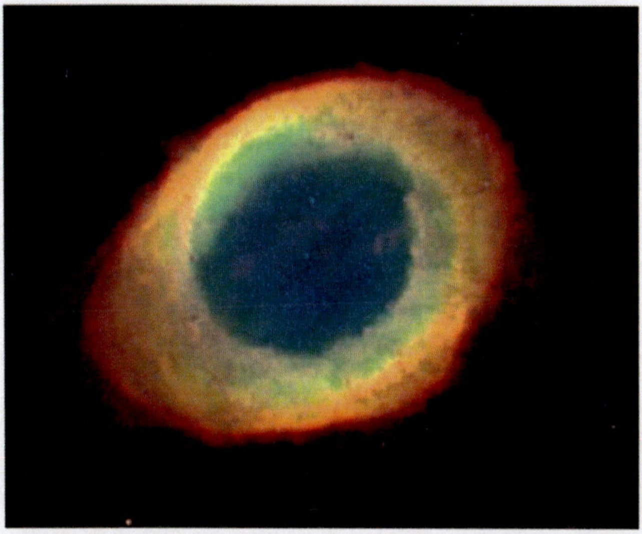

Fig. 1.4 The Ring Nebula (M57). The figure shows gas and dust that has been cast off thousand of years after the death of the star

Let us consider next some of the details of these 'life-generating' astrophysical events. Indeed, stars whose mass is similar to that of the Sun remain at the red giant stage for a few hundred million years. The last stages of burning produce an interesting anomaly: the star pushes off its outer layers forming a large shell of gas much larger than the star itself. This structure is called a planetary nebula. The star itself collapses under its own gravity, compressing its matter to a degenerate state. The laws of microscopic physics (called quantum mechanics, cf., Illustrated Glossary) eventually stabilize the collapse. This is the stage of stellar evolution called a "white dwarf ". The stellar evolution of stars more massive than the Sun is far more interesting: After the massive star has burnt out its nuclear fuel (in the previous process of nucleosynthesis of most of the lighter elements, a catastrophic explosion follows in which an enormous amount of energy and matter is released (cf., Table 1.2 and Fig. 1.5).

It is precisely these 'supernovae' explosions that are the source of enrichment of the chemical composition of the interstellar medium. This chemical phenomenon, in turn, provides new raw material for subsequent generations of star formation, which in some cases leads to the production of planets.

Table 1.2 Chemical composition of the Sun.

Element	*Symbol*	*Percentage*
Hydrogen	H	87.000
Helium	He	12.900
Oxygen	O	0.025
Nitrogen	N	0.020
Neon	Ne	0.030
Carbon	C	0.01
Magnesium	Mg	0.003
Silicon	Si	0.002
Iron	Fe	0.001
Sulfur	S	0.001
Others		0.038

Adapted from (Ponnamperuma,1989)

Fig. 1.5 The Crab Nebula is a supernova remnant in the constellation of Taurus. The nebula corresponds to a bright supernova recorded by Chinese and Arab astronomers in 1054

Several types of supernovae exist. Types I and II can be triggered in one of two ways, either turning off or suddenly turning on the production of energy through nuclear fusion. After the core of an aging massive star ceases generating energy from nuclear fusion, it may undergo sudden gravitational collapse into a neutron star or black hole, releasing gravitational potential energy that heats and expels the star's outer layers. In fact, the Chandra X-ray Observatory has discovered the youngest black hole known to exist in our neighborhood. The 30-year-old black hole, known as Supernova 1979C (SN 1979C), is about 51 million light years from Earth and occurred in galaxy M-100 (cf., Fig. 1.6). Chandra has revealed a bright source of X-rays that has remained steady during observation from 1995 to 2007. This data suggests that the object is a black hole being fed either by material falling into it from the supernova or a binary companion.

Fig. 1.6 Composite image shows a supernova within the galaxy M100

This is the nearest example where the birth of a black hole has been observed. SN 1979C is believed to have formed when a star about 20 times more massive than the sun collapsed. Many black holes are thought to form because of gamma-ray bursts (GRBs) that are flashes of gamma rays associated with extremely energetic explosions in distant galaxies. They are the most luminous electromagnetic events known to occur in the universe. However, SN 1979C is different because it belongs to a class of supernovas unlikely to be associated with a GRB. However, another interpretation of the X-ray emission is possible: the phenomenon could be due to a young, rapidly spinning neutron star (or "pulsar", cf., Illustrated Glossary). This would make the object in SN 1979C the youngest and brightest example of such a pulsar and the youngest known neutron star.

An alternative triggering mechanism for a supernova is due to a white dwarf star that may accumulate sufficient material from a stellar companion (either through accretion or via a merger) to raise its core temperature enough to ignite carbon fusion, at which point it undergoes runaway nuclear fusion, completely disrupting it (Mazzali *et al.*, 2007). Stellar cores whose furnaces have permanently gone out collapse when their masses exceed the Chandrasekhar limit, while accreting white dwarfs ignite as they approach this limit (roughly 1.38 times the mass of the Sun). In a typical Type II supernova (SNII) the newly formed neutron core has an initial temperature of about 6000 times the temperature of the Sun's core. Neutrinos carry away much of the thermal energy, allowing the formation of a stable neutron star. (The neutrons would "boil away" if this cooling did not occur.)

Late in stellar evolution stars are still poor in some of the heavier biogenic elements (such as, for instance, magnesium and phosphorus). Such elements are the product of nucleosynthesis triggered in the extreme physical conditions that occur in the supernova event itself. By this means, the newly synthesized elements are disseminated into interstellar space, becoming dust particles after a few generations of star births and deaths (cf., Chapter 2, Cosmochemistry).

1.7 Molecular clouds and circumstellar disks

As a result of several generations of stellar evolution, a large fraction of the gas within our galaxy is found in the form of clouds that originally consisted only of molecular hydrogen, but now include many of the heavier elements. The liner dimensions of these clouds can be as large as several hundred light years. The masses involved may be something in the range 10^5 to 10^6 solar masses.

It is useful to consider one example in detail. A large molecular cloud in the constellation of Orion has a few hundred thousand solar masses; its diameter is about 1500 light years. Its temperature ranges from 10 to 50 K. Within the last two million years a few stars have been born there. The most recent observations demonstrate that star formation continues there even today.

Images from the Earth-orbiting Hubble Space Telescope (HST) have shown circumstellar disks surrounding young stars. This supports the old theory of planetary formation from a primeval nebula surrounding the nascent star. In the case of our solar system this gas formation has been called the 'solar nebula'. As we shall see in the next chapter, the coplanar orbits of the planets together with the direction of their angular momentum, coinciding with the direction of the solar angular momentum, argue in favor of the hypothesis of the solar nebula.

The HST images indicate, in exact agreement with the hypothesis, the presence of opaque disks of gas and dust. What is more important, the estimated sizes of the observed circumstellar disks are compatible with the size of the Solar System. The abundance of elements in the solar nebula is an important issue that will be relevant later on. In general we may say that solar-nebula abundance should coincide to a large extent with solar values (cf., Table 1.2). There are some small differences that will be due to the nuclear-reaction processes taking place in the Sun.

In addition, there will be some differences in the abundance observed in the planets, because as the nebula evolves into a solar system the elements themselves will be partitioned according to simple physicochemical laws. In strict thermodynamic equilibrium carbon would have reacted with oxygen producing exclusively carbon monoxide (CO) in the inner solar system due to the higher thermal energy available in this region (Pollack and Atreya, 1992). The cooler parts of the outer solar nebula would have favored the formation of methane. However, the testimony of more complex forms of carbon present in comets and meteorites suggests that the parent molecular cloud had already had a condition favorable for chemical reactions that lead to the formation of hydrocarbons. Indeed, observations at radio wavelengths that can penetrate those dark molecular clouds have revealed about 100 different kinds of molecules that are formed there. This sketch of the pathway to carbon compounds and its partition during planetary formation will be very significant in the following chapters (particularly in Chapter 8): it will help us to trace the abundance of organic compounds in sites of the Solar System where life could arise. It is worthwhile to point out at this stage that in 1983 the Infrared Astronomical Satellite (IRAS) discovered some stars with excess thermal radiation not accountable by the amount that the star could emit. The result was interpreted in terms of disks of dust around these stars. The most famous example is the second brightest star of the southern celestial hemisphere constellation of "The Painter's Easel" (Pictor).

According to tradition, the star is called by the second letter of the Greek alphabet and by a name in Latin: "Beta-Pictoris" (β-Pictoris, or β-Pic, at a distance of 59 light years). More recently, a highly unusual example of one of these circumstellar disks has been obtained with an infrared camera (the Near Infrared Camera and Multi-Object Spectrometer (NICMOS) mounted on the HST (Schneider *et al.,* 1999). The disk is a ringlike structure with characteristic width of 17 astronomical units (the distance equivalent to the radius of the Earth's orbit: it is abbreviated as AU); the inner and outer radii are respectively about 55 AU and 110 AU. This suggests that the constraint that distorts the disk might be the presence of planets. This is remarkable since the star (catalogue name HR 4796A) at the center of the ring is young (8 million years, Myr), at a distance some three and a half times further away than the star β-Pic. (The precise position had been determined by HIPPARCOS, the European astrometric satellite, launched in 1989 and used for 4 years.) Both HR 4796A and β-Pictoris are blue-white stars (technically called type A) like Vega, the brightest star in the constellation Lyra, well known in the night skies of the northern hemisphere. The dust distribution has been found to be analogous to the Kuiper Belt.

1.8 Stellar and biological evolution

We have already seen that stars evolve as nuclear reactions convert mass to energy. In fact, stars such as our Sun follow a well-known pathway along a *Hertzprung-Russell* (HR) diagram (cf., Fig. 1.7). The origin of this diagram can be traced back to work that begun in 1911 by the Danish astronomer Ejnar Hertzsprung and independently by the American astronomer Henry Norris Russell.

They observed many nearby stars and found that the plot of their luminosity (i.e., the total energy of visual light radiated by the star per second), and surface temperatures, the plot exhibits a certain regularity: Indeed, the stars lie on the same curve in a diagram (the HR diagram), whose axes are the two parameters considered by the above-mentioned early 20th-century astronomers. Such stars are called *main sequence* stars that lie on a diagonal from the upper left of the HR diagram (represented by bright stars) to the lower right (gathering cool stars). The set of large, cool, stars turn up in the upper right, and the white dwarfs lie in the lower left. As can be appreciated in Fig. 1.6 the Sun lies near the middle of the main sequence. It is interesting to remark that the surface temperature of the star is indicated by its "Planckian" radiation, which is measured by its spectral type, or color index. (This important parameter - the surface temperature - is indicated by what is technically known as its 'Planck radiation', the name given in honor of the Physics Nobel Laureate Max Planck, to the distribution with wavelength of the radiation emitted by the star at various temperatures.

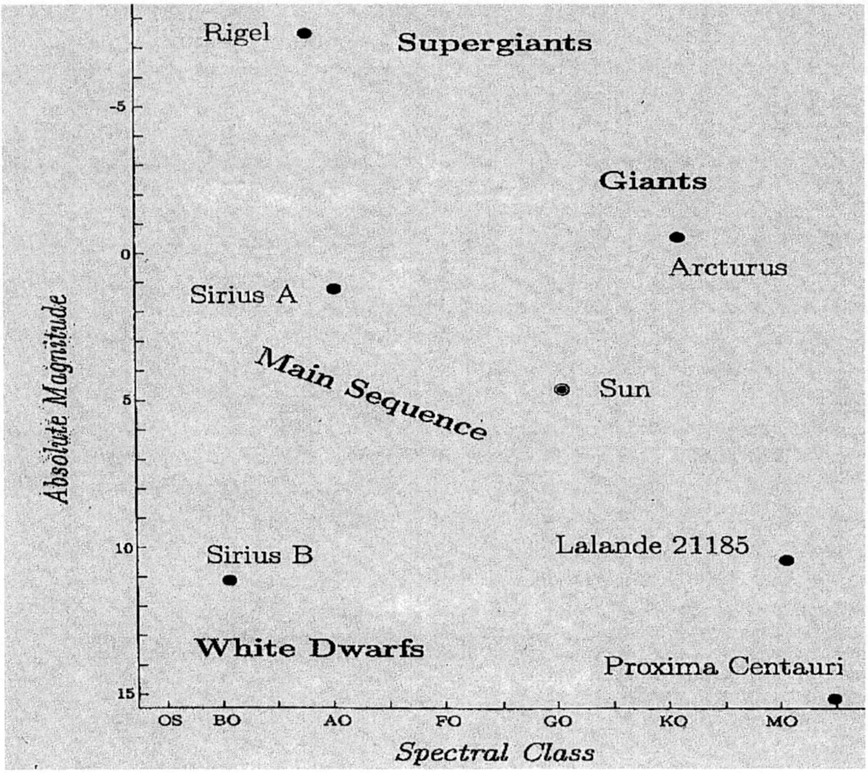

Fig. 1.7 The *Hertzprung-Russell* (HR) diagram. The coordinate axes are also proportional to luminosity (visual light with respect to the Sun and surface temperature, instead of the more astronomical terms of absolute magnitude and spectral class respectively.

We may ask: How do stars move on the HR diagram as hydrogen is burnt? Extensive calculations show that main sequence stars are funneled into the upper right hand of the HR diagram, where we find red giants of radii that may be 10 to 100 times the solar radius. In the lower part of the diagram we find dwarf stars. Stellar evolution puts a significant constraint on the future of life on Earth, since the radius of the Earth's orbit is small. (Since the eccentricity of the Earth orbit is small, we may speak to a good approximation of a circular orbit, instead of an elliptical one. On the other hand, the theory of stellar evolution tells us that the radius of the Sun is bound to increase as it evolves off the main sequence in the HR diagram.

In the case of the Sun the expected growth of the solar radius will be such that the photosphere will reach the Earth's orbit, thus eventually ending life on Earth. The current estimate is that life may continue on Earth for another 4-5 Gyr before the Earth's orbit is no longer in what we may call a 'habitable zone'. However, the implication of the remaining period of habitability of the Earth is profound for our species, as suggested in the next section.

According to the standard view of paleoanthropology, the species to which humans belong is referred to as *Homo sapiens* in order to distinguish us from an independent hominid line that became extinct some 40,000 years ago, i.e., *Homo neanderthalis.* In fact, humans have evolved in less than a few million years since the last common ancestor of the hominids. This common ancestor (technically known as a hominoid) must have lived just some 5 Myr BP. (Cf. Chapter 5, "evolution of the hominoids".) Such a fast evolutionary tempo has occurred within about 1/1000 of the geologic time that is still available for life on Earth.

Regarding our past, it is possible to link brain development and the origin of language with natural selection (cf. Chapter 5). It is remarkable that the growth of the brain of our ancestors took place in the relatively short time of a few million years. Indeed, the australopithecine lived 3-4 Myr BP, had a cranial capacity just less than 400 cm^3. Today the brain of *H. sapiens* is more than 300% of that capacity. Furthermore, according to the estimate of the previous paragraph, our brain will still be subject to evolutionary pressures for a very long time. In fact, the normal evolution of the Sun will allow life on Earth for a considerable time: life may persist for over a thousand times as long as the period in which natural selection brought the primitive brain of the earliest hominids to that of man.

Science is at present unable to extrapolate the previous growth of the human brain into the future, but this has not deterred fiction writers from a model for a future speciation event. For instance, the English science-fiction writer Herbert George Wells imagined future speciation events of humanity, which he called *eloi* and *morlock,* species of the genus *Homo* (Wells, 1996).

But, returning to the facts, not fiction, at present, from a combined consideration of cosmic, stellar and biological evolution, it is not clear how contemporary humans will continue to evolve. The role natural selection will play in the future of humanity is no longer clear. As we mentioned above, life on Earth is expected to continue for not more than four billion years. (Once again, we have to emphasize that the Earth will remain only for a finite period of time within the habitable zone of the solar system, since the Sun will continue its own evolution, according to what we know about stellar evolution, as explained above when we discussed the HR diagram.) Up to the present, however, the role of natural selection has been easier to identify, since for the major part of the existence of our species, technology has not played a significant role in our survival. Humans, which at one time would not have survived to pass on their genes to progeny, now survive due to an inevitable result of our culture: the development of medicine. It could also be argued that something has changed! Our species is overcrowding planet Earth. The absence of the proliferation of individual species to such a degree was one of the factors that led Charles Darwin to his original insight (Darwin, 1859). The future of Man is still an open problem (Pritchard, 2010).

Supplementary Reading

Cairns-Smith, A.G. (1985) *Seven Clues to the Origin of Life,* Cambridge University Press, London.

De Duve, C. (1995) *Vital Dust.Life as a Cosmic Imperative,* Basic Books, New York.

De Grasse Tyson, N. (2007) *Death by a Black Hole and Other Cosmic Quandries* W.W. Norton & Company, New York.

Delsemme, A. (1998) *Our Cosmic Origins From the Big Bang to the Emergence of Life and Intelligence,* Cambridge University Press,London.

Dyson, F. (1985) *Origins of life,* Cambridge University Press, London.

Greenberg, J.M., Mendoza-Gomez, C.X. and Pirronello, V. (Eds.) (1993) *The Chemistry of Life's Origins.* Kluwer Academic Publishers, Dordrecht, The Netherlands.
Weinberg, S. (1977) *The first three minutes,* Fontana/Collins, London.

References

Coyne, S. J., G. (1996) *Cosmology: The Universe in Evolution,* in Chela-Flores and Raulin (1996), loc. cit., pp. 35-49.

Darwin, C. (1859) *The origin of species by means of natural selection or the preservation of favored races in the struggle for life,* John Murray/Penguin Books, London, 1968, p. 455.

Freedman, W. L. and Madore, B. F. (2010) The Hubble Constant. *Annu. Rev. Astro. Asttrophys.***48**, 673-710.

Kennicutt Jr., R. C. (1996) *Nature* **381,** 555-556. Subsequent work with the HST (Wendy Freedman and co-workers) tends to confirm this value of H_0 .

Mazzali, P. A., K. Röpke, F. K., Benetti, S., Hillebrandt, W. (2007) A Common Explosion Mechanism for Type Ia Supernovae. *Science* **315**, 825–828.

Meyer, M. A. (1996) The search for life and its origins: Why NASA?, Poster 3, 8th ISSOL Meeting. 11th International Conference on the Origin of Life, Orléans, France, July 8-13, Book of Program and Abstracts, p. 63.

Pollack, J. B. and Atreya, S. K. (1992) *Giant planets: Clues on Current and Past Organic Chemistry in the Outer Solar System,* in G.C. Carle, D.E. Schwartz, and J.L. Huntington, (eds.), *Exobiology in Solar System Exploration,* NASA publication SP 512. p.96;

Ponnamperuma, C. (1989) Experimental studies in the origin of life, *J. British Interplanetary Soc.* **42,** 397-400.

Pritchard, J. K. (2010) How we are evolving. *Scientific American*, October 2010. pp. 29

Schneider, G., Smith, B. A., Becklin, E. E., Koerner, D. W., Meier, R. Hines, D. C., Lowrance, P. J., Terrile, R. J., Thomson, R. I. and Rieke, M. (1999) NICMOS Imaging of the HR 4796A Circumstellar Disk, *Astrophys. Journal* **513,** L127-L130.

Sneden C. (2001) The age of the universe *Nature* **409,** 673-674.

Wells, H. G. (1996) *The Time Machine,* Phoenix, London, Abridged edition.

2

From chemical to prebiotic evolution

We have seen in Chapter 1 that the biogenic elements were formed in stellar cores and later were expelled by the host star through stellar explosions [1] *and other processes. Subsequently, they combine in the atmospheres of evolved stars to form diatomic and triatomic molecules that are to have transcendental consequences in the subsequent stages of prebiotic and biological evolution. A few examples are:* C_2*, OH, and* H_2O*, but even larger molecules have been detected in interstellar clouds.*

The reader is advised to refer especially to the following entries in the Illustrated glossary: *Accretion, amino acids, bases, carbonaceous chondrites, hydrocarbons, interstellar dust particles, Jovian planets, photosphere, silicate, terrestrial planets.*

2.1 Organic cosmochemistry

We now know from observation that, within the range of lower temperatures that occur in circumstellar and interstellar media, some key biogenic molecules are formed, such as hydrogen, ammonia, water, formaldehyde, hydrogen cyanide, cyanacetylene, carbon monoxide and hydrogen sulfide. The chemistry of interstellar clouds sets the stage for prebiotic evolution. In other words, the nine molecules listed above are sufficient for the synthesis of some of the biomolecules of life, such as the amino acids. It is reasonable to assume that the Solar System was formed out of a disk-shaped cloud of gas and dust, which we called the solar nebula in Chapter 1. Such disk-shaped clouds have been observed about other young stars. Most of the original matter of which the solar nebula was made has since been incorporated into its planets and the central star itself, namely, the Sun. Interstellar dust is the product of the condensation of heavy elements, which were themselves produced in stellar interiors, for example, carbon, a major component of interstellar dust, magnesium, silicon and iron. Later, in interstellar space they combined with elements such as oxygen to form silicate particles measuring typically 0.1 microns. Even though a small proportion of the mass of the interstellar medium is composed of solid grains, they play a significant role on the gas itself. For instance these grains are capable of absorption of stellar radiation. Stellar formation takes place in clouds that are dark and cold—on these sites dust grains play a decisive role.

J. Chela-Flores, *The Science of Astrobiology: A Personal View on Learning to Read the Book of Life*, Cellular Origin, Life in Extreme Habitats and Astrobiology 20, DOI 10.1007/978-94-007-1627-8_2,

A very significant clue as to the nature of the original solar nebula can still be inferred from the study of comets. In fact, these small bodies formed and remained at the edges of the solar nebula. They represent the primordial conditions of the nebula. As comets pass in the vicinity of the Sun and the Earth, dust particles are released as the comets heat up. Interstellar dust particles (IDP) are a major component of the Galaxy, but IDPs are a minor component within the Solar System: they are currently identified by their preferential speed, namely their relative speed of Solar System with respect to the Galaxy. Interplanetray dust that is easily collected close to the Earth, or in the high atmosphere too, has been heavily processed due to collisions, heating and other factors. The source of IDP could be from asteroids too, while dust in comets is expected to be much less processed and, hence much closer to the original status of ISD. These simple remarks underlie the importance of collecting dust directly from comets. It is not yet possible to distinguish particles from comets from those of interstellar dust. Hence, we can appreciate the current general interest in the future direct retrieval of cometary matter. The average density of particles in space is less than one particle per cubic kilometer. Nevertheless, the total number of particles is large. One can appreciate the ubiquity of these grains by the fact that light is totally absorbed by them before it traverses a distance of 1% of the diameter of the galaxy.

2.2 Precursor biomolecules in interplanetary dust

Given the importance of even the simplest molecules that are used by living organisms (biomolecules), it is appropriate to begin with comments on the origin of some of the most important ones. They may have been synthesized in the early Earth, or transported here from space. There are also many indications that the origin of life on Earth may not exclude a strong component of extraterrestrial inventories of the precursor molecules that gave rise to the major biomolecules. About 98% of all matter in the universe is made of hydrogen and helium. In fact, besides hydrogen, the other biogenic elements C, N, O, S and P make up about 1% of cosmic matter. This is a disproportionately large amount if we recall the numerous elements of the periodic table that would be included in the remaining 1% of all cosmic matter.

Some of the more refractory material (i.e., that has gone through chemical reactions at higher temperatures) has remained in the inner solar nebula and has condensed in the form of meteorites, which are called carbonaceous chondrites; these small bodies formed in the early Solar System. (Its constituents are silicates, bound water, carbon and organic compounds, including amino acids.) These witnesses of the nature of the solar nebula also contain some of the biogenic elements (cf., Table 2.1). Some comments on Table 2.1 are needed. The main lesson to be inferred from the table can be illustrated with the case of oxygen. In spite of being the most abundant element in the Earth's crust (the case of meteorites is similar to that of the Earth), oxygen represents only about 1/1000 of the hydrogen atomic abundance in solar matter.

Cosmochemistry must explain the evolution of the inventories of oxygen and other volatiles on Earth from the solar nebula, whose composition is radically different. This is also an open problem in the case of the other terrestrial planets. (Mercury, Venus, the Earth, Mars and, in spite of being the Earth's satellite, the Moon is also included in the group of the terrestrial planets.)

• The chemistry of the solar nebula is the chemistry of the eight elements H, O, C, N, Mg, Si, Fe, and S, and to a lesser extent Al, Ca, Na and P. We leave helium and neon out of this group since they are inert.

To put meteorite composition on a convenient scale, we have adopted the standard normalization, in which an abundant element in the silicates (a 'lithophile') with high melting temperature is chosen; in the present case it is Si.

Table 2.1 Selected abundance of the elements in the solar photosphere and in a carbonaceous (C) chondrite, emphasizing biogenic elements. They are an abundant type of stony meteorite, classified with respect to an analogue-the prototypical Ivuna (I) meteorite. (Chondrites are thus denoted by 2-parameters

Atomic number	*Element*	*Abundance in a CI chondrite ($Si = 10^6$ atoms)*	*Abundance in the solar photosphere ($H = 10^{12}$ atoms)*
1	Hydrogen (H)	5.91×10^7	10^{12}
6	Carbon (C)	1.01×10^7	3.98×10^8
7	Nitrogen (N)	3.13×10^6	1×10^8
8	Oxygen (O)	2.38×10^7	10^9
12	Magnesium (Mg)	1.074×10^6	3.80×10^7
14	Silicon (Si)	1×10^6	3.55×10^7
16	Sulfur (S)	5.15×10^5	1.62×10^7
26	Iron (Fe)	9×10^5	3.24×10^7

(Fegley Jr., 1993; Grevesse, 1984)

Other possibilities are Mg or Al. In the solar photosphere the standard choice is the element hydrogen (cf., Lipschutz, and Schultz, 1999). For meteorites the data are presented on a weight basis, such as in the Elsevier Table, where the results are given in ppm by weight (cf., Lof, 1987). Alternatively, the results are given on atomic basis, as we have done in Table 2.1. For the hydrogen abundance in the CI chondrite, we have used the specific value for the Orgueil chondrite (where we have equated the carbon abundance of 34,500 ppm by weight with the atomic scale value cited in Table 2.1). The abundance of biogenic elements would suggest that the major part of the molecules in the universe would be organic. In Table 2.2 we give examples of precursor biomolecules in IDPs.

2.3 Cosmic dust and comets: their role in astrobiology

Interstellar dust particles are the predominant form of the condensable elements in the galaxy, which are not in the form of star matter. The evidence supports the view that comets are aggregates of interstellar ice and dust. In fact, out of over a hundred molecules that have been detected, either by microwave or infrared spectroscopy, 75%

are organic (Oro, 1995). Once again, we see an agreement between chemical evolution experiments and observations of the interstellar medium. Some of the molecular species detected by means of radio astronomy are precisely the same as those shown in the laboratory to be precursor biomolecules.

Table 2.2 A few precursor biomolecules in interstellar dust particles

Molecule	*Formula*
Hydrogen	H_2
Water	H_2O
Ammonia	NH_3
Carbon monoxide	CO
Formaldehyde	CH_2O
Hydrogen sulfide	H_2S
Hydrogen cyanide	HCN
Cyanacetylene	HC_3N

Adapted from (Oro, 1995)

In turn, dust grains are aggregates of inorganic matter (67% of the total mass) and organic matter as well (33% of the total mass). This insight has supported the seminal conjecture that comets are responsible for life on Earth (Oro, 1961). The grains form in interstellar clouds and in gas outflows from stars. On the other hand, interplanetary dust represents debris recently liberated from comets and asteroids of the Solar System (Brownlee and Sandford, 1992). A great deal of information on comets has been retrieved from those that have come during a period when appropriate instrumentation has been available. The results are summarized in Tables 2.3 and 2.4.

Table 2.3 Dust in the Halley Comet: The organic elements are 33% of the total mass

Elements	*Mean chemical composition in mass %*
Unsaturated hydrocarbons	16
H, C+ O	5.2
H, C+ N	4.5
H, C+ S	1.8
Water	5.5

Table 2.4 Dust in the Halley Comet: The inorganic elements are 67% of the total mass

Elements	***Mean chemical composition in mass %***
Silicates	51.5
FeS (troilite)	6
C (graphite)	3
S (sulfur)	1
Water	5

We have seen in this chapter that in our galaxy most of interstellar space is filled with gas in the form of clouds. These formations are evident since their density is higher than the surrounding space. This produces the extinction of light coming to us from distant stars, as well as from distant galaxies. These darker patches of heavens are the birthplace of stars, such as our Sun.

The abundance of hydrogen and helium (cf., Table 1.2) is not surprising, since physicists have demonstrated that within the Big Bang model, elementary subnuclear interactions are capable of generating lighter elements, once the universe is cool enough (cf. Chapter 1, where "The Origin of Elements" was discussed). But it takes the formation of stars, as we have seen, to produce the heavy elements. Before reviewing chemical evolution, we should first understand current ideas of how the Earth itself originated.

2.4 Origin of the Solar System: the Rosetta mission

The origin of the solar system took place 4.6 Gyr BP by the gravitational collapse of the solar nebula when a certain critical mass was reached. In fact, due to the original angular momentum and subsequent evolution of the original collapsing solar nebula, a natural explanation is provided for basic characteristics of our Solar System, namely the orbits of the planets and the Sun (in a central position) all are mostly coplanar; the planet's revolution about the Sun are the same direction, rotating mainly in the same direction; their rotation axes are nearly perpendicular to the orbital plane.

Thus, the protosun, the protoplanets, the comets, the parent bodies of meteorites, and other planetesimal bodies were formed as the result of this condensation of interstellar matter (Cassen and Woolum, 1999). The lunar cratering record demonstrates that during its initial stages, the solar system may have been in a chaotic state with frequent collisions of planetesimals amongst themselves as well as with other larger bodies, including the protoplanets. The composition of the solar nebula must have been analogous to that of the interstellar clouds, namely it may have consisted mostly of hydrogen, helium, as well as carbon compounds, dust and ice. This has been confirmed by astronomical studies of Jupiter, Saturn and their satellites.

To get deeper insights into the origin of the Solar System we had to wait until the launching of the Rosetta mission. The name has been taken from the famous "Rosetta Stone", presently at the British Museum in London. This stone has an inscription in three languages, ancient Egyptian hieroglyphs, Demotic Egyptian and Greek. Knowing

Greek and having some knowledge of Demotic Egyptian, Champollion was able to use this text to decipher the hieroglyphs. In an analogous manner, the proposed Rosetta mission intends to address inter-related questions: the interstellar medium, cometary material and meteorites.

The mission has one lander: Philae. It will study the nucleus of the comet 67P/Churyumov-Gerasimenko. This lander could provide tests for conflicting hypotheses for the origin of the oceans on Earth (and eventually of life). A cometary origin (Delsemme, 1992) has to take into account that the ratio of H_2O and HDO in the Earth's oceans differs by a factor of 2 with the same ratio measured in comets Halley, Hyakutake and Hale-Bopp. Instead, this ratio coincides with water in chondrites, suggesting that chondrites, instead of comets, may have been a significant source for the Earth's oceans during the heavy bombardment period. Hence the great interest and understandable excitement for the future results of the Rosetta mission is easily understood. Selected in November 1993 as a cornerstone mission of ESA's long-term science program, the Rosetta probe was launched by an Ariane 5 on 2 March 2004, on a 10-year journey to the comet. The instrumentation of the ESA mission consists of a large orbiter (cf., Fig. 2.1). It has been designed to operate for a decade.

Fig. 2.1 An artist view of the probe for the Rosetta cometary mission. The spacecraft is covered with dark thermal insulation in order to keep its warmth having kept in mind the extreme low temperatures to which the instrumentation will be exposed in the outer Solar System, inside the Jovian orbit

Philae (cf., Fig. 2.2) will carry a large set of scientific experiments in addition to those carried by Rosetta designed to complete the most detailed study of a comet ever attempted. Together they will return data on the comet's materials that are believed to be largely unchanged since the formation of the Solar System, more than 4 Gyr ago.

2.5 Origin of the terrestrial planets

From the combined information provided by the small objects of the inner Solar System and messengers from the outer Solar System (comets), we are hoping to reconstruct an outline of the main steps in the formation of the terrestrial planets.

Fig. 2.2 The Rosetta mission's target is comet 67P/Churyumov-Gerasimenko, a mountainous ball of ice, rock and dust. The mission will rendezvous with the comet in 2014. The parent Rosetta probe will go into orbit around the object and drop the lander Philae on to its surface

We expect high temperatures at the center of the solar nebula and low temperatures at its periphery. This is today an empirical observation for we can observe in our galaxy nebulae where similar processes of star formation are occurring today. As the thin interstellar medium gets concentrated in some regions in space it forms a protonebula, which eventually collapses producing a hot central condensation that becomes the star.

The collapse is coupled with rotation of the gas. A basic law of physics is the conservation of angular momentum. This means that a certain quantity (angular momentum), which depends on mass, distance and rotational velocity is conserved. In the process of collapse the mass remains the same, the distance represented by the dimension of the nebula decreases; so, as a consequence of the conservation law, the angular velocity must increase as the collapse continues. This leads to a proto-Sun with a corresponding rotating disk, which will give rise to the planets and small bodies of the Solar System. Due to the heat, only refractory materials, namely materials that vaporize at high temperatures that formed previously in the interstellar space (silicon combining with oxygen into silicate particles is an example) can survive in the inner part of the nebula. This is the first stage of a sequence that will lead to chondrites, planetesimals and eventually to terrestrial planets.

2.6 Origin of the Jovian planets

The search for the origins of life set out in the first of the four themes for Cosmic Vision 2015-2025 must begin in the Solar System. Understanding the Jovian System over a range of timescales is a main priority. Other prominent questions are how the

Galilean satellites originated and what is their structure are aspects of the same quest. Answering these and other questions involves the detailed study of mainly Europa, Ganymede, but also Io and Callisto (Fig. 2.3). ESA has already taken major initiatives elsewhere in the outer Solar System, as we shall discuss in some detail in Chapter 9, with the Huygens probe to Titan. In the inner Solar System, Mars Express was an important landmark to give answers to the remaining mysteries, but we await the results of ExoMars and further missions in the future for more definite answers (cf., Chapter 7). To sum up, the main goal should now be an in-depth exploration of one of the giant planets in the outer Solar System, of which Jupiter is the most accessible.

By the same physical principle that we referred to in the previous section, namely the conservation of angular momentum, the evolution of the matter distribution of the solar nebula will differ further out form the inner region. Gas is more abundant than IDPs, and yet the temperature will be low enough to allow the existence of water ice.

Fig. 2.3 This image is a composite of the Jovian system. It includes Jupiter on the left hand side, and the Jovian four largest moons—the Galilean satellites—Io, Europa, Ganymede and Callisto (from top to bottom). □□Europa, the smallest of the four moons, is about the size of Earth's moon, while Ganymede is the largest moon in the Solar System

Without going into the details, straightforward physical arguments suggest the reason why the terrestrial planets are denser than the Jovian planets. The reader should keep in mind that the discoveries of planets outside the Solar System may require deeper insights than those sketched in Chapter 2. In particular, the Jovian density is 1.3 gm/cm^3. This should be compared with the terrestrial density of 5.5 gm/cm^3.

At a distance of about 5-10 AU the temperature was not low enough to discriminate gases from dust (as it happened in the inner region). Consequently, the composition of Jupiter and Saturn, which were formed at these distances, are expected to reflect the composition of the original solar nebula. These plausibility arguments are subject to predictions that can be confronted with what is known about the chemical composition of the giant planets with that of the Sun. One of the reasons for sending the probe into the Jovian atmosphere, as soon as the Galileo mission approached the Jupiter system was to put to the test the foundations of this aspect of planetary science.

2.7 Origin of the moons in the outer Solar System

NASA and ESA decided at the end of the last decade to first pursue a mission to study Jupiter and its four largest moons, and plan for another mission to visit Saturn's largest moon, Titan, and Enceladus. These outer-planet flagship missions could eventually answer questions about how the satellites of our solar system formed and whether habitable conditions exist elsewhere in the Solar System. These missions that will throw additional information on the origin of the Jovian planets will be discussed in some detail in Chapters 8 and 9.

In the complex process of accretion of the Jovian planets (Pollack and Atreya, S., 1992), a further event is relevant for our eventual concern for studying environments where life could arise. When the masses of these planetary cores became large enough through the accretion of solids, they attracted gas from the surrounding solar nebula. It has been estimated that when the core reached about 10 Earth masses or more, the process of adding gas from its surroundings increased rapidly. This led to a situation in which the envelope and planetary core were about equal. Having reached this 'critical' value of the core, further accretion was dominated by the addition of a large amount of external gas. In our solar system none of the terrestrial planets reached this stage early enough before the gas of the solar nebula was trapped by the Sun or escaped into the outer regions of the nebula or into interstellar space. (Apparently this is not necessarily the case in other solar systems, as we shall see in Chapter 10.)

At the end of accretion, a combination of cooling and the force of gravity led to a process of gradual contraction. During an early part of this phase a circumplanetary disk of gas and dust, in perfect analogy with the parent disk that led to the planets themselves, developed around the Jovian planets. The disk originated from the solar nebula as well as from the outer gas envelopes that the planets had already formed.

Satellites formed from these disks, following the same principles that guided the formation of the planets around the Sun itself. For this reason the elements constituting the satellites of the Jovian planets would tend to be enriched in those elements that characterized the solar nebula itself; but we have already seen that physical and chemical processes acting in these nebulae can segregate compounds and elements from the solar nebula composition. This remark will be particularly relevant in Chapter 8 when we consider the possibility of life arising on Europa. The observed satellite sizes, distribution and fraction of the mass of the total Jovian planet are a clear demonstration that the nebular disk hypothesis works as well for the formation of satellites around planets as it does for the formation of the planets around the Sun.

Supplementary Reading

Matteucci, F. (2001) *The chemical evolution of the Galaxy*. Astrophysics and space science library, Vol. **253**, Dordrecht, Kluwer Academic Publishers, 293 pp.

McSween, H.Y. Jr and G. R. Huss (2010) *Cosmochemistry*. Cambridge, Cambridge University Press, 549 pp.

Oro, J. and Cosmovici, C. (1997) Comets and life on the primitive Earth, in C.B. Cosmovici, S. Bowyer and D. Werthimer (eds.), *Astronomical and Biochemical Origins and the Search for Life in the Universe,* Editrice Compositore, Bologna, pp. 97-120.

Owen, T.C. (1997) Mars: Was there an Ancient Eden, in C.B. Cosmovici, S. Bowyer and D. Werthimer, *Astronomical and Biochemical Origins and the Search for Life in the Universe,* Editrice Compositore, Bologna, pp. 203-218.

Winnewisser, G. (1997) Interstellar molecules of prebiotic interest, in C.B. Cosmovici, S. Bowyer and D. Werthimer (eds.), *Astronomical and Biochemical Origins and the Search for Life in the Universe,* Editrice Compositore, Bologna, pp. 5-22.

Pollack, J. B. and Atreya, S. K. (1992) *Giant planets: Clues on Current and Past Organic Chemistry in the Outer Solar System,* in G. C. Carle, D. E. Schwartz, and J. L. Huntington (eds.), *Exobiology in Solar System Exploration,* NASA publication SP 512, pp. 83-101.

References

Brownlee, D. E. and Sandford, S. A. (1992) Cosmic Dust, in G. C. Carle, D. E. Schwartz, and J. L. Huntington, (eds.), *Exobiology in Solar System Exploration,* NASA Publication SP 512, pp. 145-157.

Cassen, P. and Woolum, D. S. (1999) *The origin of the Solar System,* in P. R. Weissman, L.-A. McFadden and T. V. Johnson (eds.), *Encyclopedia of the Solar System,* Academic Press, San Diego, pp. 35-63.

Chela-Flores, J. (2010) Instrumentation for the search of habitable ecosystems in the future exploration of Europa and Ganymede. *International Journal of Astrobiology* **9**, 101-108. http://www.ictp.it/~chelaf/jcf_IJA_2010.pdf

Delsemme, A. H. (1992) *Comets: Role and importance to Exobiology,* in G. C. Carle, D. E. Schwartz, and J. L. Huntington (eds.), *Exobiology in Solar System Exploration.* NASA publication SP 512, pp. 177-197.

Fegley Jr., B. (1993) Chemistry of the Solar Nebula, in *The Chemistry of Life's Origins,* J.M. Greenberg, C.X. Mendoza-Gomez and Piranello, V. (eds.), Kluwer Academic Publishers, Dordrecht, pp. 75-147;

Grevesse, N. (1984) Accurate atomic data and solar photospheric spectroscopy, *Physica Scripta* **T8,** 49-58.

Lipschutz, M. E. and Schultz, L. (1999) Meteorites, in *Encyclopedia of the Solar System,* P. R. Weissman, L.-A. McFadden and T. V. Johnson (eds.) Academic Press, San Diego, 629-671.

Lof, P. (ed.), (1987) *Elsevier's Periodic Table of Elements*, Elsevier Science Publishers, Amsterdam.

Oro, J. (1961) Comets and the formation of biochemical compounds on the primitive earth. *Nature* **190,** 389-390.

Oro, J. (1995) Chemical synthesis of lipids and the origin of life, in Ponnamperuma, C. and Chela-Flores, J. (eds.), (1995) *Chemical Evolution: The Structure and Model of the First Cell,* Kluwer Academic Publishers, Dordrecht, pp. 135-147.

3

Sources of life's origin: A search for biomarkers

Until the commitment of the United States by President John F. Kennedy in 1961 to send a man to the Moon by the end of that decade, the only extraterrestrial samples that were available to science were meteorites. For the next four decades human presence in space was constrained to low orbital crafts After the commemoration of four decades from that event the scientific community is full of excitement for the next steps in Moon exploration by the United States, Europe, Russia, Japan China and India. On our own planet there are several sources of potential information that will help us in the search for biomarkers. Antarctica is given special attention.

The reader is advised to refer especially to the following entries in the Illustrated glossary:
Apollo Program, arnothosite, basalt, C-type asteroid, carbonaceus chondrite, SNC meteorites

3.1 The Moon and its water content as a mirror of the early Earth

The primary scientific importance of the Moon is due to the record it preserves of the early evolution of a terrestrial planet, and of the near-Earth cosmic environment in the first billion years or so of Solar System history. This record may not be preserved anywhere else (Crawford, 2004). The manned Moon project was interrupted only five years after the initiation of the Apollo Missions (cf., Table 3.1). However an enormous boost to astrobiology was given especially due to the lunar samples from six of the Apollo missions during the four-year time interval from1969 till 1972. Altogether the lunar astronauts brought back about 382 kg. In addition, three Soviet Luna Missions added a valuable 300 grams from 1970 till 1976. But as we shall see later on Antarctica has proved to be a wonderful source of meteorites, some of which have its origin in the Moon. From these sources we have a considerable insight that can be described in terms of a lunar surface that is igneous, in other words, the cooling of lava has formed the rocks. (By contrast, the most prevalent rocks exposed On the Earth's surface due to the presence of water and wind our rocks are "sedimentary". These include basalts and anorthosites. Thanks to the Moon exploration we know that Moon basalts are in the maria. In the highlands the rocks are instead mostly anorthosites. Some other rocks in these lunar regions are breccias, namely, fragments produced collisions in an earlier era and reagglomerated by subsequent impacts. In Table 3.1 we review briefly some important data related to the NASA missions.

J. Chela-Flores, *The Science of Astrobiology: A Personal View on Learning to Read the Book of Life*, Cellular Origin, Life in Extreme Habitats and Astrobiology 20, DOI 10.1007/978-94-007-1627-8_3,

Table 3.1 Statistics on the later Apollo Program that allowed the first sample-return missions.

Apollo	*Date*	*Landing site*	*Weight of the Moon rocks* (kg)
11	20 July, 1969	Mare Tranquilitatis	21.7
12	19 November, 1969	Oceanus Procellarum	34.4
14	31 January, 1971	Fra Mauro	42.9
16	21 April, 1972	Decartes	94.7
17	11 December, 1972	Taurus-Littrow	110.5

In addition, the Apollo missions returned excellent images (cf., Fig. 3.1), which have been completed with subsequent missions of Solar System exploration (cf., 3.2):

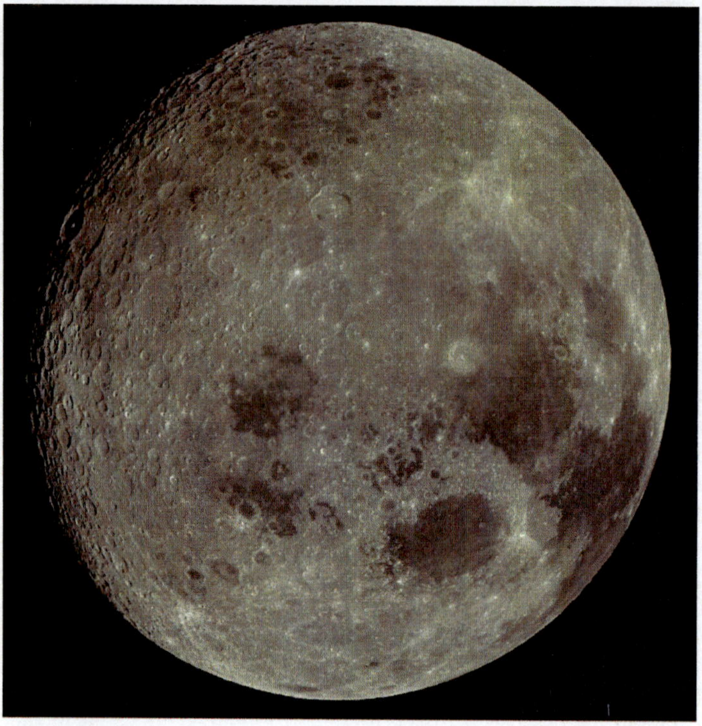

Fig. 3.1 The Moon as photographed by the crew of Apollo17 in 1972

Fig. 3.2 The Earth and Moon from a distance of over 6 million kilometers, (Galileo spacecraft, 1992). The Moon is in the foreground. The Moon was moving from left to right

Although the program of human exploration was cancelled at the beginning of the second decade of the 21st century by the USA, other leading nations have not ruled out the objective of sending astronauts to our satellite in the foreseeable future such as China and India, a principal achievement of the American Apollo Program was being the first "sample-return mission". The Apollo Missions provided Moon rocks that were tested for the presence of biogenic elements. After more than four decades of that event, the enormous achievement can be seen now in its proper perspective (Taylor, 1994) setting the Moon statistics (Table 3.2) in its proper perspective.

There were no signs of signs of life on our satellite, but the question remained whether there had been some degree of chemical evolution over the four billion years since our Moon originated. From 1969 till 1976 samples were returned from nine sites of the Moon surface. The next step in sample return missions will not take place for some years. For various technical reasons the first Mars soil retrieval mission has been postponed. Just a few months before passing away Cyril Ponnamperuma, the director of one of the laboratories chosen for the analysis of Moon samples, spoke during the Third Trieste Conference about his recollections of those very exciting days of the birth of astrobiology (Ponnamperuma, 1995).

Table 3.2 Physical parameters of the Moon

Parameters	***Values***
Mass /Earth mass	1/81
Diameter (km) / Eccentricity	1738 / 0.0549
Mean density (g/cm^3)	3.344
Mean Earth density (g/cm^3)	5.520
Albedo for the maria(%)	5 -10
Albedo or reflectance for the highlands(%)	12-18
Period of rotation on its axis (days)	27.32166
Mean distance from the Earth (Earth radii)	60.3904
Mean distance from the Earth (km)	384,400

(Ross Taylor, 1999)

The soviet 'Luna Program' was able to retrieve samples from three sites, while the Apollo Program succeeded in six different locations. The main biogenic elements that were found are listed in Table 3.3.

Table 3.3 Lunar biogenic element abundance mean value, parts per million

Element	***Soils***	***Breccias***	***Rocks***
H	40	68	2.1
C	115	78	1.5
N	82	68	3
S	850	756	1950, basalt 300, anorthosite

(Gibson and Chang, 1992)

The renewed interest in the Moon included in 1990 the Japan Space Agency JAXA with the Hiten spacecraft in orbit around the Moon. The spacecraft released a probe into lunar orbit, but the transmitter failed. In September 2007, Japan launched the SELENE spacecraft, with the objectives of obtaining data of the lunar origin and evolution and to develop the technology for the future lunar exploration.

NASA launched the Clementine mission in 1994, and Lunar Prospector in 1998. ESA launched a small lunar orbital probe called SMART 1 in 2003. SMART in order to take imagery of the lunar surface (X-ray and infrared). SMART 1 entered lunar orbit on November 15, 2004 and continued to make observations until September 3, 2006, when it was decided to crash the spacecraft into the lunar surface to retrieve information on the impact plume. More recently, however, a widespread interest in the exploration of our own natural satellite is increasing. The Lunar Reconnaissance Orbiter (LRO, cf., Fig. 3.3) spacecraft has orbited the Moon on a low 50 km polar mapping mission simultaneously with the Lunar Crater Observation and Sensing Satellite (LCROSS) (a robotic spacecraft that succeeded in discovering water in the southern lunar crater Cabeus).

Fig. 3.3 The Lunar Reconnaissance Orbiter, a NASA robotic on a low Moon orbit

LRO is a precursor to future manned missions to the moon by NASA and other space agencies: ESA is expected to launch a lander near the Moon's south pole around 2018. Some new technologies will be tested for future exploration. The difficult terrain of the south-pole region is still attractive due to access to solar power and water ice. The aim of ESA's proposed precursor mission to visit the Moon's south polar region is to probe the moonscape's unknowns and test new technology to prepare for future human landings. To these efforts we should add the Chinese and Indian space agencies successes (cf., Fig 3.4).

Fig. 3.4 The ESA proposed Moon lander

The China National Space Administration (CNSA) undertook the Lunar Exploration Program of the Moon. It uses Chang'e lunar orbiters, rovers and soil return spacecraft. The first spacecraft of the program, Chang'e 1, an unmanned lunar orbiter was successfully launched on October 24, 2007. Chandrayaan-1 was India's first unmanned lunar probe, launched by the Indian Space Research Organisation (ISRO) in 2008 (cf., Fig. 3.5):

Fig. 3.5 The ISRO lunar orbiter Chandrayaan-1. The spacecraft carried three European experiments on board which are direct descendents of ESA's SMART-1 to study the mineralogy and the chemical composition of the lunar surface and the interaction between the lunar surface and the solar wind. It also carried the successful NASA Moon Mineralogy Mapper M^3 as discussed in the text

The mission included a lunar orbiter and an impactor. The probe impacted near Shackleton Crater at 20:31 ejecting underground soil that could be analyzed for the presence of lunar water ice. It was intended to survey the lunar surface to produce a complete map of its chemical characteristics and three-dimensional topography. Chandrayaan operated for 312 days with its suite of 11 scientific instruments. Among its successes was the discovery of the widespread presence of water molecules in lunar soil due to the excellent response of NASA's mapping tool M^3, one of the probe's five foreign-built instruments (Pieters *et al.*, 2009). M^3 was designed to map the lunar surface in unprecedented detail in visible and near-infrared light, even with the low-resolution data we have from the first phase, we have several new and completely unexpected discoveries. In particular, M^3 detected absorption features near 2.8 to 3.0 micrometers on the surface of the Moon. For silicate bodies, such features are typically attributed to hydroxyl- and/or water-bearing materials. This suggests that the formation and retention of hydroxyl and water are ongoing surficial processes that could make the lunar regolith a candidate source of volatiles for human exploration.

3.2 Biogenic elements on asteroids

There are approximately 200 asteroids that have already been identified between the orbits of Mars and Jupiter, where there are many more still to be studied in detail. The largest amongst them are even a few hundred kilometers across (cf., Table 3.4).

Table 3.4 Statistics of the largest C-type asteroids

Asteroid	*Semi-major orbital axis (AUs)*	*Diameter (Km)*
1 Ceres	2.767	940
10 Hygeia	3.144	430
511 Davida	3.178	324
52 Europa	3.097	292

(Britt and Lebofsky, 1999)

Some asteroids have been shown to possess small satellites (cf., Fig. 3.6).

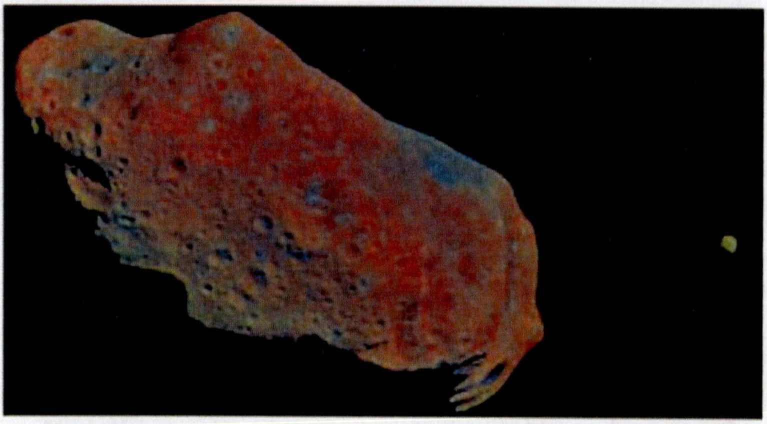

Fig. 3.6 The asteroid Ida and its satellite Dactyl

Collisions amongst asteroids produce smaller pieces that are thrown into orbits intercepted by the Earth. If they survive the transit through our atmosphere these small bodies are called 'meteorites'. We shall devote considerable space to these small bodies from the outer solar system, whose chemical composition can give us a great deal of information on the origin of elements that were to trigger off the process of chemical evolution on Earth and elsewhere, particularly in Titan, the Saturn moon.

However, there is another aspect of the chemistry of asteroids that is common to several other small bodies of the solar system, namely satellites and comets. In fact, since the 1970s various experimental techniques, amongst them infrared observations, have demonstrated that a variety of the small objects contain very dark material that is either black, or red in color. These small bodies are called asteroids of C-type. Normally the term C-type asteroid is reserved for a dark, carbonaceous asteroid in a classification according to the spectra of reflected sunlight. Such asteroids reflect light poorly; we say that their reflectance (or *albedo*) is low. We recall that albedo is defined as the percentage of incoming visible radiation reflected by the surface.

3.3 Biogenic elements on comets

Comets are important indicators of the early Solar System due to their orbits as well as to their small size. This is a fast-moving field and at the time of going to press five cometary nuclei have been imaged by spacecraft ending for the present with images of Comet Hartley that reveal icy jets triggered by carbon dioxide. Evidence has been gathering that some comet nuclei have similar composition to the C-type asteroids (dark carbonaceous and silicate dust). We begin with the small objects that are easily accessible on Earth. The spectra of several comets are similar (in the light and near ultraviolet range). This suggests that the composition of a comet nucleus is similar. The composition contains a great deal of information that has been is brought within reasonable distance form the Earth to allow useful information to be gathered. Perhaps one of the most impressive comets left an unprecedented record of images towards the end of the 20th century: it was with the Shoemaker-Levy comet (Eugene and Carolyn Shoemaker and David Levy) that collided with Jupiter. The colliding fragments of the comet were separated by an average of several hours, each fragment leaving significant interactions with Jupiter's atmosphere (cf., Fig. 3.7).

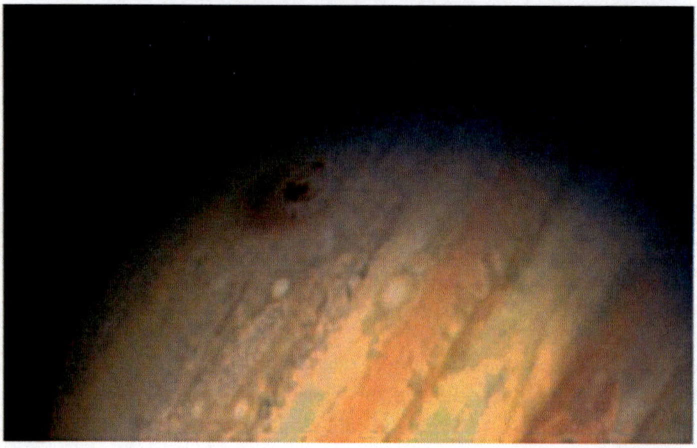

Fig. 3.7 The image shows sites of impact of fragments of the comet Shoemaker-Levy 9

The American astronomer Fred L. Whipple proposed in 1950 a model for comets that has been influential: known as the "dirty snowball model". Wipple assumed that the nucleus is essentially an icy conglomerate with dust is incrusted by water ice, but even by icy forms of volatile molecules. He remarked the stability of its structure allowing it to approach the Sun at relatively close distances before continuing towards the outer solar system along its orbit. This stability pointed out by Wipple has allowed some remarkable estimates of their composition, as illustrated in Table 3.5.

Table 3.5 Estimated volatile abundance in comet nuclei

Molecule in comet nuclei	***Abundance of ices***
H_2O	0.85
CO	0.04
CO_2	0.03
H_2CO	0.02
CH_3OH (methyl alcohol)	0.02
N_2	0.01
H_2S hydrogen cyanide, ammonia and hydrocarbons	Less than 0.03

(Boice and Huebner, 1999)

Comets are a source of CNOPS elements, which develop gaseous envelopes when they are in the close vicinity of the Sun. The comets expel gases, carrying some of the dust particles. On the other hand, the Earth's biosphere is that part of this planet where life can survive; it extends from a few kilometers into the atmosphere to the deep-sea vents of the ocean, as well as into the crust of the Earth itself. The Delsemme model is based on the assumption of the cometary origin of the biosphere (Delsemme, 2000). According to this viewpoint, an intense bombardment of comets has brought to the Earth most of the volatile gases present in our atmosphere and most of the carbon extant in the carbonate sediments, as well as in the organic biomolecules. With the measuring ability that was available throughout last century, molecules have been identified by their spectra in a wide range of wavelengths, and even by *in situ* mass spectrometry during spacecraft flybys of comet P/Halley. These measurements provided composition of parent volatiles and dust, properties of the nuclei and physical parameters of the coma. Besides the Halley comet already mentioned, two other comets were particularly useful objects for remote sensing measurements: comet Hale-Bopp, and comet Hyakutake (Campins, 2000). These comets led to the detection of HCN, methane, ethane, carbon monoxide, water, as well as a variety of biogenic compounds.

Between the orbits of Mars and Jupiter there is an important belt of asteroids. In fact, the majority of asteroids in our solar system belong to that belt. The origin of the asteroids could be found in the nuclei of extinct comets, which themselves may have been formed in the Kuiper belt outside the orbit of Pluto and extending for 500 AU.

Other comets, with longer orbital periods (longer than 200 years), may have been formed in the Oort Cloud, which lies beyond the Kuiper belt). Its extension is truly colossal: it goes as far as halfway to the nearest star. An interesting still unanswered question is whether these two largely unknown components of our Solar System are present in other solar systems (cf., Chapter 10 where we will discuss the discovery of other solar systems elsewhere in the universe).

3.4 Biogenic elements on meteorites

A large number of meteorites are of the stony kind, the so-called chondrites, as they contain chondrules (mineral-rich blobs). Amongst the chondrites the most ancient ones, contemporary with the formation of the Solar System, are rich in carbon and are, therefore, called carbonaceous chondrites.

The isotopic variations in elements within meteorites (oxygen, neon, xenon and titanium) can be rationalized by assuming that nuclear processes that predated the formation of the Solar System cause these variations. The interstellar grains that were the carriers of these isotope anomalies were probably formed in stellar atmospheres and preserved the signatures of the isotope-formation processes of certain stars. There is a very important fraction of all the meteorites that have been collected on the surface of the Earth, whose origin has been estimated to be from the early Solar System. They may be referred to as basaltic meteorites.

In particular a few meteorites are certainly Martian in origin. This assertion was finally proved when it was found that several of these bodies contained trapped gases having a composition identical to that of the Martian atmosphere as we learnt to recognize from the measurements by the Viking landers that were searching for biogenic signatures. The rocks may have been ejected from the Martian surface by large impacts, subsequently going into solar orbit for several million years before falling on Earth in the various places indicated in Table 3.6:

Table 3.6 Some important meteorites. The SNC meteorite will be considered in detail in the next section. The initials S, N and C stand for the locations where these meteorites were first retrieved (shergottites, nackhlites and chassignites, respectively). The Murchison meteorite will also be discussed below in a separate section

Place where the meteorite fell	*Type*	*Weight*
Grootfontein, Namibia	Iron (90% Fe, 10% Ni)	60 tons
Jilin, China	Stony	1.7 tons
Murchison, Australia	Carbonaceous chondrite	Many fragments
Allan Hills, Antarctica	SNC meteorite (Martian origin)	2 kg

(McSween and Stolper, 1980)

A large sample of meteorites has been found in Antarctica. From the point of view of astrobiology, the advantages are being far removed from densely populated regions and the low temperatures that are recorded. The latter factor helps in the preservation of those valuable extraterrestrial records. For some time now the composition of the Martian atmosphere has been known, (cf., Table 3.7). However, inside some of the meteorites found in Antarctica there is some gas that has been trapped when the minerals were laid down. Such meteorites are called SNC, an acronym that stands for Shergotty, Nakhla, Chassigny, which are the names of the places where the meteorites fell. Studies of these meteorites have shown that the gas that they contained has the same composition as that shown in Table 3.7.

Table 3.7 Composition of the Martian atmosphere.

Chemical component	*Fraction by weight (%)*
Carbon dioxide, CO_2	95
Molecular nitrogen, N_2	1.5
Water	0.03

Hence their origin is assumed to be from Mars. In particular, the Shergotty meteorites are named after the city in India where the prototypical meteorite fell in 1865. Further support for the Martian origin of the SNC meteorites comes from a comparison of the mineral content of a Shergotty meteorite and that of Martian soil, as we show in Table 3.8.

Table 3.8 Comparison of minerals in shergottites and Martian soil

Mineral	*Shergotty meteorite (%)*	*Presence in the Martian soil (%)*
Silicon dioxide, SiO_2	50.4	53.9
Ferrous oxide, FeO	19.3	19.7
Calcium oxide, CaO	9.6	6.7
Magnesium oxide, MgO	9.3	10.0
Aluminum oxide, Al_2O	7.0	6.8
Titanium dioxide, TiO_2	0.9	1.0
Potassium oxide, K_2O	0.3	0.1

Adapted from (McSween and Stolper, 1980)

The SNC meteorites were driven off the surface of Mars by large impacts produced by large objects. In time geophysicists working in Antarctica have retrieved a few of these meteorites. Another important aspect of these meteorites is that their mineral composition implies that they were at one time in water, one of the indicators that Mars may have had sometime in its past a 'clement period', with conditions appropriate for the origin of life.

3.5 The Murchison meteorite

Exploring the consequences of the laws of physics and chemistry may give us some insights on the question whether or not we are alone in the cosmos. The concept of the constraint on chance (cf., Chapter 5) militates against the older criterion of Monod who, during his lifetime, had less information than we have accumulated now on the mechanisms that are behind the origin and evolution of life on Earth. Indeed, chance (rate of mutations) and necessity (natural selection) do not imply that life elsewhere in the cosmos is unlikely (Monod, 1972). Thus, the prospect for detecting life in the newly discovered exoplanets remains a possibility, either by detecting life-supporting volatile elements, or by directly detecting their radio messages.

It has been clear for some time now that the extraterrestrial option should not be ruled out. Firstly, a significant illustration of the plausibility of an extraterrestrial origin of the precursors of the biomolecules is based on the meteorite that fell in the town of Murchison in southeastern Australia on September 28, 1969 (cf., Fig. 3.8).

Fig. 3.8 The image shows a fragment of the Murchison meteorite

Since the meteorite exploded in mid-air, many fragments were retrieved near the town. This meteorite is classified as a CM2 meteorite, as carbonaceous chondrites are classified into 9 classes according to a two-parameter system as follows: The abbreviations CI, CM, CV, CO refer to the analogues of the prototypical meteorites:

- Ivuna,
- Mighei,
- Vigarano and
- Ornans

(There are five more classes of chondrites that have been omitted).

The laboratory of Cyril Ponnamperuma was able to obtain a piece of the Murchison meteorite. At that time they were preparing for the first analysis of the lunar rocks. They were able to get the first conclusive evidence of extraterrestrial amino acids. These results and others have demonstrated the universality of the formation of some organic compounds that are essential for life today (Wolman *et al.*, 1972). The result of the analysis is shown in Table 3.9.

Table 3.9 Amino acids from an extraterrestrial source and from chemical evolution experiments. The relative abundance of the amino acids in the experiments that imitate the paleoatmosphere as a mixture of CH_2, NH_3, H_2O and H_2. In the first column we show the amino acids found in the Murchison meteorite. The relative amounts are denoted by dots.

Amino acid	***Murchison meteorite***	***Experiments***
Glycine	• • • •	• • • •
Alanine	• • • •	• • • •
Valine	• • •	• •
Proline	• • •	•
Aspartic acid	• • •	• • •
Glutamic acid	• • •	• •

(Oro *et al.*, 1992; Orgel, 1994).

However, the above chemical evolution sketch of the origin of life is incomplete, a fact that is illustrated, for instance, with the question of the gases that were present in the early Earth atmosphere, a topic which is not settled (Kasting, 1993). Perhaps one of the most important aspects of the analysis of the Murchison meteorite is the fact that in that meteorite a detailed analysis of the relative abundance of the precursors of the biomolecules is known (cf., Table 3.10).

On the other hand, carbon dioxide must have been sufficiently abundant to prevent the Archean Earth from freezing under the Sun at a lower level in the main sequence of the Hertzprung-Russell (HR) diagram (cf., Fig. 1.8). In simpler language without reference to the HR diagram, we may say that at the time that life first appeared on Earth, the Sun was some 30% less luminous than it is today. This is referred to as the "*faint young Sun paradox*". Through a greenhouse effect there would have been the appropriate temperatures for producing the bacterial associations that we know must have been in existence some 3.5 Gyr BP (cf., Chapter 2). Besides, it also remains to be clarified what was the relative importance of the precursors of life that were brought to

Earth by comets, meteorites, and micrometeorites compared with the inventories that were part of the Earth as it formed out of our own solar nebula.

Table 3.10 Selected precursors of the biomolecules in the Murchison meteorite. The total carbon content is 2.0-2.5%. PPM denotes parts per million

Precursors of the key biomolecules	***Abundance***
Carbonate and CO_2	0.1-0.5 %
Hydrocarbons	
aliphatic	12-35 PPM
aromatic	15-28 PPM
Amino acids	10-20 PPM
N-heterocycles	
purines	~ 1 PPM
pyrimidines	~ 0.05 PPM

(Wolman et al., 1972)

3.6 Biogenic elements in extreme terrestrial environments

In most habitats the available energy is due to the Sun; in other words, the food chain begins when plants trap the energy they need by means of the process of photosynthesis. Both plants and some bacteria convert energy from the Sun into chemical energy, which is used to produce carbohydrates, a class of organic compounds of great biological importance both structurally and as energy stores.

However, there are alternative energy sources to the Sun at the bottom of the oceans and other water environments, including the Gulf of Mexico that harbors a rich fauna supported by underlying fields of oil and gas. Mainly geologic processes, including volcanic activity, fuel the vent environments. We illustrate this point with hot springs. Indeed, ever since deep-sea hydrothermal vents were discovered in 1977, they have been found subsequently in several tectonically active areas in the ocean floor. John Corliss and others have been defending the thesis that life may have originated at hydrothermal vents.

This conjecture raises an interesting point regarding our understanding of the possible distribution of life in the solar system. In fact, with our present incomplete understanding of life's origins, it is not at all evident that living organisms will evolve exclusively within, or close to what has traditionally known as a "habitable zone", namely the region that is neither too near (and hence too hot, i.e., Mercury) nor too far from the Sun (and hence too cool, i.e., Jupiter), in order to allow life to emerge. Effectively the complete extension of the Solar System is a possible site for the evolution of life, if the hypothesis that life could emerge at hydrothermal vents is correct. This hypothesis, however, must await complete experimental confirmation, a point to which we shall return in Chapter 8, when we discuss the possibility that microscopic life could exist in the Jovian satellite Europa. Mars, the most promising candidate, had the most favorable period for the origin of life during the so-called 'clement period', or Eden-like era, which was contemporary with our own Archean Eon. During that time the conditions on Mars were favorable to life. We shall return to this question later on.

On the other hand, current parallel evolution based on a totally primary production based on chemosynthetic bacteria cannot be excluded from the outer solar system satellites, where we know that there exists frozen water, or water in hydrated silicates. As we mentioned above, the most prominent candidate is the Jovian satellite Europa.

Close to the warm vents there exist dense invertebrate communities, in which the chemosynthetic bacteria are in symbiosis with organisms in the ecosystem. For instance, bivalves and gastropods live in symbiosis with their bacteria in their gill tissues. Vestimentiferans, polychaetous annelids have been observed with a very broad distribution throughout the ocean floor.

In the bottom of the Gulf of Mexico, over 500 meters from the surface there is an underwater lake of brine so dense that it remains in a depression of the seafloor. A vast salt deposit produces the gradual seeping through of gases that autotrophic bacteria can turn into foodstuff, and these bacteria become the first stage in the food chain of clams, mussels and tubeworms. Because they are so foreign to our everyday experience and also far removed by catastrophes that may exterminate other ecosystems, the hot-spring environments have been assumed to be a sort of refuge against evolution, as it is the background of the elimination of species that new species evolve. It has also been observed by *in situ* submarine investigation that hot-spring communities of animals are remarkably similar throughout the world. This 'refuge' concept has prevailed even in astrobiology: There has been some speculation on the possible origin of life at the bottom of the Europan putative ocean, who have only considered that Archea may have also evolved in the benthic regions (deep sea), which were heavily dependent on bacterial chemosynthesis. However, scientists at the Urals branch of the Russian Academy have identified fossils from the earliest hydrothermal-vent community dating from the late Silurian Period over 400 Myr BP. This particular community has its own case of species extinction, as the fossils have been identified and correspond to lampshells and snail-like organisms (cf., Little *et al.*, 1997, where additional references to the literature may be found).

This discovery has the profound implications when specific biological experiments shall be designed for testing for the presence of life in Europa, as we cannot maintain today that the analogous conditions that may exist there will induce the appearance of Archean-like organisms. Given the common origin of all the Solar System, the antiquity of the favorable conditions in Europa may have been conducive not only to the first steps in evolution. On the other hand, the source of life (hydrothermal vents at the bottom of the Europan ocean) cannot be considered a permanent refuge only for Archaea, but the transition to a complex cell, or even beyond into metazoans, has first to be ruled out by experimental tests (cf., Chapter 12 for a discussion). The effort is worthwhile, as it was the most momentous step in the evolution towards intelligent behavior on Earth. The theoretical basis arguing in favor of common basic cellular plans for life in the universe have been argued in the past. The proposal for a space mission has been put forward by several space agencies (cf., Chapter 8).

3.7 A terrestrial analog to Europa in Antarctica: The dry valley lakes

Microorganisms constitute the major part of all life on Earth, and they are mainly organized in microbial mats and into biofilms. For this reason the description of different aspects of microbial mats that are fundamental for our deeper understanding of a major cross-section of microbiology in general, and especially for astrobiology. For a much broader review of the topics considered in this and subsequent sections of this chapter, we refer the reader to the work done in collaboration with Aranya B. Bhattacherjee and Suman Dudeja (Dudeja *et al.*, 2010). Amongst the multiple implications of the study of microbial mats emerges the understanding of the early Earth, before multicellularity evolved. Microbial mats may help us to understand the possibility of life elsewhere in our own solar system, such as on Europa, Mars, Enceladus and Titan.

Indeed, microbial mats are ubiquitous in extreme environments: at high and low temperatures; in hypersaline bodies of water such as the Dead Sea; in hot springs, where they not only survive, but thrive as exemplified by the startling colored microbial mats that live in Yellowstone National Park. Microbial mats are also present in volcanic vents on the ocean floor, called black smokers. Other environments suitable for microbial mats are deserts and, specifically the Dry Valleys of Antarctica in the McMurdo region that is traversed by striking glaciers. The British explorer Sir Robert Scott discovered the Dry Valleys in 1905 (cf., Table 3.11 taken from Doran *et al.,* 1994; Wharton *et al.*, 1983; Parker *et al.*, 1982).

Table 3.11 Statistics of the Dry Valleys lakes in Antarctica

Lake or pond	***Maximum depth (meters)***	***Elevation (meters above sea level)***	***Lake type***
Lake Fryxell	18	17	Perennial ice cover; liquid water
Lake Hoare	34	73	Perennial ice cover; liquid water
Lake Vanda	69	123	Perennial ice cover; liquid water
Lake Joyce	37	1677	Perennial ice cover; liquid water

Some of the most interesting lakes in this region are permanently covered by ice. These extraordinary environments present us with an ideal window to glance at significant events that are relevant for ancient life, and even for paleolimnology that is suggestive of the possible perseverance of life on Mars in an earlier Eden-like epoch. The ice-covered lakes of Antarctica's McMurdo Dry Valleys have long been of interest to astrobiology. These environments harbor unique microbial ecosystems that could orient us how to plan our experiments on Europa. Lake Joyce is of special interest to NASA as it is ice covered year-round: Its icy surface is 6 meters deep. Yet even the few percent of light that penetrates through the ice are enough to support an algal ecosystem in the lake. Many of the structures on the lake bottom look like what we see in the

Archean rock record from about 3 Gyr BP, because it's waters harbor carbonate structures known as microbialites. These unique structures are formed with layers of cyanobacteria. The research team is interested in how these organisms are able to grow in the dark, cold waters of Lake Joyce (cf., Fig. 3.9):

In these environments the extremophiles that are trapped in microbial mats may also be living under the Taylor Glacier in the Taylor Valley, a region that is bounded by the Ferrar Glacier and the Asgard Range. These microbes probably lived in the ocean at one time, but when the floor of the Dry Valleys rose more than a million years ago, the glacier covered seawater when it advanced and trapped the microorganisms in pockets of water.

Fig. 3.9 Lake Joyce is a small lake in Pearse Valley (S77° 43.138', E161° 35.829'), close to the Taylor glacier. It is 4.8 km long and 0.9 km wide

We have detailed knowledge of how microbes survive in Antarctica (cf., Table 3.12).

Table 3.12 Microorganisms living in the Dry Valleys lakes, Antarctica.

Organism	***Domain***	***Habitat***
Cyanobacteria	Bacteria	Lakes Chad, Fryxell and Vanda
Leptothrix	Bacteria	Lakes Fryxell and Hoare
Achronema	Bacteria	Lakes Fryxell and Hoare
Clostridium	Bacteria	Lakes Fryxell and Hoare
Chlamydomonas subcaudata (Phylum Chlorophyta)	Eucarya	Lakes Bonney (east lobe) and Hoare
Diatoms (Phylum Bacillariophyta)	Eucarya	Lakes Bonney, Chad, Fryxell, Hoare and Vanda
Bryum (a moss)	Eucarya	Lake Vanda

An intriguing feature, the Blood Falls, suggests the presence of microbial mats underneath the Taylor glacier. The name is due to the resemblance with a blood-red color waterfall at the glacier's extreme end. This coloring is analogous to the colored microbial mats that live in the hot springs of the Yellowstone National Park. Isotopic measurements of sulfate, water, carbonate, ferrous iron and gene analyses imply that a microbial consortium facilitates a catalytic sulfur cycle analogous to the metabolic events that may sustain life elsewhere in the Solar System (Mikucki *et al.,* 2009). This is especially relevant to the icy satellites of the outer Solar System, including Europa, where the Galileo Mission discovered sulfur patches (1995-2003). These stains on the icy surface of the Jovian satellite are suggestive of chemosynthetic products of metabolism

From the point of view of geology and microbiology, some of the best studied in frozen lakes are in the Taylor Valley, namely Lake Fryxell and Lake Hoare. Further north, in the Wright Valley, Lake Vanda is also remarkable for its biota. Amongst the microbial mats that are permanently thriving in the frozen lakes there are examples of both prokaryotes and eukaryotes. Besides, some of the most interesting geologic paleoindicators for reconstructing the history of these lakes are stromatolites. In the Dry Valleys these structures consist of various species of cyanobacteria, such as *Phormidium frigidum* Fritsch, a prokaryote that forms the matrix of most mat types (Wharton *et al,* 1983). Modern organisms analogous to ancient life are to be found in the Dry Valley lakes. What is most significant is that single-celled eukaryotes are amply represented in this Antarctic biota (cf., Table 3.13).

Table 3.13 A few examples of eukaryotes present in Antarctica

Organism	***Domain***	***Habitat***
Diatom shells	Eucarya (Bacillariophyta)	Lake Vostok (ice core, at depth of 2375m)
Caloneis ventricosa	Eucarya (Bacillariophyta)	Lakes Chad, Fryxell, Hoare and Vanda
Navicula cryptocephala	Eucarya (Bacillariophyta)	Lakes Bonney, Fryxell, Hoare and Vanda
Chlamydomonas subcaudata	Eucarya (Chlorophyta)	Lakes Bonney and Hoare
Tetracystis sp.	Eucarya (Chlorophyta)	Lakes Fryxell, Hoare and Vanda
Yeast	Eucarya (Ascomycota)	Lake Vostok (ice core)

Amongst the related paleoindicators that have been found are diatom frustules, cyst-like structures, most likely of crysophycean origin have also been identified. These intriguing lakes contain various taxa of planktonic and benthic microorganisms. These environments are dominated by lower life forms inviting us to search for biomarkers of an earlier biota since grazing, for instance, is totally absent (Doran *et al.,* 1994). Microbial mats in lake Bonney, Chad, Fryxell, Hoare and Vanda have been thoroughly documented, especially since the 1980s. For instance, in these environments microbial mats are known to include heterotrophic bacteria, eukaryotic algae (mainly diatoms) and fungi (Baublis *et al.,* 1991) besides the above-mentioned cyanobacteria. There are some dinoflagellates *Gymnodinium* and *Glenodinium* in Lake Fryxell, where in addition protozoan taxa are associated with the algal mats (Cathey *et al.,* 1981). The existence of these permanently frozen lakes adds an extra bonus to our model of the Europan Ocean. Modern organisms analogous to the early Earth biota are found in the Dry Valley lakes. Single celled eukaryotes are represented. In Tables 3.12-3.13 we summarize the names of some of the organisms that inhabit in these lakes (Doran *et al*., 1994; Wharton *et al*., 1983; Parker *et al*., 1981, 1982; Ellis-Evans and Wynn-Williams, 1996).

3.8 A second terrestrial analog to Europa near the South Pole: Lake Vostok

From the point of view of the possibility of the existence of life on Europa (cf., Chapter 8), we should consider a lake called Vostok, which is the largest of about 80 subglacial lakes in Antarctica. Its surface is of approximately 14,000 km^2 and its volume is 1,800 km^3. Indeed this Ontario-sized lake in Eastern Antarctica is also deep, with a maximum depth of 670 m. On the other hand, from the point of view of microbiology, the habitat-analogue provided by Lake Vostok for the Europa environment seems appropriate. At the time of writing the ice above the lake has been cored to a depth of over 3,600 m, stopping just over 100 m over the surface of the lake itself. This work has revealed great diversity of single-celled organisms: yeast, actinomycetes, mycelian fungi (which remain viable for almost 40,000 years), the alga *Crucigenia tetrapodia,* diatoms, and most interestingly, 200,000 year old bacteria. Besides it appears that water temperatures do not drop too far below zero centigrade, with the possibility of geothermal heating raising the temperatures above this level. Extrapolation of data retrieved from work deep in the ice core to the lake itself, implies that Lake Vostok may support a microbial population, in spite of the fact that that large volume of water has been isolated from the atmosphere for over one million years.

Bacterial density is found to be two to seven-fold higher in accretion ice than in the overlying glacial ice. This implies that Lake Vostok is a source of bacterial carbon beneath the ice sheath. Phylogenetic analysis of the amplified small subunit ribosomal ribonucleic acid (rRNA) gene sequences in this accretion ice has revealed the presence of *alphaproteobacteria*, *betaproteobacteria* and *gammaproteobacteria* (Christner *et al.,* 2006). Lake Vostok and its relevance for astrobiology have been extensively reviewed (Christner *et al.,* 2006; Priscu *et al*., 2003). It has been estimated that the youngest water is at least 400,000 years old. It is a window into life forms and climates of primordial eras. Lake Vostok is the largest of more than 140 subglacial lakes (Siegert *et al*., 2005). The zone of ice layer up to 3,309 m (referred to as I), and the layer between 3,310 to 3,509 m (zone II) provide detailed information about the paleoclimate record spanning during the last 420,000 years.

The basal portion of the ice core from 3,539 to 3,623 m has many features differing from overlying glacial ice and its geochemical composition indicates that it represents actual lake water that has accreted (i.e., frozen) to the underneath of ice sheet. In spite of extremely cold air temperatures above the ice (an average of -55 °C), liquid water is stable in the lake owing to the combined effect of background geothermal heating, the insulating properties of the overlying icy sheet, (Siegert *et al*., 2003).

Lake Vostok appears to be harboring hydrothermal vents beneath the water surface. This is suggestive of what may be occurring on Europa. The circulation of pure water in

Lake Vostok will be driven by the differences between the density of meltwater and lake water. Geothermal heating will warm the bottom water to a temperature higher than that of the upper layers. The water density will decrease with increasing temperature resulting in an unstable water column. This leads to vertical convective circulation in the lake, in which cold meltwater sinks down the water column and water warmed by geothermal heat ascends up the water column (Siegert *et al.*, 2001). Similarly, Europa may also have geothermally-heated warm water under its ice-crust. Processes of the type that occur in Lake Vostok may be taking place on Europa, where biogenic sulfur may be reaching the surface.

In the near future, scientists from the Vostok Base plan to enter the lake (Schiermeier, 2011). This would be done by sending what we once called a "hydrobot" into the lake to collect water samples and sediments from the bottom (cf., Chapter 8, Sec. 8.6 "Technological challenges for reading further into the Book of Life"). However, even though Lake Vostok is what we have called in this section: "A second terrestrial analog to Europa near the South Pole", within the foreseeable future the concept of a hydrobot is suitable for the search of life on Earth—in Lake Vostok—but the costs would be prohibitive in the Europa Ocean itself.

3.9 A third terrestrial analog to Europa in the Canadian Arctic

Besides the biogenic sulfur-related traffic through the icy surfaces of the lakes that is well understood in the dry valley lakes lying on the western shore of McMurdo Sound (cf., Sec. 3.7 and Chela-Flores, 2006), there is another major Europan analog in North America. Once again, we are dealing with a valley that has been given the name of "Borup Fiord Pass", where minerals accumulate on glacial ice. This site is a geologically relevant feature that lies on Ellesmere Island in northern Canada. The valley lies over saline springs that are rich in sulfide and sulfate that make their way all the way to the icy patchy surface. This additional traffic of surficial sulfur provides a terrestrial laboratory to test the instrumentation that may be used later on the next mission to Europa (cf., Chapter 8).

The biosignatures that have been studied in this environment are not related to the sulfur isotopes, as we have done in our above-mentioned work of 2006. Rather, what has been followed up in Ellesmere Island is the process of biomineralization (Gleeson *et al.*, 2010). This interesting biogenic process takes place when biominerals are generated as a result of interactions of microbial life and its environment. In chemosynthesis this is a well-understood process (Konhauser, 2007).

Supplementary Reading

Chadha, M. S. and Phondke, B. (1994) *Life in the universe*, Publications and Information Directorate, New Delhi.

Doran, P. T., Berry Lyons, W. and McKnight, D. M. (2010) *Life in Antarctic Deserts and Other Cold Dry Environments*. Cambridge.

Feinberg, G. and Shapiro, R. (1980) *Life beyond Earth: The intelligent Earthling's Guide to Life in the Universe.* William Morrow and Co., New York.

MacDonald, I.R. and Fisher, C. (1996) *Life without light.* National Geographic Magazine, October, pp. 87-97.

McSween, H.Y and Stolper, E.M. (1980) Basaltic meteorites, *Scientific American* **242,** Number 6, pp. 44-53.

Prieur, D., Erauso, G. and Jeanthon, C. (1995) *Hyperthermophilic life at deep-sea hydrothermal vents. Planet Space Sci.* **43,** 115-122.

Raulin, F. (1994) *La vie dans le cosmos.* Flammarion, Paris.

Shapiro, R. (1994) *L'origine de la vie.* Flammarion, Paris.

References

Baublis J. A, Wharton, R. A. Jr., Volz P. A. (1991) Diversity of micro-fungi in an Antarctic dry valley, *J Basic Microbiol.*, **31**(1), 3-12.

Boice, D.C. and Huebner, W. (1999) Physics and Chemistry of Comets, in P. R. Weissman, L.-A. McFadden and T. V. Johnson (eds.), *Encyclopedia of the Solar System,* Academic Press, San Diego, pp. 519-536.

Britt, D. T. and Lebofsky, L. A. (1999) Asteroids, in P. R. Weissman, L.-A. McFadden and T. V. Johnson (eds.), *Encyclopedia of the Solar System,* Academic Press, San Diego, pp. 585-605.

Campins, H. (2000) The chemical composition of comets, In: Chela-Flores, J., Lemarchand, G. A. and Oro, J., (eds.), Astrobiology: Origins from the Big Bang to Civilization, Kluwer Academic Publishers: Dordrecht, The Netherlands, pp. 163-176.

Cathey, D. D., B. C. Parker, G. M. Simmons Jr., W. H. Yongue Jr. and M. R. Van Brunt (1981) The microfauna of algal mats and artificial substrates in Southern Victoria Land lakes of Antarctica, *Hydrobiologia* **85**, 3–15.

Crawford, I. (2004) The scientific case for renewed human activities on the Moon. *Science Policy* **20**, 91-97.

Chela-Flores, J. (2006) The sulphur dilemma: Are there biosignatures on Europa's icy and patchy surface? *International Journal of Astrobiology,* **5**, pp. 17-22. http://www.ictp.it/~chelaf/ss64.html

Christner, B.C., Roysto-Bishop, G., Foreman, C.M., Arnold, B.R., Tranter, M., Welh, K.A., Lyons, W. B., Tspain A.I., Studinger M., and Priscu J.C. (2006) Limnological conditions in subglacial Lake Vostok. *Antarctica. Limnology and Occeanography* **51**, 2485-2501.

Delsemme, A. H. (2000) Cometary Origin of the Biosphere (The 1999 Kuiper Prize Lecture). *Icarus*, **146,** 313-325.

Doran, P. T., Wharton, Jr., R. A. and Berry Lyons, W. (1994) Paleolimnology of the McMurdo Dry Valleys, Antarctica, *J. Paleolimnology* **10**, 85-114.

Dudeja, S., Bhattacherjee, A. B. and Chela-Flores, J. (2010) Microbial mats in Antarctica as models for the search of life on the Jovian moon Europa. In: J. Seckbach and A. Oren (eds.) *Microbial Mats. Modern and Ancient Microorganisms in Stratified Systems.* In the COLE series, Springer. pp. 543-561. http://www.ictp.it/~chelaf/Dudeja.pdf

Ellis-Evans, J. C. and Wynn-Williams, D. (1996) A great lake under the ice, *Nature* **381**, 644-646.

Gibson,, E. K. and Chang, S. (1992) The Moon: Biogenic elements, in G. C. Carle, D. E. Schwartz and J. L. Huntington (eds.), *Exobiology in Solar System Exploration,* NASA SP 512, pp. 29-43.

Gleeson, D., Pappalardo, R. T., Grasby, S. E., Anderson M. S., Beauchamp, B., Castano, R., Chien, S., Doggett, T., Mandrake, L., and Wagstaff, K. (2010) Characterization of a sulfur-rich, Arctic spring site and field analog o Europa using hyperspectral data. Remote sensing of Environment, 114, 1297.1311.

Kasting, J. F. (1993) Earth's Early Atmosphere, *Science* **259**, 920-926.

Konhauser, K. (2007) *Introduction to geomicrobiology*. Blackwell Publishing, Oxford, 425 pp.

Little, C. T. S., Herrington, R. J., Maslennikov, V. V., Morris, N. J. and Zaykov, V. V. (1997) *Nature* **385**, 146-148.

McSween, H. Y and Stolper, E. M. (1980) *Basaltic meteorites, Scientific American* **242**, Number 6, pp. 44-53.

Mikucki, J. A. Pearson, A., Johnston, D. T., Turchyn, A. V. Farquhar, J., Schrag, D. P., Anbar, A. D., Priscu, J. C. and Lee, P. A. (2009) A Contemporary Microbially Maintained Subglacial Ferrous "Ocean", *Science,* **324**, 397 – 400.

Monod, J. (1972) *Chance and Necessity An Essay on the Natural Philosophy of Modern Biology,* Collins, London.

Orgel, L. (1994) The origin of life on Earth, *Scientific American* **271**, No. 4, 53-61.

Oro, J. Squyres, S. W., Reynolds, R. T., and Mills, T. M. (1992) Europa: Prospects for an ocean and exobiological implications, in G. C. Carle, D. E. Schwartz and J. L. Huntington (eds.), *Exobiology in Solar System Exploration,* NASA SP 512, pp. 103-125.

Parker, B. C., Simmons, Jr., G. M., Gordon Love, F., Wharton, Jr., R. A. and Seaburg, K. G. (1981) Modern Stromatolites in Antarctic Dry Valley Lakes, *BioScience* **31**, 656-661.

Parker, B. C., Simmons, Jr., G. M., Wharton, Jr., R. A., Seaburg, K. G. and Gordon Love, F. (1982) Removal of organic and inorganic matter from Antarctic lakes by aerial escape of bluegreen algal mats, *J. Phycol.* **18**, 72-78.

Pieters, C. M., Goswami, J. N., Clark, R. N., Annadurai, M., Boardman, J., Buratti, B., Combe, J. -P., Dyar, M. D., Green, R., Head, J. W., Hibbitts, C., Hicks, M., Isaacson, P., Klima, R., Kramer, G., Kumar, S., Livo, E., Lundeen, S., Malaret, E., McCord, T., Mustard, J., Nettles, J., Petro, N., Runyon, C., Staid, M., Sunshine, J., Taylor, L. A., Tompkins, S. and Varanasi, P. (2009) Character and Spatial Distribution of OH/H_2O on the Surface of the Moon Seen by M^3 on Chandrayaan-1. *Science* **326**, 568 - 572.

Ponnamperuma C. (1995) The origin of the cell from Oparin to the present day, in Ponnamperuma, C. and Chela-Flores, J. (eds.), (1995) *Chemical Evolution: The Structure and Model of the First Cell,* Kluwer Academic Publishers, Dordrecht, pp. 3-9.

Priscu, J.C., Bell, R.E., Bulat, S.A., Ellis-Evans, C.J., Kennicutt, M.C., Lukin, V.V., Petit, J.-R., Powell, R.D., Siegert, M.J., and Tabacco, I. (2003) An international plan for Antarctica subglacial lake exploration. *Polar Geogr.* **27**: 69-83.

Ross Taylor, S. (1999) *The Moon,* in P. R. Weissman, L.-A. McFadden and T. V. Johnson (eds.), *Encyclopedia of the Solar System,* Academic Press, San Diego, pp. 247-275.

Taylor, G.J. (1994) The scientific legacy of Apollo. *Scientific American,* **271** (1) pp. 26–33.

Schiermeier, Q. (2011) Race against time for raiders of the lost lake. *Nature* **469**, 275 (2011) doi:10.1038/469275a

Siegert, M.J., Ellis-Evans, J.C., Tranter, M., Mayer, C., Petit, J.R., Salamatin, A., and Priscu, J.C. (2001) Physical, chemical and biological processes in Lake Vostok and other Antarctic subglacial lakes. *Nature* **414**, 603-609.

Siegert, M. J., Tranter M., Ellis-Evans J.C., Priscu, J.C., and Lyons, W.B. (2003) The hydrochemistry of Lake Vostok and the potential for life in Antarctic subglacial lakes. *Hydrol. Processes* **17**, 795-814.

Siegert, M.J., Carter, S., Tabacco, I., Popov S., and Blankenship, D.D. (2005) A revised inventory of Antarctic subglacial lakes. *Antarctic Sci.* **17**, 453-460.

Wharton, R. A. Jr., Parker, B. C. and Simmons G. M. Jr (1983) Distribution, species composition and morphology of algal mats in Antarctic Dry Valley lakes, *Phycologia* **22**, 355–365.

Wolman, Y., Haverlard, W. J., and Miller, S. L. (1972) Non-protein amino acids from spark discharges and their comparison with Murchison meteorite amino acids, *Proc. Natl. Acad. Sci. USA* **69**, 809-811.

4

From prebiotic evolution to the emergence of single cells

In our studies of the origin of life we will encounter the major macromolecules of life: proteins, nucleic acids, polysaccharides and lipids. We shall learn that there is great unity in all of biochemistry. This important remark will be illustrated with two stunning examples: the analogous asymmetry of all the main molecules of life and secondly, the universality of the genetic code.

The reader is advised to refer especially to the following entries in the Illustrated glossary: *Abdus Salam, bilayer, Cabibbo, CP violation, electroweak interactions, enantiomer, genetic code, glycerol, Lee, lipid, molecular clouds, neutron star, Pasteur, Smirnov, standard model, supernova, taxonomy, Yang, Wigner, Wolfenstein, Wu.*

4.1 Which are the macromolecules of life?

To understand the next great transition in the ascent of life from organic chemistry to the living cell, we must first comment on certain molecules that have played a key role in that ascent. In some cases we will be considering large macromolecules that take part in the all-important process of producing multiples copies of them, in order to drive the living process. They are, on the one hand, amino acids and the polymers they form, namely the proteins. On the other hand, we have the bases and the polymers they form, the so-called nucleic acids. Later on, when we consider our earliest common ancestor, we shall have to refer to other molecules of life that serve to cover up into a sac-like structure the proteins and nucleic acids already mentioned that drive the basic process of life, that of information storage (in the nucleic acids) and the translation of such information into useful proteins (the combined action of nucleic acids and proteins). A third major group of macromolecules that are relevant consists of molecules, the lipids, which due to their response when they are in contact with water have become essential in the formation of membranes around the cells that contain proteins and nucleic acids.

Let us consider the main molecules that we have already referred to above:

• Amino acids are any of twenty organic compounds that are the building blocks of proteins, which are synthesized at one of many small cellular bodies called ribosomes. In meteorites over 70 amino acids have been detected, but the proteins of living organisms only make use of 20 amino acids. Their three-letter abbreviations are given

J. Chela-Flores, *The Science of Astrobiology: A Personal View on Learning to Read the Book of Life*, Cellular Origin, Life in Extreme Habitats and Astrobiology 20, DOI 10.1007/978-94-007-1627-8_4,

in Table 4.1. (The precise chemical formula of the amino acids is not essential for following the arguments in this book.)

Table 4.1 The 20 amino acids and their three-letter abbreviations

Ala: Alanine	Gly: Glycine	Pro: Proline
Arg: Arginine	His: Histidine	Ser: Serine
Asn: Asparagine	Ile: Isoleucine	Thr: Threonine
Asp: Aspartic acid	Leu: Leucine	Try: Tryptophan
Cys: Cysteine	Lys: Lysine	Tyr: Tyrosine
Gln: Glutamine	Met: Methionine	Val: Valine
Glu: Glutamic acid	Phe: Phenylalanine	

• Proteins are organic compounds, which are essential biomolecules of all living organisms. Their elements are: hydrogen, carbon, oxygen and nitrogen and sulfur. They are made up of a series of amino acids. (A medium-sized protein may contain 600 amino acids.)

• The nitrogenous base compounds U, C, G, A and T are ring compounds which are constituents of nucleic acids. These letters represent the initial of the corresponding base; for instance, U denotes uracil. The full names of the other bases are: cytosine, guanine, adenine and thymine. To follow the arguments presented in this book the detailed chemical formulae are unnecessary.

• Nucleic acids are organic molecular structures consisting of five-carbon sugars, a phosphate and mainly, although not exclusively, one of the five bases already mentioned above.

• Lipids are a wide group of organic compounds having in common their solubility in organic solvents, such as alcohol. They are important in biology, as they are constituents of the cell membrane and have a multitude of other important roles.

4.2 The primitive Earth

There are many reasons why we cannot be certain of the exact conditions in which life evolved on Earth. One important factor is what has been called the heavy bombardment period in the early solar system. Wherever conditions may have prevailed in the atmosphere, oceans and lithosphere, they must have been altered significantly when a large number of large bodies collided with the early Earth. So much so that the Moon itself is believed to have been the product of a massive collision of the Earth and a Mars-like object. Entire oceans may have boiled off down to a depth of several kilometers.

Nevertheless, there is another body in our solar system that is within reach of experimental probing. Indeed, a very successful mission reached Saturn's satellite Titan in 2005. We shall discuss this success story in the exploration of the Solar System in Chapter 9; suffice it to say at the moment that Titan's atmosphere is dense and lacking in oxygen. This gives some weight to the earliest experiments searching for mechanisms that may have led to the biomolecules of life which have been enumerated in the previous section: the amino acids, nucleic acids and lipids.

As we learnt in the Introduction Stanley Miller, after some significant previous experiments by Melvin Calvin, attempted to create a model for the early Earth, in which it was postulated that the main components of the atmosphere were methane (as in the case of Titan), ammonia, hydrogen and water. His experimental set up included a flask of water that was boiled to induce circulation of the gases, which at the same time was capable of trapping any volatile water-soluble products that were formed during the experiment. An electric spark acted on the gases for a period of time. After suspending the electric discharge the water was found to contain several small organic compounds, two of which are found in all proteins: the amino acids glycine and alanine.

Since the Miller experiment was concluded half way through last century, the debate has continued regarding the nature of the primitive atmosphere. Today, some arguments lead us to think that the original gases may not have coincided with those of the Miller experiment. However, that is not the main point of the revolutionary trend that Miller started. What is more significant is that oxygen was missing. Miller's tutor was the Chemistry Nobel Laureate Harold Urey. If he had advised the young student to add oxygen to his gas mixture, the result of the experiment would have been that no amino acids would have formed. This will be rationalized later on in this book, when we shall mention that the generation of oxygen was mainly due to life itself.

4.3 The origin of the first cell

We are about to enter the most exciting phase of our ascent form the cosmic creation of the basic molecules of life to the stage of encapsulating them into a living cell with a membrane of lipids. In the interior of the cell the chemistry of life will take place. We still do not have the final answer about the origin of the first cell, although much research has gone in this direction. Two proposals have been associated with the earliest work in the area. The Russian chemist Alexander Oparin suggested that a mixture of organic polymers when heated are able to assemble themselves into a membrane, which in principle could contain the rest of the constituents of the living cell (cf., Introduction Sec. I.2). Since this early 20th century contribution was published, the Oparin structures—the coacervate droplets—can properly be said that they could contain the genetic material, the biopolymers DNA and RNA made up out of the four bases that were mentioned above. (The T base is a constituent of DNA that is systematically replaced in RNA by the U base.)

Much impressed by this work, an American scientist, Sidney Fox suggested that small, spherical membranes are able to assemble spontaneously by heating amino acids in water (cf., Introduction Sec. I.2). Fox called these structures 'proteinoid microspheres'. He persisted with his work for half a century with numerous colleagues. We have a wonderful memory of Sidney Fox's lecture in Trieste where he delivered a lecture co-authored by twenty colleagues who collaborated with him over a good fraction of his career.

The essential contribution of Oparin, Fox and others was to demonstrate that the entrapment of biomonomers and biopolymers within a membrane was possible in conditions that may have resembled those of the early Earth. However, it is also true that it is not evident how these proto-membranes, the forerunners of the modern cell, may have given rise to the membranes that we see today.

Independent experiments by a group of younger scientists that began their work after Fox, have demonstrated that under primitive Earth-like conditions the real components of the modern cell, the lipids may indeed be synthesized. Joan Oro, Cyril Ponnamperuma and many other organic chemists have led the way to a more comprehensive picture, which were mentioned amongst the pioneers of astrobiology in the Introduction.

4.4 A biochemical relic of the earliest stages of life

Independent of the exact details of the pathway that may have led to life on Earth, today all living organisms have extraordinary relics of the early stages of life. We shall complete this necessarily sketchy chapter by focusing attention on two of them. The simplest is a universal handedness that is observed from bacteria to humans, a certain asymmetry of all the biomolecules, although for simplicity we shall refer to amino acids. The second will be an almost universal code that translates sequences of genes into proteins. In order to do so we must first cover a certain number of physical and chemical properties of matter in general. Indeed, many chemical substances, when they are extracted form a liquid environment, in which they are dissolved, assume a definite crystalline form.

Generally, crystals are regular in form. For instance, they may be divided into two symmetrical halves by a plane. But this is not always the case. Sometimes there is an asymmetry that has been understood since the time of the great French scientist Louis Pasteur. A most remarkable aspect of the origin of life on Earth is the unity of biochemistry. At the lowest level an asymmetry makes that unity evident. The key biomolecules have the same "handedness". This phenomenon occurs when molecules are asymmetric, in such a manner that they are able to exist in two configurations, which like a pair of gloves, mirror each other's shape; this is referred to by saying that both partners (or 'stereo-isomers') are mirror images of each other, or that such stereo-isomers have the same handedness. For example, amino acids all come in two versions, which are technically referred to as "two stereo-isomers".

These molecules are optically active, just like many three-dimensional structures. As we said above, we owe to Pasteur our understanding of the relationship between molecular asymmetry and response to the light going through a liquid in which they are dissolved. Pasteur in 1848 devoted his studies to the effect on light of mixtures of certain crystals. Thus, the key macromolecules of life have the same handedness, or, more often these molecules are said to have the same "chirality" or, preferably that they are "homochiral". (These words are taken from the Greek language, as *cheir* means hand.) Finally, a mixture of equal quantities of the left- and right-handed forms of an optically active compound is called a "racemic mixture". In simpler terms, when amino acids are created in the laboratory by means of a chiral device, they are said to be racemic (they contain equal numbers of left- and right-handed molecules).

Chiral molecules have non-superposable three-dimensional mirror image structures or 'enantiomers' (once again, these are words derived from the Greek *enantios morphe*, whose meaning is 'opposite shape'). Molecules that respond to beams of light in the above-mentioned manner are said to be optically active. The monomers of proteins (amino acids) are examples of single-handed molecules that will concern us in this review. Surprisingly, these amino acids are exclusively left-handed.

4.5 Symmetry in the cosmos

Discrete symmetries are formal concepts that are close to those symmetries that will be relevant in biology. In general, symmetries describe non-continuous changes in a system. For example, a square possesses discrete rotational symmetry, as only rotations by multiples of right angles will preserve the square's original appearance. Discrete symmetries sometimes involve some type of exchange called reflections.

The standard model of fundamental forces (discussed in Chapter 15) has three related natural near-symmetries. These state that the actual universe about us is indistinguishable from one where:

• Every particle is replaced with its antiparticle. This is C-symmetry (charge symmetry).
• Everything appears as if reflected in a mirror. This is P-symmetry (parity symmetry first formulated by Eugene P. Wigner). This conservation of parity meant that it not possible to distinguish right from left and clockwise from counterclockwise (mirror reflection was meant to be an intrinsic property of nature).
• The direction of time is reversed. This is T-symmetry (time symmetry).

These symmetries are approximate symmetries because each one is broken (i.e., they are not of universal validity) in the present-day universe. The discovery of parity violation was a fundamental contribution. Some elementary particles decay through the weak interaction, such as electrons ejected along with antineutrinos from certain cobalt nuclei in processes that are called beta decay. In these cases Wu and collaborators in 1957 demonstrated that weak interactions are predominantly left-handed (cf., Fig. 4.1).

Fig. 4.1 Chien-Shiung Wu, first on the right in the front row during her participation in the 1980 American Physical Society Workshop, Baddeck, Canada

These experiments were motivated by Cheng Ning Yang and T. D. Lee with a discovery a year earlier in relation with the decay of subnuclear particles called K-mesons (cf., Fig. 4.2).

Fig. 4.2 Left: Tsung-Dao Lee lecturing at the 1968 Ettore Majorana School. Together with C. N. Yang they demonstrated that parity is violated when particles decay through the electroweak interactions **Right:** C. N. Yang during his 1980 visit to Universidad Simon Bolivar, Caracas.

However, the Standard Model predicts that the combination of the three 'discrete' symmetries (that is, the simultaneous application of all three symmetries) must itself be a new symmetry called CPT symmetry. Yet the violation of the combination of C- and P-symmetry (CP violation) is necessary for the presence of significant amounts of baryonic matter in the universe and thus is a prerequisite for the existence of life. CP violation is a fruitful area of current research in physics.

Lincoln Wolfenstein called interaction that is responsible for CP violation the "superweak" force. This new force, which is much weaker than the nuclear weak force, assumes in a more modern version (due to two Japanese scientists Kobayashi and Maskawa) that certain microscopic effects between quarks is the source of CP violation. The attractive aspect of the original Wolfenstein model was that it used only one variable—the size of the force—in order to explain this violation of discrete symmetries. The superweak force also had the merit that in the early stages of the study of CP violation we were able to orient ourselves in terms of simple phenomenology to understand the main parameters that began to be necessary from the work that derived from accelerators of that period (Glashow, 1967, Chela-Flores, 1968; 1969).

The violation of CP symmetry in the light of the validity of the CPT symmetry raises the possibility to make an absolute distinction between matter and antimatter, which may have profound implications for cosmology. (This subject and its implications in astrobiology will be discussed in Chapter 15.) One of the mysteries that remain in physics is why the observable universe is made chiefly of matter. (Largely unknown sectors of the universe still have to be understood.) It can be argued that the observed matter-antimatter ratio may have been produced by the occurrence of CP violation in the first seconds after the big bang, but we leave this topic at this stage, not to overload the reader with too many details, before we discuss a significant aspect of astrobiology in the next section.

4.6 Symmetry in the life sciences

Certain animals, particularly most sponges and protozoans lack symmetry. The vast majority of animals, however, exhibit a definite symmetrical form. Four such patterns of symmetry occur among animals: spherical, radial, biradial, and bilateral. For instance, in spherical symmetry can be illustrated by the protozoan groups, whose body plan body has the shape of a sphere (Radiolaria and Heliozoia). Such an animal has no ends or sides. The spherical type of symmetry is possible only in minute animals of simple internal construction.

ASYMMETRY IN THE ANIMAL WORLD

Asymmetry in organisms is fairly common due to how cells divide in organisms. Louis Pasteur suggested that biological molecules are asymmetric because the cosmic forces that contribute to their synthesis are asymmetric. It is known that there are fundamental physical asymmetries as we discussed in the previous section. (Left-right symmetry violation—parity violation—is a well-known feature of subnuclear physics). Asymmetry and important evolutionary traits can be illustrated by the handedness in traits that are usually symmetric. A classical example is the following: Over a wide world distribution, especially in the Americas, there are burrowing crabs—*Uca pugnax*—with an extraordinary left-right asymmetry. In the male out of two claws, one is enlarged and is frequently held in a position that reminds us how musicians hold their fiddle. For this reason the asymmetrical crabs of the genus *Uca* are called "fiddler crabs".

ASYMMETRY IN THE MICROSCOPIC WORLD: CHIRALITY

In Table 4.2 we highlight, now at a microscopic level rather than at the above macroscopic level of the fiddler crab, some of the chemical reactions that lead from the precursors to the macromolecules of life themselves that will exhibit asymmetries as well. Altogether, about 12 molecular species that occur spontaneously in interstellar clouds have been shown to be precursors of the main biomolecules. Such research in astrobiology extends right to the present. Inevitably, this effort to learn more about our origins will continue its robust progress in the future.

Phospholipids are biomolecules of the cell membrane having a stable ion formed from phosphoric acid H_3PO_4 (a phosphate group) and one or more molecules with two distinct regions reacting differently with water, one highly soluble, called hydrophilic, the other being water insoluble fatty acids.

Table 4.2 Reactions relevant to chemical evolution

Precursor molecule	***Macromolecule of life***
Formaldehyde CH_2O	Ribose, glycerol
Carbon monoxide + hydrogen $CO + H_2$	Fatty acids
Hydrogen cyanide HCN	Purines (adenine, guanine)
Cyanamide H_2NCN	Peptides and phospholipids

(Oró, 1995)

These remarks argue in favor of the readily formation of lipid bilayers—the basis of cell membranes. These are also 'chiral' molecules in the same sense as amino acids, as we explained above. We may say that they have biomarkers related with their handedness, as we shall proceed to explain: In phospholipids two of the -OH groups of a constituent molecule (glycerol) are linked to fatty acids.In fact, glycerol is a colorless, sweet-tasting viscous liquid, widely distributed in all living organisms. (At the molecular level its atomic formula is that of an alcohol.) Indeed, glycerol illustrates a very important property of the cell membrane of many microbes.

We will see repeatedly that at the cellular level we may group microorganisms into three large groups or taxons called domains, cf., for example Table 6.1. In fact, two of them contain exclusively right-handed-glycerol (Kandler, 1995). "Bacteria" is the group, or *domain* of bacteria, the so-called eubacteria. On the other hand, Eucarya is a domain that encompasses the animals, the ciliates, the green plants, the fungi, the flagellates and the microsporidia. From this point of view the Archea are exceptional and have left-handed glycerol in the phospholipids of their cell membrane instead. (They were previously called archaebacteria and form the third domain, which contains microorganisms living in extreme conditions of temperature, pressure or salinity.)

HOMOCHIRALITY AS A BIOMARKER OF LIFE

Molecular asymmetry ('homochirality') may be considered as a biomarker of life. Due to the progress in research and the availability of a whole series of forthcoming space missions currently being planned beyond the first decade of this century, the problem of chemical evolution has been virtually transferred from the laboratories of organic chemistry to space exploration of the Solar System (an evident example is the chemical evolution that is currently taking place on Titan, Saturn's largest moon, cf., Chapter 9). In this context we may recall intriguing ideas which suggest, independently, that circularly polarized light from neutron stars on passing clouds of interstellar dust may selectively eliminate one or the other enantiomer in the dust mantle if mirror image molecules are originally present (Bonner, 1991; Greenberg *et al.,* 1995). (Circular polarization of a light wave refers to the orientation of its oscillating electric field, which rotates 360 degrees clockwise or anticlockwise during each cycle.). According to some recent work another possible astrophysical source of homochirality may be supernovae (Cline *et al.*, 1995). The subject has been reviewed extensively (Guijarro, and Yus, 2009).

ABDUS SALAM'S VISION OF SYMMETRY IN NATURE

Abdus Salam was fascinated with the possibility of linking up the symmetry of nature in the large domain of the macromolecules of life with the symmetry that takes place in the subnuclear domain. He developed a theory in which the chiral symmetry of the electroweak interaction was probed as a possible source of the chirality of the amino acids (Salam, 1991). At the time of his publication, in view of the large gap of knowledge separating the life scientists from physicists, he wished his theory to be explained in simpler terms more accessible to biologists (Chela-Flores, 1991). Salam throughout his life was led by an intuition expressed early (Salam 1966 and Fig. 4.3):

Allied with the wonder of God's creation all explanation we have ever found is based on symmetry concepts.

Fig. 4.3 Abdus Salam who focused his attention on the question of the chiral bias of the macromolecules of life underlying the relevance of symmetry principles in science

Salam persistently requested his collaborators to be able to appeal not only to knowledge in the scientific domain, but even in the frontier of science and the humanities (Isham, 2008). But in spite of Salam's intuition and the support that some of his close associates gave him at the time of the 1991 paper (Fraser, 2008), the question of the sources of homochirality from the point of view of subnuclear physics still remains as a challenge. Following Salam's insistence of paying special attention to the frontier of science and the humanities, we shall discuss the largely ignored frontier of astrobiology and the humanities in Chapters 13 to 16.

On the other hand, the discovery in 1969 of the Murchison meteorite was a landmark in our understanding of chirality in terrestrial macromolecules that was not completely cleared up until well after Salam's contribution (cf., Section 3.5 and Cronin and Pizzarello, 1997). For the Murchison meteorite contained non-biological amino acids that were not present in the proteins of terrestrial organisms. In fact, of the amino acids that have been identified in the Murchison meteorite about 50 are non-protein amino acids (Oro *et al.,* 1990). These numbers give us a considerable insight into the origin of life, when we recall that the protein constituents are chosen from a very restricted set of twenty amino acids (cf., Sec. 4.1). This would indicate a cosmic origin of the homochirality in living systems (proteins, DNA and RNA). The electroweak interaction breaks chiral symmetry, but a robust process is needed. One possibility is that of supernova explosions (SNII, cf., Chapter 1); antineutrinos could provide this mechanism in the solar cloud (Cline, 2004, Cline et al., 1995).

THE ORIGIN OF CHIRALITY AND THE MURCHISON METEORITE

Analysis of the Murchison meteorite has been notoriously difficult ever since it fell in Australia, while the main chemical evolution laboratories around the world were preparing themselves for the analysis of the Moon samples returned by the Apollo 11 mission (cf., Sec. 3.5 and Ponnamperuma, 1995). All the preliminary experiments searching for a chiral bias reported negative results. However, it has been observed that the endogenous amino acids contained in the meteorite coincide in part with those that constitute proteins. There is a direct approach to detect deviations of racemisation (i.e., deviations from equal numbers of left-handed and right-handed amino acids). With this objective only those amino acids that are not biogenic have been studied, including those that are rare on Earth. In this manner, the difficulty is removed of detecting a chiral bias, as well as the uncertainty of whether the positive result is due to biogenic contamination. Following this strategy, some rare amino acids in the Murchison meteorite have been shown to have a small excess of left-handed versions of four amino acids ranging up to 10%.

Homochirality of the macromolecules of life is valid for all organisms, but some care most be taken with the concept of homochirality as we consider the highest taxa: Exceptionally, in the domain Bacteria (encompassing the flavobacteria and relatives, the cyanobacteria, the purple bacteria, the Gram-positive bacteria, and the green non sulfur bacteria), cell walls may contain right-handed amino acids, as in the case mentioned in the Introduction of *Lactobacillus arabinosus*. A second exception should be stressed. It concerns the domain Archaea, which, for instance, includes the genus *Pyrodictium* and the genus *Thermoproteus;* such microorganisms are capable of producing methane as a by-product of the reduction of carbon dioxide. These microorganisms are known to contain left-handed-glycerol in their membrane phospholipids, instead of the standard right-handed-glycerol as we mentioned above, which is characteristic of the phospholipids of the cellular membranes of the other two domains, Eucarya and Bacteria.

The differences that we have pointed out in this section are relevant, since it is important at all stages in the study of the origin and evolution of life on Earth to be aware that we can recognize the degree of evolution of microorganisms by noticing wherever possible the 'biomarkers' that are characteristic of each of the highest classification groups (or taxons). For example, evolution from the simple Archaea with characteristic left-handed glycerol in their cell membranes differ from the more evolved nucleated cells of higher organisms that have right-handed glycerols in their cell-membrane phospholipids. In the future campaign of exploration of the Solar System, it is necessary to distinguish the degree of evolution of putative living microorganisms in environments that may be favorable to life.

4.7 An analogy with languages: the genetic code

The analogy with languages will allow us to make a comfortable summary of the major facts that have been learnt regarding another major insight in our understanding of the origin of life. We are referring to the language of proteins and nucleic acids. It concerns the establishment of a 'genetic code'. This may be considered as a dictionary, which translates from the language of nucleic acids (i.e., DNA and RNA) to the language of proteins. Both of these classes of biomolecules are basic for all the living cells on Earth. It is the bases and amino acids that constitute the genetic alphabet. The language of the DNA consists of words of four letters corresponding to the four bases of the nucleic acids that we have mentioned above (T, C, A, G). The language of the proteins consists of words of 20 letters corresponding to the 20 amino acids (cf., Table 4.1) that take part in the structure of proteins.

Nucleic acids, in particular RNA molecules, are capable of serving as a first step in the implementation of the transfer of the information. This transfer takes place from triplets of bases that code for a given amino acid (for this reason they are called 'codons'), according to the standard genetic code (cf., Table 4.3).

In other words, since U is a constituent base of RNA and T replaces U as a constituent base of DNA, only four letters of the (RNA) nucleic acid dictionary (U, C, A and G) are used in the genetic code. On the other hand the codon denoted as "stop" needs some explanation: It is sometimes called a *termination codon* and serves, within the genetic machinery, to signal the cutting of the nascent protein that is in the process of synthesis. In other words, the stop codon indicates that the corresponding triplet terminates the polypeptide chain that is being synthesized on any of the ribosomes (Bonitz *et al.*, 1980). One example should suffice to grasp how the code is read and interpreted: To begin with, consider the bottom right-hand corner of the code shown in Table 4.3. We learn that both 'codons' GGU and GGC code for the amino acid glycine (cf., Table 4.1); we can think of phenylalanine as the first monomer in a polymerization process that will end when the complete protein will be synthesized on the ribosome. The carrier of the genetic information is a transcript of the original DNA, which is an RNA molecule. This molecule has the capacity of traveling from the nucleus to the site where the synthesis will be carried out-the ribosome. For this reason it is given the reasonable name of "messenger-RNA", and is abbreviated as mRNA. A set of enzymes is responsible on the ribosome for the joining of the amino acids into the full protein. The reader will appreciate at this stage the enormous step that the discovery of the genetic code meant for the science of molecular biology. For the first time one was able to comprehend how the sequence of bases on the DNA that carries the genetic information is related with the sequence of amino acids in protein synthesis Robert W. Holley, Har Gobind Khorana and Marshall W. Nirenberg received the 1968 Nobel Prize in Medicine and Physiology.

4.8 The origin of cellular organelles

The standard genetic code diverges at organelles that are sites of the cell's energy production; these organelles are called mitochondria (singular: mitochondrion). By a divergence we mean that in some cases the codons will not coincide with the assignments shown in the standard code (cf., Table 4.3).

Table 4.3 The standard Genetic Code. The notation for the twenty amino acids follows the standard three-letter notation.

UUU	Phe	UCU	Ser	UAU	Tyr	UGU	Cys
UUC		UCC		UCC		UGC	
UUA	Leu	UCA	Ser	UAA	Stop	UGA	Stop
UUG		UCG		UAG	Stop	UGG	Trp
CUU	Leu	CCU	Pro	CAU	His	CGU	Arg
CUC		CCC		CAC		CGC	
CUA	Leu	CCA	Pro	CAA	Gln	CGA	Arg
CUG		CCG		CCG		CGG	
AUU	Ile	ACU	Thr	AAU	Asn	AGU	Ser
AUC		ACC		AAC		AGC	
AUA	Ile	ACA	Thr	AAA	Lys	AGA	Arg
AUG	Met	ACG		AAG		AGG	
GUU	Val	GCU	Ala	GAU	Asp	GGU	Gly
GUC		GCC		GAC		GGC	
GUA	Val	GCA	Ala	GAA	Glu	GGA	Gly
GUG		GCG		GAG		GGG	

At least three examples are known in which the mitochondrion genetic code diverges from the standard code: yeast, *Drosophila* and human mitochondria (Bonitz *et al.*, 1980; Bruijn, 1983; Barrel *et al.,* 1979). Symbiosis is a process by means of which individuals of different species interact. It has been assumed that the cellular organelles, such as the mitochondrion and the chloroplast, originated by means of an obligatory symbiosis, in analogy with what happens with the well-known obligatory symbiosis between an alga and a fungus, the lichen (Margulis, 1993).

One possible rationalization for the mitochondrion genetic code deviations is that symbiosis may have occurred very early in evolution, the possible scenario being that of a cellular predator invading a larger cell, such as *Thermoplasma,* thus presumably preceding the appearance of the eukaryotic cells themselves (Margulis and Sagan, 1987) most of which have adopted the standard genetic code. However, there remains the fact that some bacteria and protists do have the standard genetic code (cf., Table 4.4). Hence, it is not evident how to establish a time chronology in the evolution of the genetic codes of the contemporary living cells.

4.9 From chemical to cellular evolution

We have only sketched two of the major steps (homochirality and the universal genetic code), which are significant in the pathway towards life during chemical evolution on Earth. These steps should have taken place from 4.6 - 3.9 Gyr BP, the preliminary interval of geologic time, which is known as the Hadean Subera.

It should be noticed that impacts by large asteroids onto the early Earth do not necessarily exclude the possibility that the period of chemical evolution may have been considerably shorter. Indeed, it should not be ruled out that the Earth might have been continuously habitable by non-photosynthetic ecosystems from a very remote date, possibly over 4 Gyr BP (Sleep *et al*., 1989). Some evidence is provided by the content and the ratios of the two long-lived isotopes of reduced organic carbon in some of the earliest sediments (retrieved from the Isua peninsula, Greenland, some 3.8 Gyr BP) may convey a signal of biological carbon fixation expressed in terms of standard geochemistry, as described in the Introduction (Schidlowski, 1995).

This reinforces the expectation that chemical evolution may have occurred in a brief fraction of the Hadean Subera (4.6 - 3.9 Gyr BP, cf., Fig. 5.1), in spite of the considerable destructive potential of large asteroid impacts which took place during the same geologic interval in all the terrestrial planets, already referred to as the heavy bombardment period.

In subsequent suberas of the Archean (3.9 - 2.5 Gyr BP), life, as we know it, was present. This is well represented by fossils of the domain Bacteria, which is well documented by many species of cyanobacteria (Schopf, 1993). We may not exclude from the geochemical data earlier dates for the first prokaryotes. Indeed, the possible origin of life on Earth may have occurred immediately after the end of the Hadean Subera, some 3.9 Gyr BP.

4.10 The influence of the ancient Sun on the early evolution of life

The early Earth was much more dynamic geologically and most of the records of large impacts were deleted. The same geological activity of the early Earth was most likely responsible for partial outgassing of a secondary atmosphere, the exact nature of which can be inferred from the isotopic composition of the noble gases: It has been shown that comets are capable by themselves of providing noble gases in the correct proportions provided that the laboratory experiments duplicate the conditions for cometary formation (Owen and Bar-Nun, 1995). Besides the temperatures had descended to about 100^0 C, or below, by about 4.4 Gyr before the present (BP). This scenario for planetary origin allows the possibility of an early origin and evolution of life on Earth. However, it should be remembered that the lunar record demonstrates that some difficulties may arise in this scenario since the Imbrium basin on the Moon was formed by a large impact as late as 3.8 Gyr BP. This implies the persistence of catastrophic impacts for life on Earth, since our planet has a larger effective cross section than our satellite (Sleep *et al.,* 1989).

Yet, in this harsh environment photosynthesis of prokaryotes did arise. It is evident from fossils of the stromatolitic-forming cyanobacteria. But the best evidence comes from geochemical analyses of the ancient rocks that militate in favor of the presence of bacterial ecosystems in the period that we are discussing in this section, namely 3.8-3.9 Gyr BP. The question of the metamorphism to which the Isua samples have been subjected remains controversial (Schidlowski, 1988). Stromatolites consist of laminated columns and domes, essentially layered rocks. Prokaryotic cells called cyanobacteria form them.

In addition, they are users of chlorophyll-a to capture the light energy that will drive the photosynthetic process. We remind the reader that chlorophyll is a pigment that is present in chloroplasts that captures the light energy necessary for photosynthesis. Chlorophyll-a is the most common of five such pigments absorbing well at a wavelength of about 400-450 nm and at 650-700 nm. The reason that there are so many pigments in photosynthesis is that each pigment absorbs light more efficiently in a different part of the spectrum. Cyanobacteria are mat-building communities. Right back into ancient times such mats covered some undermat formation of green sulfur and purple bacteria. Such underlying microorganisms are (and were) anaerobes that can actually use the light that impinges on the mat above them by using bacteriochlorophylls that absorb wavelengths of light that pass through the mat above them (Schopf, 1999). Stromatolites have persevered practically without changes for over 3 billion years. The exact date for the earliest stromatolitic fossils is at present under discussion (Brasier *et al*., 2002; Schopf *et al*., 2002). They have been dated at around 3.5 Gyr BP (Schopf, 1993). If the fossils are accepted, life's origin must be in the Archean, or even earlier, considering the complexity of a cyanobacterium itself.

Another signature of the early Sun is provided by isotopic fractionation of the five stable noble gas elements, namely, He, Ne, Ar, Kr, and Xe. The early atmosphere arose from collisions during the accretion period, the so-called heavy bombardment of the surface of the Earth. Planetesimal impacts increase the surface temperature affecting the formation of either a proto-atmosphere or a proto-hydrosphere by degassing of volatiles (Matsui and Abe, 1986).

Life emerged on Earth, during the Archean (3.8 - 2.5 Gyr BP). From the point of view of space weather (SpW), we should first of all appreciate the magnitude of the ionizing radiation that may have been present at that time. According to some theoretical arguments the origin of life may be traced back to the most remote times of this eon (3.8 Gyr BP). Indeed, isotopic and geologic evidence suggest that photosynthesis may have been already viable by analysis of the biogeochemical parameter delta 13 C (Schidlowski, 1988). Besides in the Archean the atmosphere was to a large extent anoxic. As a result the abundance of ozone would not have acted as a UV defense mechanism for the potential emergence of life. UVB (280–315 nm) radiation as well as UVC (190–280 nm) radiation could have penetrated to the Earth's surface with their associated biological consequences (Cockell, 1998; Elster, 1999).

If the distribution of life in the Solar System took place by transfer of microorganisms between planets, or satellites, knowledge of SpW becomes fundamental during the early stages of its evolution, to have some constraints on the possible transfer of microorganisms, as investigated extensively (Horneck and Cockell, 2001). The most radiation resistant organism known at present exhibits a remarkable capacity to resist the lethal effects of ionizing radiation. The specific microorganism is a non-spore forming extremophile found in a small family known as the Deinococcaceae. In fact, *Deinococcus radiodurans* is a Gram-positive, red-pigmented, non-motile bacterium. It is resistant to ionizing and UV radiation. Various groups have studied these microorganisms (Battista, 1997; Daly et al., 2004; Levin-Zaidman *et al.*, 2003). Members of this bacteria taxon can grow under large doses of radiation [up to 50 grays (Gy) per hour]. They are also known to recover from acute doses of gamma-radiation greater than 10,000 Gy without loss of viability. Such microorganisms demonstrate that life could have survived at earlier times, when the Earth surface was more exposed to solar radiation.

4.11 Eventual deeper insights into the origin of life from solar missions

Space climate and space weather are also fundamental for understanding the early evolution of life. This section and the previous section are based on earlier work that may be consulted for more comprehensive account in the context of space weather (Messerotti and Chela-Flores, 2009, and earlier references contained therein.)

The DNA repair mechanisms that extremophilic organisms evolved in response to a continuously changing space environment. The Sun changed considerably in time its temperature and luminosity, key factors for astrobiology that require better understanding, so that we would be in a position to predict with a certain degree of confidence what were the conditions like during the first Gyr of the evolution of the Solar System. The violent eruptions of solar flares during its earliest stage (T-Tauri), soon gave way to a very broad range of physical conditions that have to be clarified by further research.

There are several reasons to raise the question: Why do we need improved understanding, and predictions of solar activity? One reason arises from theoretical modeling of the earliest organisms. Predictions range broadly in their claims. Improved understanding of solar activity in the first billion years of the Earth would provide essential clues. Additional aspects of astrobiology, such as the question of the distribution of life in the Solar System, also depend on further research outside the frontier of astrobiology, namely by acquiring improved understanding of solar activity, such as the preliminary information that has been possible to retrieve from the International Space Station (ISS). Much progress has been achieved since the 1990s missions for studying the Sun. Firstly, remarkable progress has been possible with the Solar and Heliospheric Observatory (SOHO, cf., Fig. 4.4), a joint NASA-ESA spacecraft. SOHO is concerned with the physical processes that form and heat the Sun's corona, maintain it and give rise to the expanding solar wind. A second mission launched in the last decade of the 20th century is Ulysses with measurements of the Sun from a polar orbit.

Fig. 4.4 Artist's impression of the SOHO telescope

It is also dedicated to interplanetary research especially with interplanetary-physics investigations, including close (1992) and distant (2004) Jupiter encounters.

More recently, the Transition Region and Coronal Explorer (TRACE) is giving us information on the three-dimensional magnetic structures that emerge through the photosphere, defining both the geometry and dynamics of the upper solar atmosphere. In this respect, another significant contribution to the exploration of the Solar System is due to the Solar Terrestrial Relations Observatory (STEREO, cf., Fig. 4.5).

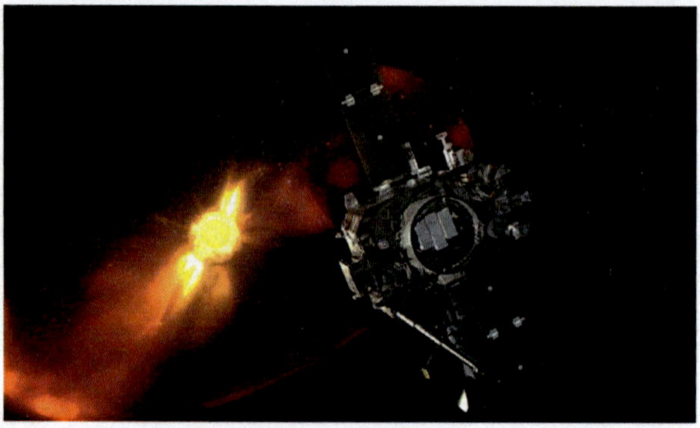

Fig. 4.5 An artist's depiction of one of the STEREO probes

STEREO was launched in February 2006 to study the nature of coronal mass ejections (CME) that in spite their significant effects on the Earth, their origin, evolution, or extent in interplanetary space still remains as a challenge. Persevering with the solar missions like Ulysses has been useful, especially when addressing an important question: whether in the past 2 - 3 Gyr similar events may have produced space climate and weather harmful, or beneficial for the evolution of life in the solar system.

There was another reason for the success of the Ulysses mission: The focus of SpW from the point of view of astrobiology gives us a hint. Jupiter's moon Io is emitting volcanic particles at passing spacecraft. The dominant source of the Jovian dust streams is Io's volcanoes (Graps *et al.,* 2000, cf. Fig. 4.6).

Fig. 4.6 Ulysses spacecraft shown in flight configuration in orbit close to the Jovian system

In September 2004 the Jovian satellite Io emitted dust particles, whose impact rate was recorded by the Cosmic Dust Analyzer (CDA) on board of Ulysses on its way to Saturn. This instrument was meant to measure the size, speed, and direction of tiny dust grains near Saturn. Compared to the corresponding CDA aboard the Ulysses spacecraft, the instrument on Cassini is significantly more advanced. The discovery of Io's dust particles dates back to 1992 when, a stream of volcano dust hit Ulysses, as it approached within 1 AU from Jupiter (cf., Fig. 4.7 and Grun *et al*., 1993).

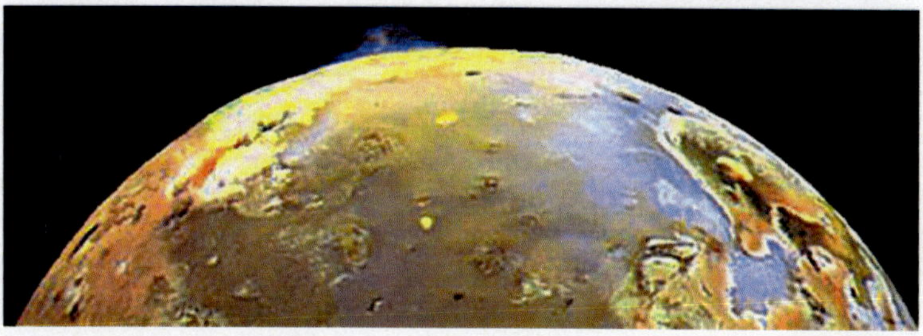

Fig. 4.7 A Galileo Mission image of one of the Io volcanoes during its eruption (top and centre of the image)

The Cassini-Huygens Mission dust detector on board of the Cassini probe was more capable than the instrumentation on Ulysses when faced with a similar event (Srama *et al.,* 2000). In addition to mass, speed, charge and trajectory, Cassini measured elemental composition finding sulfur, silicon, sodium and potassium, whose origin is volcanic. This discovery emphasizes the relevance of the study of dust particles, even if their source is not the Sun, for there could be information that is relevant to astrobiology by better appreciation of the extent of influence of the Jovian satellites.

In this chapter we have attempted only a preliminary comprehensive discussion of how research in the conditions of the early Sun may combine with observations in several disciplines to give us insights into the factors that lead to the emergence of life in a given solar system (biogeochemistry, lunar science, micropaleontology and chemical evolution), especially the early evolution of life on Earth.

These considerations are necessary to approach the conditions that will allow life to emerge in a given solar system anywhere in the universe. Due to the relevance of getting deeper insights into the close relationship between a star and its suite of young terrestrial planets potentially carrying the ingredients of life, we should increase our understanding of astrobiology with the new solar probes, such as Solar Probe Plus, is to be launched no later than 2018 (cf., Fig. 4.8).

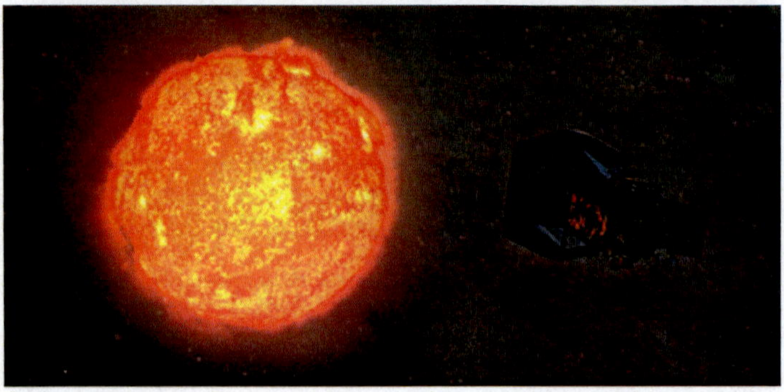

Fig. 4.8 An Artistic interpretation of the Solar Probe Plus future approach to the Sun. It will be provided with a thermal protection system consisting of a carbon heat shield to protect the probe form the solar heat

The spacecraft will approach the Sun's atmosphere to an approximate distance of less than 6 million kilometers of the surface. This is equivalent to eight times closer than any of the previous missions mentioned so far in the present chapter. It will explore several science objectives to solve two significant questions of solar physics: Why is the Sun's outer atmosphere so much hotter than the Sun's visible surface and What propels the solar wind that affects Earth and the Solar System?

To answer these questions the following instrumentation has been considered:

Solar Wind Electron Alphas and Protons Investigation (SWEAP) to measure the electrons protons and helium ions in the solar wind.

Instruments to record 3D images of the solar corona, measure electric and magnetic fields and shock waves, and compile an inventory of elements in the Sun's atmosphere.

As the spacecraft approaches the Sun, its heat shield must withstand temperatures exceeding 2,550 degrees Fahrenheit and blasts of intense radiation protected by a carbon-composite heat shield. The spacecraft will have a close and view of the Sun enabling deeper insights into space weather so crucial for human exploration of the Solar System. Mysteries are plentiful in solar physics: Why is the solar wind subject to transitions from subsonic to supersonic speeds? The phenomenon of high temperature of the solar corona has accompanied us for a long time. We recall that the corona is the outermost region of the Sun's atmosphere, consisting of hot ionized gas. It extends more than 13,000,000 kilometers from the solar surface, or "photosphere". It is affected by the Sun's magnetic field. Indeed, emission lines in the corona spectrum show ionized iron atoms, which in turn imply that its temperature should be about 200 times hotter than the 5.500 ºC photosphere. Although some theoretical explanations have been put forward, the direct data that Solar Plus would provide would give us necessary insights. We would have to wait after six flybys of Venus before reaching its closest approach towards the end of the year 2024. Eventually the Solar Probe will make a high-resolution 360º images of the corona. Most relevant of all from the point of view of this book will be stronger basis for understanding the influence of the Sun on the first steps of the origin of life on Earth. By probing the Sun we will be in a position to understand our nearest star and thereby get deeper insights into stellar behavior.

There are three other missions for solar observation that are worth recalling, as they might give us further insights into the early Sun and its influence on the origin and evolution of life on Earth.

• The Hinode (Solar-B). This is a Japan Aerospace Exploration Agency (JAXA) mission launched on 22 September 2006. It consists of a remarkable observational satellite with three solar telescopes of unprecedented resolution. Its solar optical telescope (SOT) focuses on the solar magnetic fields that are capable of resolving a feature with the size of 50 cm, if it was observing the terrestrial surface. Its X-ray telescope (XRT) has a resolution of three times as high as the previous Japanese instrument *Yohkoh* (A Solar observatory spacecraft of the Institute of Space and Astronautical Science (Japan) with United States and United Kingdom collaboration. It was launched into Earth orbit August 30, 1991 and ending ten years later). Finally, the EUV imaging spectrometer (EIS) has sensitivity superior to the SOHO instrument. Hinode is addressing the following questions: Why does a hot corona exist above the cool atmosphere? What drives explosive events such as solar flares? What creates the Sun's magnetic fields?

• The Solar Dynamics Observatory (SDO). This NASA mission has been observing the Sun since February 2010 (cf., Fig. 4.9).

Fig. 4.9 The Solar Dynamics Observatory (SDO)

It is expected to end its mission in 2015. NASA has an ongoing program called "Living With a Star" (LWS) in which SDO has been inserted with the objective of developing information that is useful to understand the Sun–Earth system especially those items that are most relevant to life on Earth. For instance, the Sun's influence on Earth and near-Earth space by means of a deeper understanding of the solar atmosphere. Like Solar-B, SDO will retrieve further insights into the Sun's magnetic field: Its generation, how magnetic energy is transferred into the heliosphere by a variety of means including the solar wind.

The main components of SDO from the point of view of instrumentation are: The Helioseismic and Magnetic Imager (HMI), which studies solar variability and characterize the Sun's interior and the various components of magnetic activity. The Extreme Ultraviolet Variability Experiment (EVE) to measure the Sun's extreme ultraviolet irradiance with precision over preceding measurements made for instance by SOHO. The instrument will contribute to our understanding of the relationship between solar extreme ultraviolet radiation (EUV)—a high-energy ultraviolet radiation spanning wavelengths from 120 nm down to 10 nm—and magnetic variation changes in the Sun. Finally, the Atmospheric Imaging Assembly (AIA), will provide the Sun's images at high spatial and temporal resolution.

• The Solar Orbiter (SOLO). If approved by ESA SOLO will be devoted to observing the Sun. Its launch is expected to be early in 2017. It is expected to obtain measurements of the inner heliosphere and the early stages of the solar wind. In addition, it will observe the polar regions of the Sun, hoping to address questions such as how does the Sun create and control the heliosphere? SOLO will approach the Sun up to 60 solar radii (R_S), almost 70% closer to the Sun that the terrestrial surface.

Supplementary Reading

Barron, L. D. (2004) *Molecular light scattering and optical activity*. Cambridge: Cambridge University Press, 443 pp.

Bertola, F., Calvani, M. and Curi U. (eds.), (1994) *Origini: l'universo, la vita, l'intelligenza.* Il Poligrafo, Padua.

Brack, A. and Raulin, F. (1991) *L'evolution chemique et les origines de la vie*, Masson,Paris.

Greenberg, J.M., Mendoza-Gomez, C.X. and Pirronello, V. (eds.), (1993) *The Chemistry of Life's Origins,* Kluwer Academic Publishers, Dordrecht, The Netherlands.

Lilensten, J. (ed.) *Space Weather. Research towards Applications in Europe.* Springer, Dordrecht, The Netherlands, Astrophysics and Space Science Library (ASSL) Series, Vol. **344**.

Moldwin, M. (2008) An Introduction to Space Weather. Cambridge University Press. 134 pp.

Schopf, William ed. (2002). *Life's Origin The Beginnings of Biological Evolution.* University of California Press, Berkeley (2002).

References

Battista, J.R. (1997) Against all odds: The survival strategies of *Deinococcus radiodurans. Ann. Rev. Microbiol.* **51,** 203-224.

Barrel, B. G., Bankier, A.T. and Drouin, J. (1979) A different genetic code in human mitochondria, *Nature* **282,** 189-194.

Bonitz, S. G., Berlani, R., Corruzzi, G., Li, M., Macino, G., Nobrega, F. G., Nobrega, M. P., Thalenfeld, B. E., and Tzagoloff, A. (1980) Codon recognition rules in yeast mitochondria, *Proc. Natl. Acad. Sci. USA* **77**, 3167-3170.

Bonner, W. A. (1991) The origin and amplification of biomolecular chirality, *Origins of Life and the Evolution of the Biosphere* **21**, 59-111.

Brasier, M.D., Green, O.W., Jephcoat, A.P., Kleppe, A.K., Van Kranendonk, M.J., Lindsay, J.F., Steele, A. and Grassineau, N.V. (2002) Questioning the evidence for Earth's oldest fossils. *Nature* **416**, 76-81.

Bruijn, M. H. L. (1983) *Drosophila melanogaser* mitochondrial DNA, a novel organization and genetic code, *Nature* **304**, 234-241.

Chela-Flores, J. (1968) Relation between CP violating parameters. *Nucl. Phys.* **B7,** 409- 412.

Chela-Flores (1969) *Symmetry in the Weak Interactions of hadrons and leptons.* Ph.D. Thesis. University of London (1969). 115 pp.

Chela-Flores, J. (1991) Comments on a Novel Approach to the Role of Chirality in the Origin of Life. *Chirality* **3**, 389-392. http://www.ictp.it/~chelaf/Chirality1.pdf

Cline, D. B. (2004) Supernova antineutrino interactions cause chiral symmetry breaking and possibly homochiral biomaterials for life. *Mendeleev Communications* **14,** 301-304.

Cline, D., Liu, Y. and Wang, H. (1995) Effect of a chiral impulse on the weak interaction induced handedness in a prebiotic medium, *Origins of Life and the Evolution of the Biosphere* **25**, 201-209.

Cockell, C.S. (1998) Biological Effects of High Ultraviolet Radiation on Early Earth - A Theoretical Evaluation. *J. theor. Biol.* **193**, 717-729.

Cronin, J. R. and Pizzarello, S. (1997) *Enantiomeric excesses in meteoritic amino acids, Science* **275**, 951-955.

Daly, M. J., Gaidamakova, E. K., Matrosova, V. Y., Vasilenko, A., Zhai, M., Venkateswaran, A., Hess, M., Omelchenko, M. V., Kostandarithes, H. M., Makarova, K. S., Wackett, L. P., Fredrickso, J. K. and Ghosal, D. (2004) Accumulation of Mn(II) in *Deinococcus radiodurans* Facilitates Gamma-Radiation Resistance. *Science,* **306**, 1025-1028.

Elster, J. (1999) Algal versatility in various extreme environments, in *Enigmatic Microorganisms and Life in Extreme Environments.* Ed. J. Seckbach, Kluwer Academic Publishers, Dordrecht, The Netherlands, pp. 217-227.

Fraser, G. (2008) *Cosmic Anger Abdus Salam-the first Muslim Nobel scientist.* Oxford University Press, Chapter 13, pp. 233-246.

Glashow, S. L. (1967) Kaon decay and pion phase shifts. *Phys. Rev. Letters* **18**, 524-526.

Graps, A. L., Grun, E., Svedhem, H., Kruger, H., Horanyi, M., Heck, A. and Lammers, S. (2000) Io as a source of the jovian dust streams. *Nature* **405**, 48 - 50.

Greenberg, J. M., Kouchi, A., Niessen, W., Irth, H., van Pardijs, J., de Groot, M., and Hermsen, W. (1995) *Interstellar dust, chirality, comets and the origins of life: Life from dead stars?,* in Ponnamperuma, C. and Chela-Flores, J. (eds.), (1995) *Chemical Evolution: The Structure and Model of the First Cell,* Kluwer Academic Publishers, Dordrecht, pp. 61-70.

Grün, E., Zook, H. A., Baguhl, M., Balogh, A., Bame, S. J., Fechtig, H., Forsyth, R., Hanner, M. S., Horányi, M., Khurana, K. K., Kissel, J., Kivelson,M., Lindblad, B. A., Linkert, D., Linkert, G., Mann, I., McDonnell, J. A. M., Morfill, G. E., Phillips, J. L., Polanskey, C., Schwehm, G., Siddique, N., Staubach, P., Svestka, J., and Taylor, A. (1993) Discovery of Jovian dust streams and interstellar grains by the Ulysses spacecraft. *Nature* **362**, 428 - 430.

Guijarro, A. and Yus, M. (2009) *The origin of chirality in the molecules of life: a revision from awareness to the current theories and perspectives of this unsolved problem.* Royal Society of Chemistry, Cambridge, 150 p.

Horneck, G. and Cockell, C.S. (2001) The History of the UV Radiation Climate of the Earth—Theoretical and Space-based Observations. *Photochemistry and Photobiology*, **73**, 447–451.

Isham, C. (2008) Memories of working with Abdus Salam. In: *Salam +50*. M. Duff (ed.) Imperial College Press, London, pp. 21-27.

Kandler, O. (1995) Cell wall biochemistry in Archaea and its phylogenetic implications, in Ponnamperuma, C. and Chela-Flores, J. (eds.), (1995) *Chemical Evolution: The structure and model of the first cell,* Kluwer Academic Publishers, Dordrecht, pp. 165-169.

Levin-Zaidman S, Englander J, Shimoni E, Sharma AK, Minton KW, Minsky, A. (2003) Ringlike structure of the *Deinococcus radiodurans* genome: a key to radioresistance? *Science* **299**, 254-256.

Margulis, L. and Sagan, D. (1987) *Microcosm,* Allen & Unwin, London, p. 132.

Margulis, L. (1993) *Symbiosis in Cell Evolution,* Freeman & Co., San Francisco.

Messerotti, M. and Chela-Flores, J. (2009) Solar Activity and Life. A Review. *Acta Geophysica* **57** (1), 64-74.
http://www.ictp.it/~chelaf/MesserottiJCF.pdf

Oró, J. (1995) *Chemical synthesis of lipids and the origin of life,* in Ponnamperuma, C. and Chela-Flores, J. (eds.), (1995) *Chemical Evolution: The Structure and Model of the First Cell,* Kluwer Academic Publishers, Dordrecht, pp. 135-147.

Oro, J., Miller, S. L., and Lazcano, A. (1990) The origin and early evolution of life on Earth *Ann. Rev. Earth Planet Sci.* **18**, 317-356.

Owen T., Bar-Nun, A. (1995) Comets, impacts and atmospheres. *Icarus* **116**, 215-226.

Ponnamperuma C. (1995) *The origin of the cell from Oparin to the present day,* in Ponnamperuma, C. and Chela-Flores, J. (eds.), (1995), *Chemical Evolution: The Structure and Model of the First Cell,* Kluwer Academic Publishers, Dordrecht, pp. 3-9.

Salam, Abdus (1966) Symmetry Concepts in Modern Physics. Atomic Energy Centre, Lahore, Pakistan, p. 54.

Salam, Abdus (1991) The role of chirality in the origin of life. *J. Mol. Evol.* **33**, 105-113.
http://www.ictp.it/~chelaf/Salam.pdf

Srama, T., Ahrens, J., Altobelli, N., Auer, S., Bradley, J. G., Burton, M., Dikarev, V. V., Economou, T., Fechtig, H., Görlich, M., Grande, M., Graps, A., Grün, E., Havnes, O., Helfert, S., Horanyi, M., Igenbergs, E., Jessberger, E. K., Johnson, T. V., Kempf, S., Krivov, A. V., Krüger, H., Mocker-Ahlreep, A., Moragas-Klostermeyer, G., Lamy, P., Landgraf, M., Linkert, D., Linkert, G., Lura, F., McDonnell, J. A. M., Möhlmann, D., Morfill, G. E., Müller, M., Roy, M., Schäfer, G., Schlotzhauer, G., Schwehm, G. H., Spahn, F., Stübig, M., Svestka, J., Tschernjawski, V., Tuzzolino, A. J., Wäsch, R., Zook, H. A. (2004) The Cassini Cosmic Dust Analyzer. *Space Science Reviews,* Vol. 114, 465–518.

Schidlowski, M. (1988) A 3.800-million-year isotopic record of life from carbon in sedimentary rocks, *Nature* **333**, 313-318.

Schidlowski, M. (1995) Early terrestrial life: Problems of the oldest record, in Chela-Flores, J., Chadha, M. Negron-Mendoza, A. and Oshima, T. (eds.), *Chemical Evolution: Self-Organization of the Macromolecules of Life,* A. Deepak Publishing: Hampton, Virginia, USA. pp. 65-80

Schopf, J. W. (1993) Microfossils of the Early Archean Apex Chert: New Evidence of the Antiquity of Life, *Science* **260**, 640-646.

Schopf, J.W. (1999) *Cradle of Life: The Discovery of Earth's Earliest Fossils*, Princeton University Press, Princeton, New Jersey, pp. 186-190.

Schopf, J. W., Kudryavtsev, A. B., Agresti, D. G., Wdowiak, T. J. and Czaja, A. D. (2002) Laser-Raman imagery of Earth's earliest fossils. *Nature* **416**, 73-76.

Sleep, N. H., Zahnle, K.J., Kasting, J. F., and Morowitz, H. J. (1989) Annihilation of ecosystems by large asteroid impacts on the early Earth *Nature* **342**, 139-142.

THE BOOK OF LIFE

PART 2: EVOLUTION OF LIFE IN THE UNIVERSE

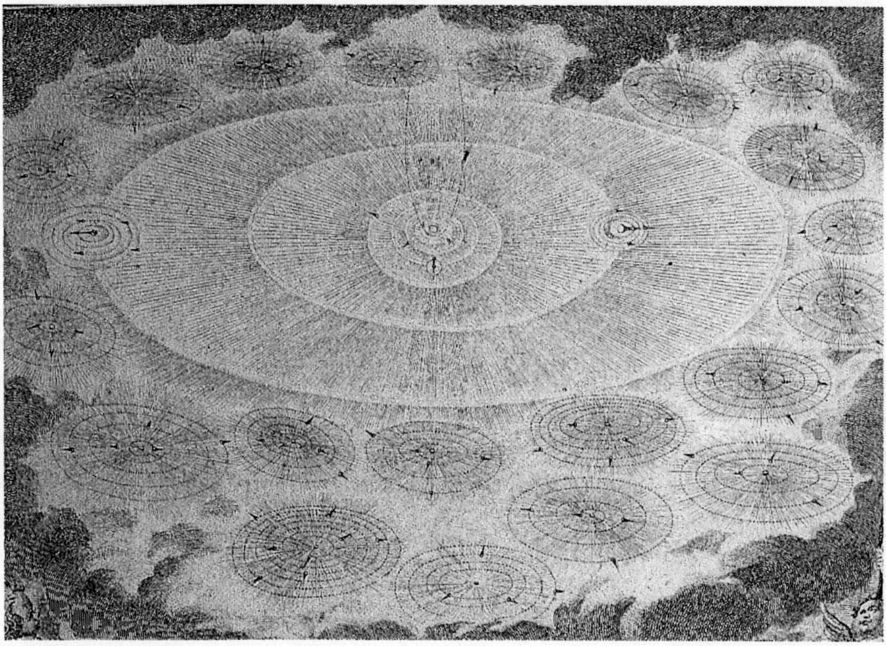

A 17th century representation of the Solar System, in which both planets and some of its satellites are present; but also neighboring stars are drawn with other worlds circling around them (Bertola, 1995). This image was to a certain extent a consequence of Bruno's thinking in his Italian Dialogues and in the debates at the University of Oxford: He stated his Third Dialogue that there were innumerable suns with earths around them, populated by living beings. This view of Bruno was a 16th century anticipation of conjecturing that the observed evolution of life on Earth can be extended to the whole universe, the central theme of our Book of Life, Part 2, in which we are attempting in the 21st century to integrate with what we know form the life sciences.

5

From the age of prokaryotes to the emergence of eukaryotes

Our subsequent discussion is based on the idea that evolution has taken place on Earth. Darwin's major thesis was that evolutionary change is due to the production of variation in a population and the survival and reproductive success of some of these variants. In this chapter we shall concentrate on the first stages of the story of life on Earth guided by Darwin's ideas. We pick up the story once the earliest and simplest living cell has already formed. As we saw in Chapter 4, it is sometimes called the progenote or, alternatively, the "cenancestor".

The reader is advised to refer especially to the following entries in the Illustrated glossary: *Banded-iron formation, chloroplast, chromatin, chromomere, chromosome, contingency, diagenesis, gene expression, genetic drift, histone, interphase, Lyell, Mesozoic, natural selection, nucleosome, nucleus (cell), organelle, Paleozoic, Phanerozoic, progenote, Proterozoic, red bed, symbiosis.*

5.1 Could prokaryotes have emerged on the early Earth 4,4 Gyr ago?

A window to the earliest stages in the geologic evolution of our planet is possible from analysis of small grains of zircon. These are minerals of great resistance to high pressure and temperature. For this reasons they are indicators of the state of the early Earth, even though it was going through the "heavy bombardment period", in which the collision of meteorites and larger bodies was more frequent than now. Specific analysis of the isotopes of elements (uranium and lead) present in a grain of zircon from the Narryer Gneiss Complex in Western Australia are consistent with the presence of continental crust and even liquid water between 4,400 and 4,300 million years before the present (Mojzsis *et al.*, 2001; Halliday, 2001).

The oxygen isotopic composition of the zircon in question was analyzed. According to the standard method it should tell us something about the magma (molten rock) from which the zircons crystallized. By implication the nature of the rocks that gave rise to the magma can be inferred. It is known that heavy oxygen is produced by the interaction between rock and liquid water, a process that should occur at sufficiently low temperatures (to allow for the liquid state of water). The presence of the heavy isotope of oxygen in the zircons form the Narryer Gneiss Complex suggest that the

J. Chela-Flores, *The Science of Astrobiology: A Personal View on Learning to Read the Book of Life*, Cellular Origin, Life in Extreme Habitats and Astrobiology 20, DOI 10.1007/978-94-007-1627-8_5,

magma had been produced by an episode of high temperature (or heavy pressure) on the surface of the early Earth, on which there was liquid water.

Given the significant implications for the first appearance of life on the surface of the early Earth these results will require further research. If such controls are done by independent teams, the first remarkable aspect for the science of astrobiology is that the mechanisms of chemical evolution and its transition to the first appearance of a single cell occur in a temporal scale that might be much shorter than previously suspected, since the occurrence of liquid water could then be traced to just over 100 million years after the formation of the planet itself.

The reader should observe that another ingredient for the origin of life, besides liquid water is expected to be a source of organic material (cf., Chapter 3, where we discussed the input form the incoming meteorites and comets). Finally, during the early geologic evolution of our planet, it has been evident since the early experiment of Stanley Miller that the third ingredient for the generation of life on Earth was the presence of sources of energy. These would be present in a variety of ways, including electric activity such as lightning, volcanic activity either on the Earth's crust or at hydrothermal vents at the bottom of oceans.

5.2 The dawn of unicellular organisms

In spite of the fundamental work of Darwin that gave rise to modern biology, the fact remained that his contemporaries were still dividing the Earth biota into animals and plants. It was only in the 1930s when taxonomy shifted its emphasis from the multicellular dominated classification to one more oriented towards basic cellular structure (Rizzotti, 2000). The encapsulation of chromosomes in nuclei was clearly absent in bacteria. This remark led to a division of all living organisms into two groups that went beyond the animal/plant dichotomy. As we have seen repeatedly, the new groups were the cells that lacked a nucleus (*karyon* in Greek) thus called prokaryotic and those that had nuclei, which were called eukaryotic (i.e., truly nucleated). However, the tree of life could not yet be constructed, since amongst bacteria it was remarked that rapid and random exchange of genes occur. This randomness deterred the use of sequences of the biomolecules, in order to construct a tree of life (phylogenetic tree). Yet, Linus Pauling and Emil Zuckerkandl had pointed out that a "molecular clock" might be identified from the slow mutation rates of some biomolecules (Zukerkandl and Pauling, 1965). The question then is to identify the proper molecule that would not be affected as evolution proceeds. A special kind of RNA together with some protein constituents makes up ribosomes. (This form of RNA is called ribosomal RNA or rRNA.) Extensive work with these molecular chronometers has led to taxonomy, according to which all life on Earth that is deeper than the dichotomy prokaryote/ eukaryote. A classification in which the highest taxa are called 'domains', instead of Kingdoms, has been suggested that will be considered in more detail in Chapter 6, Fig. 6.1 (Woese, 1983). In this approach there are three 'branches' in the tree of life:

• Archaea's most striking molecular characteristic is that the cellular membrane. It differs from the cellular membrane of the other two domains.

• Bacteria encompass all bacteria; their cellular membrane is more similar to that of the eukaryotes than that of the Archea.

• Eucarya includes all the truly nucleated cells.

This formulation of a proper taxonomy allows the discussion not only of the branches of the tree, but also of the trunk itself: the universal common ancestor (alternatively called in this text a 'progenote' or a 'cenancestor'), which since the formulation of Darwin's Theory of Common Descent has remained an important problem in evolutionary biology. As with all analogies, the tree of life with a common trunk from which the three main branches developed early in the evolution of the Earth, is not perfect. Recent molecular analysis suggests a somewhat more complicated picture (Woese, 1998). Independently of the features of the phylogenetic tree itself, several aspects of the dawn of cellular life are emerging slowly.

5.3 An early emergence of life makes the Moon a major target for astrobiology

An early mode of nutrition developed, in which microorganisms were able to use organic and inorganic substances as energy sources: these organisms are called chemotrophs. In particular, they are organisms that still live to this date that were able to use, as a source of their carbon, simply carbon dioxide present in the early Earth environment. Evidence is accumulating that supports the possibility that organisms live and prosper in the deep terrestrial subsurface, at depths as great as several kilometers below the surface. Such environments may have been favored at the dawn of cellular life on Earth. In the following chapter we shall learn that biogeochemical data suggests that the origin of life may have occurred even before the period of heavy bombardment was over. This means that practically before the time when the solar system had completed the incorporation of planetesimals into the suite of planets and small bodies that we observe today, the living process may have already working its way to possible niches.

Up to that time, soon after about 4 Gyr BP, planetesimals were still drifting into the path of the Earth orbit. These collisions are well documented since the airless and geologically inert surface of the Moon testifies with its well-preserved record of craters, the colossal collisions that the surface of the Earth was suffering. As the process of natural selection favors microorganisms that can accommodate their life styles to extreme conditions, the Moon emerges from this argument as a possible target for astrobiologists. Since we cannot exclude a very early date for the emergence of prokaryotes on Earth, the heavy bombardment period must have led to a two-way exchange of material between the Earth and its satellite. In principle this traffic would allow early terrestrial microbes to reach the Moon. The current and coming fleet of probes for the Moon can help to elucidate the question of the origin of life. In this context the penetrator technology (cf., Chapter 12) could be significant.

"Extremophile" is a term used sometimes for any of the many microorganisms that are capable of distinguishing different degrees of adaptability to extreme ranges of conditions. Many such 'extremophilic' microorganisms are known: which have been isolated from deep-sea hydrothermal vents (Prieur *et al.,* 1995). Besides, today hydrogen sulfide and methane are abundant in the fluids of the deep-sea. Chemothrophs obtain the energy required to fix carbon dioxide and produce the organic matter that they need. Such environments and microorganisms, as we have sketched in this section, are likely candidates for having given rise to the dawn of cellular life on Earth.

5.4 A unique event lost somewhere deep in the Archean

Although we follow Darwin, his ideas were developed in a different context from the origin of life. It was clear to Darwin that the tree of life had been evolving for a long period, incompatible with the estimates arising form a literal reading of the Bible. And like all trees it must have a root, which in the previous chapter we have called a cenancestor. The evident unified nature of the tree of life speaks for itself, otherwise there would be growing a whole forest of life on Earth, a fact that contradicts everything that is known to naturalists. For Darwin the origin of life was a unique event lost somewhere deep in the Archean. Darwin avoided the controversy that was raging in Victorian England between atheists and creationists. His main insight was to keep origins out of his written work (cf., Introduction).

He was content with the thesis that the origin of life on Earth was inaccessible to 19th century science. As a naturalist Darwin was only concerned with life's subsequent development. The real question in his monumental *The Origin of Species* was whether the different species of the organisms he was very familiar with had a common ancestor. His expertise on a particular organism, the barnacle, was generally accepted. He had attempted to study the whole barnacle group, and to produce a definitive text, which included fossil barnacles as well.

As in all cases of good scientific approach to intricate questions, by narrowing down the problem, Darwin was able to document his case in favor of his two theories: natural selection as the principal force in evolution, and the *Theory of Common Descent* of all species on Earth. One particular question that this narrow approach to the problem allowed him was to keep away form ideological issues. We shall also attempt to avoid the ideological trap. It is nevertheless interesting to recall the essential arguments in direct opposition to Darwin, even by some of his most illustrious friends (Desmond and Moore, 1991).

We do this at this stage in anticipation of the philosophical aspects of astrobiology that we shall discuss in Chapter 14. Sir Charles Lyell supported the view that the distant past is to be explained only by forces that we experience today (the doctrine known as uniformitarianism). It emphasized uniform processes of change in nature. What worried Lyell regarding the work of his friend Darwin was that Man would loose his special place in creation. Lyell was protecting what in his view would be "radical degradation". Although Darwin's definite strategy in the narrow approach was to stick to species and stay away from the problem of creation, Lyell argued against Darwin's 'ugly facts'. Following Darwin's example, our limited objective will be to learn from the geological history of the Earth regarding the transition from the cenancestor to more complex cells. We do not attempt to follow up the detailed chemistry underlying the genesis of our common ancestor.

5.5 The evolution of the Earth's atmosphere: The Great Oxidation Event

About 3.4 billion years ago, nitrogen was the major part of the then stable early atmosphere. An influence of life has to be taken into account rather soon in the history of the atmosphere, since hints of early life forms are to be found as early as 3.5 billion years ago.

The evolution of this early atmosphere is marked by a transition from an early atmosphere with high nitrogen content and very low oxygen content to one with increasing oxygen content up to a few percent of the present atmospheric level. In the geologic record this can be traced to the late Archean, when the oxygen-containing atmosphere began to develop. It is likely that the source of the increment was biogenic due to a large extent to photosynthesis of cyanobacteria. This can be well documented with stromatolite fossils from the main geologic archives from the two cratons in Australia and South Africa (cf., Introduction, where this topic is discussed at length).

Placing time constraints on this transition is of interest because it identifies the time when oxidative weathering became efficient, when ocean chemistry was transformed by delivery of oxygen and sulfate, and when a large part of Earth's ecology changed from anaerobic to aerobic (Holland, 2006). The oxygenation of the atmosphere and oceans are now fairly well understood. Some difficulties are outstanding with respect to the oxygenation of the oceans.

Several lines of geological and geochemical evidence indicate that the level of atmospheric oxygen was extremely low before 2.45 Gyr ago, and that it had reached considerable levels by 2.22 Gyr ago. This period has been referred to as "the Great Oxidation Event" and is significant for the rise of eukaryotes and multicellar organisms. It has been confirmed that syngenetic pyrite is present in organic-rich shales of the 2.32-Gyr-old Rooihoogte and Timeball Hill formations, South Africa. The range of the isotopic composition of sulfur in this pyrite is large and shows no evidence of mass-independent fractionation (cf., Introduction, I.8 New fingerprints of early life: mass-independent fractionations), indicating that atmospheric oxygen was present at significant levels (that is, greater than 10^{-5} PAL) during the deposition of these units (Bekker *et al.*, 2004).

There are Archean rock formations —up to 2 Gyr BP—that are significant in the evolution of life (cf., Section 5.1). These are compounds of dioxide of silicon (silica) and iron, which occur in layers. In reference to their stratified structure they are called "banded iron formations" (BIFs). The period in which the BIFs were laid out ended some 1.8 Gyr BP. In the anoxic atmosphere of the Archean, iron compounds could have been dispersed over the continental crust.

They could have absorbed some oxygen, thereby protecting photosynthesizers that could not tolerate oxygen. Such microorganisms in turn produced oxygen that combined with their environment to produce iron oxide (for example, hematite Fe_2O_3), which makes up the BIFs. In strata dating prior to 2.3 Gyr BP it has been observed that there is an abundance of the easily oxidized mineral form of uranium (IV) oxide (urininite, for example the well-known variety *pitchblende).* This argument supports the conclusion that we had to wait until about 2 Gyr BP for a substantial presence of free O_2. Another recent line of investigation throws some light on the Great Oxidation event. We discussed in the Introduction that the recent discovery of mass independent fractionation (MIF) in sulfur isotopes has provided a new tool for tracing changes in the oxygen content of the atmosphere. Sulfur of sulfides and sulfates from sedimentary units older than 2.47 Gyr has values of MIF (expressed in terms of $\Delta^{33}S$ ranging from -2.5‰ to +8.1‰, whereas all sulfides and sulfates younger than 1.9 Gyr have $\Delta^{33}S$ values <10.4 ‰ (cf., Bekker *et al.*, 2004 for references). The only known mechanism for producing MIF in sulfur isotopes is photodissociation in the gas phase (Farquhar *et al.*, 2001) that has been observed in the atmosphere (Romero, and Thiemens, 2002).

Preservation of large MIF signals in the Archean record is probably related to the lack of an ozone shield in the atmosphere, allowing deep penetration of high-energy ultraviolet and photochemical dissociation of SO_2 into elemental and water-soluble S species. In an atmosphere with an oxygen content larger than ~ 10^{-5} times PAL, sulfur species are oxidized to sulfate, exchange, and lose most of their MIF signal. The change from an anoxic to an oxygenated atmosphere therefore explains the absence of $\Delta^{33}S$ values > 0.4 ‰ in the S isotopes of sulfides and sulfates during the past 1.9 Gyr, a period during which the atmospheric partial pressure of oxygen (p_{O2}) has been much greater than 10^{-5} PAL. The work of Bekker et al. with early diagenetic pyrites in black shales of South Africa suggest that by 2.32 Gyr ago atmospheric p_{O2} was already $>10^{-5}$ PAL.

5.6 The role of oxygen and iron in eukaryogenesis

Several lines of research suggest the absence of current values of oxygen, O_2, for a major part of the history of the Earth. Some arguments militate in favor of Archean atmospheres with values of the partial pressure of atmospheric oxygen (O_2) about 10^{-12} of the present atmospheric level (PAL). We have already seen in the Introduction that the growth of atmospheric oxygen was due to the evolutionary success of cyanobacteria, which were able to extract the hydrogen they needed for their photosynthesis directly from water. One of the chief indicators of the growth of atmospheric oxygen is shale—a rock that has played a role in our understanding of biological evolution. The onset of atmospheric oxygen is demonstrated by the presence in the geologic record of red shale, colored by ferric oxide. The age of such 'red beds' is estimated to be about 2 Gyr. At that time oxygen levels may have reached 1-2% PAL, sufficient for the development of a moderate ozone (O_2) protection from ultraviolet (UV) radiation for microorganisms from the Proterozoic era (cf., Fig. 5.1), which is the time ranging form 2.5 billion years before the present (Gyr BP) to 570 million before the present (Myr BP).

In fact, UV radiation is able to split the O_2 molecule into the unstable O-atom, which, in turn, reacts with O_2 to produce O_3. This new molecule is known as ozone and is an efficient filter for the UV radiation. The paleontological record suggests that the origin of eukaryotes occurred earlier than 1.5 Gyr BP. Some algae may even date from 2.1 Gyr BP, a period comparable to the first onset of red beds: Except for the nearly 2-Gyr–old coil-shaped fossil *Grypania spiralis,* which may have been eukaryotic, evidence for morphological and taxonomic biodiversification of macroorganisms only occurs towards the beginning of the Mesoproterozoic era (1.6–1.0□Gyr, cf., Fig. 5.1).

This is still rather late, compared to the earliest available prokaryotic fossils of some 3.5 Gyr BP (Schopf, 1993). However, if we keep in mind certain affinities between eukaryotes and Archaea (such as homologous factors in protein synthesis), we may argue that Archaea and the stem group of eukaryotes may have diverged at about the same time (Runnegar, 1992).

This conjecture, combined with the lightest carbon isotope ratios from organic matter (cf., Introduction), implies that bacteria capable of oxidizing methane (CH_4) ('methylotrophs') may have been using methane produced by Archaea that were able to produce it as a by-product of their metabolism. From a careful analysis of such fossils, an age of 2.7 Gyr BP is assigned to the biological transition into eukaryogenesis.

5.7 First appearance of metazoans in the Paleoproterozoic

Once the eukaryotes enter the fossil record, their organization into multicellular organisms followed in a relatively short period (in a geological time scale). Metazoans are assumed to have arisen as part of a major eukaryotic radiation in the Riphean Period of the Late Proterozoic, some 800-1,000 Myr BP when the level of atmospheric O_2 had reached 4-8% PAL (cf., Fig. 5.1).

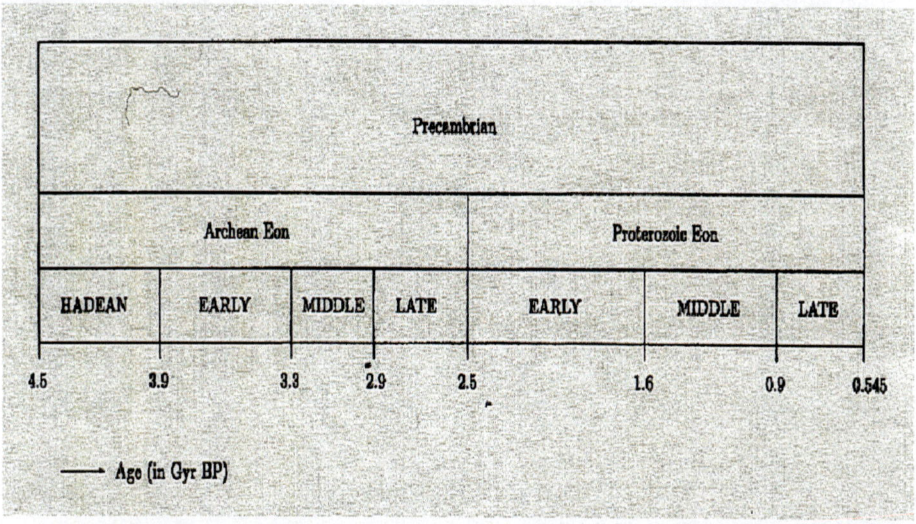

Fig. 5.1 Stratigraphic classification. The Precambrian: spans a period of time divided into several eons, ranging from the formation of the Earth around 4500 Myr BP to the beginning of the Cambrian Period, approximately 545 Myr BP

The centimeter-sized structures discovered in Gabon, West Africa, date back to 2.1 Gyr BP and may represent ancient signs of multicellular life (El Albani *et al.,* 2010). The geographic location is in the Francevillian Group, is well characterized geologically outcropping across 35,000□km^2 in Gabon. This 2.1 Gyr Formation sandstone of the Franceville Basin in Gabon hosts the Oklo natural fission reactors. It has been discovered to contain abundant Palaeoproterozoic oil-bearing fluid inclusions inside which syngenetic biomarkers are preserved. This discovery leads to the conclusion that these Gabon structures are biogenic. The presence of abundant organic matter has been confirmed in at least two separate research programs in the above-mentioned Gabon Formations (Cortial *et al.,* 1990; Mossman *et al.,* 2001), including steranes of eukaryotic origin (Dutkiewicz *et al.,* 2007) is consistent with this interpretation. Besides, contemporary with these fossils there are also large eukaryotic algae from the 2.1 Gyr Negaunee iron-formation in Michigan, USA (Han and Runnegar, 1992).

In addition, there is some evidence in the Late Proterozoic during the Vendian Period, for the existence of early diploblastic grades (Ediacaran faunas). These organisms were early metazoans with two germ layers, such as the modern coelenterates (jellyfish, corals, and sea anemones). Later on, when the level of atmospheric oxygen had reached values in excess of 10% PAL, these groups of primitive animals were overtaken in numbers by more complex groups, triploblastic phyla (cf., Illustrated glossary), as the level of atmospheric oxygen had reached 40% PAL. These organisms are called the Cambrian faunas, i.e., Early Paleozoic faunas, some 500 million years before the present (Myr BP, cf., Fig. 5.2), which were mainly metazoans with three germ layers. They constitute at present the greater majority of multicellular organisms, including *H. sapiens*.

We may obtain further insights from paleontology: acceleration in the evolutionary tempo is observed after the onset of eukaryogenesis, as it is clearly demonstrated by the microfossils of algae from the Late Proterozoic (Knoll, 1994) and by the macrofossils of the Early Phanerozoic (Conway-Morris, 1993). Such evolutionary changes within the first billion years of atmospheric oxygen raised the simple prokaryotes to eukaryotes, animals and plants.

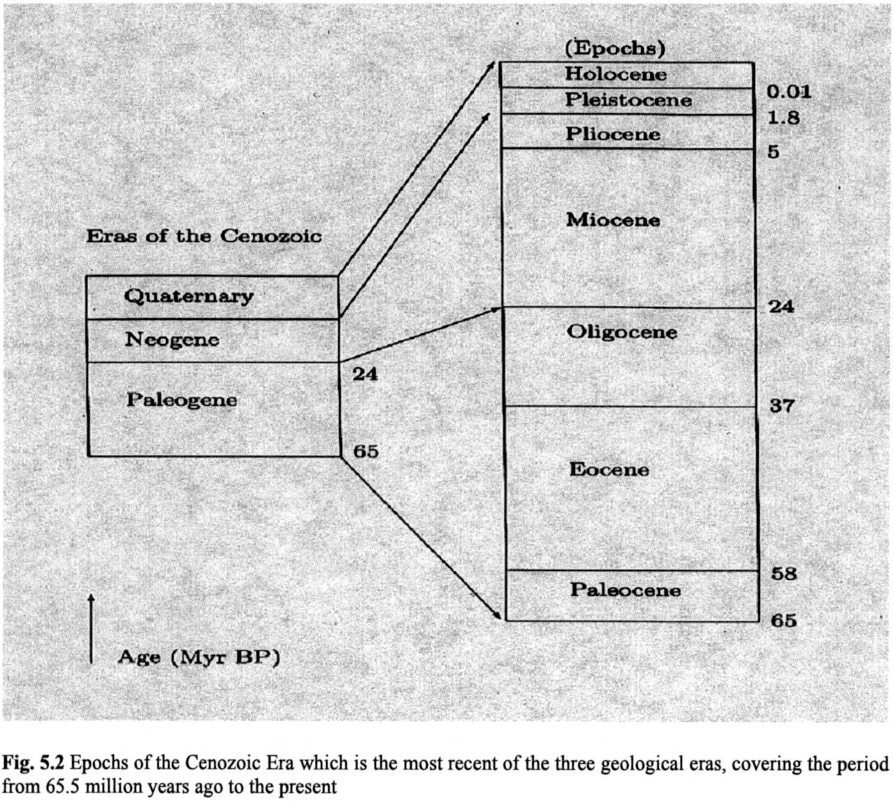

Fig. 5.2 Epochs of the Cenozoic Era which is the most recent of the three geological eras, covering the period from 65.5 million years ago to the present

5.8 The origin of chromosomes

The most remarkable period subsequent to the establishment of the genetic code, or more precisely to the establishment of the translation mechanism of the genetic message, concerns the period in which the chromosomes originated at the very beginning of the evolution of cells. It is widely believed that the genes were part of the first cell. This view implies that the genes of the most advanced higher organisms may reflect the primitive genome structure more clearly than the genes of prokaryotes (Brenner, 1994). This idea is persuasive enough for looking at the contemporary eukaryotic cell as a potential source of information on the structure and function of the first cell. This viewpoint allows the possibility of facing even some of the most difficult questions concerning the primordial cell.

Perhaps amongst such difficult questions stands out the origin and evolution of chromosomes. One general guiding line in this important aspect of the origin of the progenote (Maynard Smith and Szathmary, 1993) is to explain the past in terms of processes that may currently be present in cells. It seems likely that changes in a protocell that led to the eukaryotic cell were in some way related to the control of gene expression (Maynard Smith, 1993). Unfortunately, there is yet no general understanding of this aspect of the eukaryotic cell. One way of approaching this intricate problem is to extract from the known phenomena of genetics some general aspects of the relations between the main relevant parameters and to use such knowledge to sketch possible aspects of the earliest chromosomes.

5.9 The Molecular Clock hypothesis

As we have seen in Section 5.2 molecular changes during evolution may be used in order to reconstruct phylogeny. This possibility is of considerable importance in phylogeny and was referred to as the 'molecular-clock' hypothesis. The molecular changes of sequences of nucleic acids that take part in the universal functions of the cell have proved to be particularly fruitful. These studies are based mainly on some of the ribosome molecules. (These studies are based on the RNA molecule of one of the two subunits that make up the ribosome; specifically, the small one.). These ribosome molecules have a key to phylogeny of the entire span of living cells. Such RNA molecules also allow a natural classification of single-cell organisms.

This approach may be particularly useful in cases where morphology may be insufficient. A clear example is the difficult phylogenetic position of onychophorans (velvet worms). These animals share features of annelids (segmented worms) and arthropods (jointed-foot invertebrates); they resemble mollusks and, like mammals, have placentas. On morphological grounds it has been assumed that these animals belong to a separate phylum, but recent evidence from certain ribosomal RNA sequences suggests rather that onychophorans belong to the phylum Arthropoda (Ballard *et al.,* 1992).

The fact that the eukaryotic cell is a multiple-genome organism has been brought up in discussions of the molecular-clock hypothesis (in which one gene is taken to represent a whole organism). However, the concept of phylogeny is meaningful as long as a clear majority of the essential genes in a genome share a common heritage (Olsen and Woese, 1993). Nevertheless, some care is needed for introducing the domains Achaea, Bacteria and Eucarya beyond the well-established five-Kingdom classification (Margulis and Guerrero, 1991).

It is possible that taxa above the kingdom level may obscure the fundamental division of the biota into organisms with a single ancestry and those with multiple genomes evolving by symbiosis. Besides, the question remains of whether certain sets of molecular sequences should take precedence over other sets. This point may be underlined by the fact that plant cells have normally three genomes, while certain algae have four genomes; for example, the cryptomonads. In this case the fourth genome is in an organelle called the nucleomorph that is bounded by a double membrane, possibly the remains of the nucleus of a symbiotic eukaryote (Douglas *et al.,* 1991). Such cases are referred to as "secondary symbioses".

5.10 A transition in the pathway towards intelligence

Having discussed the origin of chromosomes, one of the questions we have to face is eukaryogenesis. This is perhaps the most significant event in the diversification of life and, according to the fossil evidence, it may have occurred not later than in the Proterozoic about 1.8 - 2.0 Gyr BP. Eukaryotes have their DNA linked in chromatin; the main organelles, such as mitochondria and chloroplasts are normally in their cytoplasm. However, protozoans may provide examples of mitochondrion-less eukaryotes. There are even phyla of amitochondrial protozoa; for instance, the microsporidia (Cavalier-Smith, 1987). Besides, in two out of three Eucarya kingdoms of multicellular organisms (i.e., Animalia and Fungi) chloroplasts are absent. The origin of these two types of organelles, mitochondria and chloroplasts, in the eukaryotes is to be found, according to the serial endosymbiosis hypothesis, in separately evolved organisms. Thus, at least the origin of the red alga chloroplast may be traced back to cyanobacteria. This may be illustrated with the analysis of ribosomal RNA of the unicellular marine red alga *Porphyridium* (Bonen and Doolittle, 1976).

On the other hand, mitochondria may be linked with purple bacteria resembling *Paracoccus denitrificans.* This prokaryote is suggested to be a plausible ancestor, because when all various biochemical parameters are taken into account, *P. denitrificans* resembles a mitochondrion much more closely than other aerobic bacteria (John and Whatley, 1975).

In fact, symbiosis implies that the ancestral prokaryote may have been taken up by a chloroplast-free amoeboid proto-eukaryote; eventually the symbiont may have lost autonomy, possibly by horizontal gene transfer between the proto-mitochondrion and the host's nucleus. These events, which are different from natural selection, were factors that would have led to the evolution of a single-cell organism that had already integrated the metabolism of the partners in symbiosis. Further support for symbiosis is that both eukaryotic organelles we have just mentioned have their own separate mechanisms of translation of the genetic message:

In Chapter 4 we have seen that separate genetic codes are known for each organelle. The origin of the nucleus does not seem to be explained by symbiosis. The main reason for sustaining the thesis of direct differentiation has been summarized briefly elsewhere (Margulis, 1993). This sets some limitation on the extent to which symbiosis may have shaped the first eukaryotic cell, leaving the question of the nature of the earliest eukaryote as an open problem. Primitive eukaryotes have been studied in sufficient detail to allow the study of their origins. One such taxon is the family Cyanidiophyceae. This group of organisms is rhodophytes, commonly known as red algae (Seckbach, 1972).

5.11 In which environments can extremophiles survive in our Solar System?

It is useful to question in which atmospheres extremophiles may survive, as we now know of several Solar System planets and satellites that have atmospheres, not all similar to our own. For example, four large planetary satellites in the outer Solar System are known to have atmospheres (Hall *et al.*, 1995): Europa (Chapter 8) and Titan (Chapter 9).

Mars currently has a CO_2 dominated atmosphere. Besides, looking elsewhere in our Solar System, Ganymede (a Jovian satellite) is known to have a tenuous oxygen atmosphere (Noll *et al.,* 1996); two satellites of Saturn, Rhea and Dione, which orbit within its magnetosphere, have accumulated oxygen and ozone, since their icy surfaces have been exposed to ion irradiation (Noll *et al.,* 1997). Although Mars atmosphere is not typical in our Solar System by comparison with the systems of Jupiter and Saturn, the possibility of extending the biosphere deep into the silicate crust of Mars has been discussed with its suggestive implications. This question seems pertinent to astrobiology and will be discussed in some detail in Chapter 7.

In the meantime looking at our experience on Earth for insights, rather than Mars, we cannot exclude the possibility that organisms, which have been found to inhabit deep in the silicate crust of the Earth, may been deposited with the original sediment, and survived over geologic time. This question has been considered in some detail from the point of view of geophysics (McKay, 1996).

5.12 Mechanisms of evolution beyond natural selection

Before we consider the evolution of man it is worth delaying the discussion in order to appreciate some of the reasons in biodiversity that we may appreciate once eukaryogenesis took place on Earth. We have learnt in the previous chapter ("The dawn of unicellular organisms") that an important genetic mechanism beyond natural selection has been called horizontal gene transfer, HGT (Amabile-Cuevas and Chicurel, 1993; Chela-Flores, 1995). HGT is a process by means of which genetic information may be implanted into

- A target species from a donor species, or
- Intracellularly between organelles, or
- Between organelles and the cell nucleus. Some genes may have been exchanged between the chloroplast and the nucleus.
- The complete sequence of the chloroplast genome of a bryophyte [*Marchantia polymorpha*, a liverwort] has been established. Its DNA has some genes that are not detected in the chloroplast of a more complex plant species [tobacco].

On the other hand, some genes may then have been exchanged between the nucleus of the host plant cell and the residual genome of the corresponding chloroplast during the evolution of angiosperms. Such HGTs may have occurred during the time span separating the first appearance of bryophytes (Paleozoic), from the first appearance of the angiosperms in the late Mesozoic.

It is instructive to appreciate that there are several constraints on chance. Some examples are relevant to the question of whether life elsewhere might follow pathways analogous to the transition from prokaryotes to eukaryotes as known to us from the only example that we know, namely the Earth biota. From various examples of constraints on chance enumerated elsewhere (De Duve, 1995), we select three and provide an additional example:

1. Not all genes are equally significant targets for evolution. The genes involved in significant evolutionary steps are few in number; they are the so-called regulatory genes. In these cases mutations may lead to unviable organisms and are not fixed.

2. Once a given evolutionary change has been retained by natural selection, future changes are severely constrained; for example, once a multicellular body plan has been introduced, future changes are not totally random, as the organisms' viability narrows down the possibilities. For instance, once the body plan of mammals has been adopted, mutations such as those that are observed in *Drosophila*, which exchange major parts of their body, are excluded. Such fruit-fly mutations are impossible in the more advanced mammalian body plan.

3. Not every genetic change retained by natural selection is equally decisive. They may lead more to increasing biodiversity, rather than contributing to a significant change in the course of evolution. This may be illustrated with the following example:

Within the Solanaceae family one tomato chromosome has a region between its center and end that consists of a row of segments in which DNA is compacted into tight masses, largely inactive in *Petunia* transcription, in spite of being another genus of the same family, the abundance of these genetic structures is not preserved. A tiny bit of more compact chromosomes may be superficially indistinguishable from a eukaryotic counterpart. These two genera illustrate how quickly evolution can induce rearrangements, while preserving general chromosome structure. This mutation has not contributed any significant change in the course of evolution.

Besides the constraints on chance just mentioned, we should recall the eternal confrontation deep in the fabric of evolutionary theory, brought to popular attention by Jacques Monod in *Chance and Necessity*. Indeed, implicit in Darwin's work we have *chance* represented by the randomness of mutations in the genetic patrimony, and their *necessary* filtering by natural selection. However, astrobiology forces upon us to accept that randomness is built into the fabric of the living process. Yet, contingency, represented by the large number of possibilities for evolutionary pathways, is limited by a series of constraints as mentioned before. What is more significant for astrobiology is to recognize that natural selection necessarily seeks solutions for the adaptation of evolving organisms to a relatively limited number of possible environments.

We have seen in cosmochemistry (Chapter 2) that the elements used by the macromolecules of life are ubiquitous in the cosmos. The Jovian planets discovered in our 'cosmic neighborhood' can bear satellites in which conditions similar to those in the satellite Europa may be replicated (cf., Chapter 10, exomoons).

To sum up, the finite number of environments forces upon natural selection a limited number of options for the evolution of organisms. From these remarks we expect convergent evolution to occur repeatedly, wherever life arises. It will make sense, therefore, to search for the analogues of the attributes that we have learnt to recognize in our own particular planet. We postpone the discussion till the end of Chapter 11, where we discuss the central question of astrobiology: is life exclusively linked to the Earth?

5.13 The adaptation of life to extreme environments

Life on Earth is ubiquitous. Most of the organisms that we know thrive in normal environments that we consider to be ambient habitats. For additional references we recommend the readers to refer to the work done in collaboration with Joseph Seckbach (Seckbach and Chela-Flores, 2011), on which this section is based, Extremophiles are among the microorganisms living on the edge of life under severe conditions. In recent years microorganisms have been discovered living in extreme environments, such as very high temperature (up to 115° C), and also at very low temperature (~ minus 20° C). In addition, they can also withstand a variety of stresses; amongst them we mention both ends of the pH range; very strong acidity vs. high alkalinity; saturated salt solutions and high hydrostatic pressure. Astrobiology considers the possibility that extraterrestrial civilizations may be present in some exoplanets in the large suite that has been discovered so far. The instruments of research are radio telescopes. Astrobiology also raises the possibility of life elsewhere in the Solar System. The most promising examples are Mars (Chapter 7), Europa (Chapter 8), and possibly Titan and Enceladus (Chapter 9). We know that life exists on Earth in almost every ecological niche. One of the prerequisites for life is the availability of liquid water, sources of energy and a reasonable supply of organic molecules. From our experience with the Earth biota, wherever there is water, there is a good opportunity of finding living organisms.

The search for extraterrestrial life is encouraged by a comparison between organisms living in severe environmental conditions on Earth and the physical and chemical conditions that exist on some Solar System bodies. The extremophiles that could tolerate more that one factor of harsh conditions are called poly-extremophiles. There are unicellular and even multicellular organisms that are classified as hyperthermophiles (heat lovers), psychrophiles (cold lovers), halophiles (salt lovers), barophiles (living under high pressures), acidophiles (living in media of the lower scale of pH). At the other end of the pH scale they are called alkaliphiles (namely, microbes that live at the higher range of the pH scale). Thermo-acidophilic microbes thrive in elevated thermo-environments with acidic levels that exist ubiquitously in hot acidic springs.

Cyanidium caldarium is a classical example of an acido-thermophilic red alga that thrives in places such as hot springs (< 57 ^{oC} and in the range 0.2 - 4 pH). This algal group shows a higher growth rate (expressed as number of cells and higher oxygen production when cultured with a stream of pure CO_2, rather than when bubbled with a stream of air. It has been reported that *Cyanidium* cells resisted being submerged in sulfuric acid. This is a practical method for purifying cultures in the laboratory and eliminating other microbial contamination (Allen, 1959). The psychrophiles thrive in cold environments, such as within the territories found in the Siberian permafrost, around the North Pole in Arctic soils, and they may also grow in Antarctica. Barophilic microorganisms can tolerate a pressure of 1,000 atmospheres on the seafloor, while other barophilic microorganisms have been detected in the subsurface of dry land. In hypersaline areas (such as the Dead Sea, Israel) we find halophilic bacteria (Arahal *et al.,* 1999) and algae that can balance the osmotic pressure of hypotonic external solutions (Oren, 1988).

Recently, the segmented microscopic animals tardigrades, (0.1 – 1.5 mm) have been under investigation (Goldstein and Blaxter, 2002; Horikawa, 2008). These "water bears" are polyextremophilic, and are able to tolerate a temperature range from about 0 °C up to + 151°C (much more that other known microbial prokaryotic extremophiles, Bertolani *et al.,* 2004). But even low Earth orbit extreme temperatures are possible: tardigrades can survive being heated for a few minutes to 151°C, or being chilled for days at -200 °C, or for a few minutes at -272°C, 1° warmer than absolute zero (Jönsson *et al.,* 2008). These extraordinary temperatures were discovered by an ESA project of research into the fundamental physiology of the tardigrade, named TARDIS. Tardigrades are also known to resist high radiation, vacuum, and anhydrous condition for a decade in a dehydrated stage and can tolerate a pressure of up to 6,000 atmospheres. These aquatic creatures are ideal candidates for extraterrestrial life and for withstanding long periods in space. They have already been used in space and have survived such stress.

Supplementary Reading

Attenborough, D. (1981) *Life on Earth.* Fontana, London.

Coyne, J. A. (2009) *Why evolution is true*. Oxford, Oxford University Press, London.

Margulis, L. and Fester, R. (eds.), (1991) *Symbiosis as a Source of Evolutionary Innovation.* The MIT Press, Cambridge, Mass.

Margulis, L. and Sagan, D. (1987) *Microcosm-* Allen & Unwin, London.

References

Allen, M.B. (1959) "Studies with *Cyanidium caldarium*, an anomalously pigmented chlorophyte," *Arch. Mikrobiol.* (Berlin, Heidelberg) **32,** 270–277.

Arahal, D. R., Marquex, M. C., Volcani, B. E., Schleifer K. H. and Ventosa, A. (1999) *Bacillus marismortui* sp. nov., a new moderately halophilic species from the Dead Sea," *Int. J. Syst. Evol. Microbiol.* **49,**521-530.

Amabile-Cuevas, C. F. and Chicurel, M. E. (1993) Horizontal Gene Transfer, *American Scientist* **81**, 332-341.

Ballard, J. W O., Olsen, G. J., Faith, D. P., Odgers, W. A., Rowell, D. M., and Atkinson, P. W. (1992) Evidence from 12S ribosomal RNA sequences that onychophorans are modified arthropods, *Science* **258**, 1345-1348.

Bekker, A., Holland, H. D., Wang, P.-L., Rumble, III, D., Stein, H. J., Hannah, J. L., Coetzee, L. L. and Beukes, N. J. (2004) Dating the rise of atmospheric oxygen. *Nature* **427,** 117–120.

Bertolani, R., Guidetti, R., Jönsson, K.I., Altiero, T., Boschini, D., and Rebecchi, L. (2004) Experiences with dormancy in tardigrades, *Journal of Limnology* **63** (Suppl 1),16-25.

Bonen, L and Doolittle, W. F. (1976) Partial sequences of 16S rRNA and the phylogeny of blue-green algae and chloroplasts, *Nature* **261**, 669-673.

Brenner, S. (1994) The ancient molecule, *Nature* **367**, 228-229.

Cavalier-Smith, T. (1987) Eukaryotes with no mitochondria. *Nature* **326**, 332-333.

Chela-Flores, J. (1995) Molecular relics from chemical evolution and the origin of life in J. Chela-Flores, M. Chadha, A. Negron-Mendoza, and T. Oshima (eds.), *Chemical Evolution: Self-Organization of the Macromolecules of Life (A Cyril Ponnamperuma Festschrift),* A. Deepak Publishing, Hampton, Virginia, pp. 185-200.

Conway-Morris, S. (1993) The fossil record and the early evolution of the Metazoa, *Nature* **361**, 219-225.

Cortial, F., Gauthier-Lafaye, F., Lacrampe-Couloume, G., Oberlin, A. and Weber, F. (1990) Characterization of organic matter associated with uranium deposits in the Francevillian formation of Gabon (lower proterozoic). *Org. Geochem.* **15**, 73–85.

De Duve, C. (1995) *Vital Dust. Life as a Cosmic Imperative,* Basic Books, New York, pp. 294-296.

Desmond, A. and Moore, J. (1991) *Darwin.* Michael Joseph, London, pp. 412-413.

Douglas, S. E., Murphy, C. A., Spenser, D. F., and Gray, M. W. (1991) Cryptomonad algae are evolutionary chimaeras of two phylogenetically distinct unicellular eukaryotes, *Nature* **350**, 148-151.

Dutkiewicz, A., George, S. C., Mossman, D. J., Ridley, J. & Volk, H. (2007) Oil and its biomarkers associated with the Palaeoproterozoic Oklo natural fission reactors, Gabon. *Chem. Geol.* **244**, 130–154.

Goldstein B. and Blaxter, M. (2002) Quick Guide: Tardigrades. *Current Biology* **12**, R475.

El Albani, A., Bengtson, S., Canfield, D. E., Bekker, A., Macchiarelli, R., Mazurier, A. E., Hammarlund, U., Boulvais, P., Dupuy, J.-J., Fontaine, C., Fürsich, F. T., Gauthier-Lafaye, F., Janvier, P, Javaux, E., Ossa Ossa, F., Pierson-Wickmann, A.-C., Riboulleau, A., Sardini, P., Vachard, D, Whitehouse, M. and Meunier, A. (2010) Signs of evolution in the Archean rock formations. *Nature* **466**, 100-104.

Hall, D. T., Strobel, D. F., Feldman, P. D., McGrath, M. A. and Weaver, H. A. (1995) Detection of an oxygen atmosphere on Jupiter's moon Europa, *Nature* **373**, 677-679.

Halliday, A.N. (2001) In the beginning..., *Nature* **409,** 144-145.

Han, T.-M. and Runnegar, B. (1992) Megascopic eukaryotic algae from the 2.1-billion-year-old Negaunee iron-formation, Michigan, *Science* **257**, 232-235.

Holland, H. D. (2006) The oxygenation of the atmosphere and oceans. *Phil. Trans. R. Soc. B* **361**, 903–915.

Horikawa, D. D. (2008) The Tradigrade Ramazzottium varieornatus as a model animal for astrobiological studies, *Biol. Sci. in Space* **22(3)**, 93-98.

John, P. and Whatley, F. R. (1975) *Paracoccus denitrificans* and the evolutionary origin of the mitochondrion, *Nature* **254,** 495-498.

Jönsson, K. I., Rabbow, E., Schill, R. O., Harms-Ringdahl. .M. and Rettberg, P. (2008) Tardigrades survive exposure to space in low Earth orbit, *Current Biology*: R729-R731.

Knoll, A. H. (1994) Proterozoic and Early Cambrian protists: Evidence for accelerating evolutionary tempo, *Proc. Natl. Acad. Sci. USA* **91**, 6743-6750.

Margulis, L. (1993) *Symbiosis in Cell Evolution,* W.H. Freeman & Co., San Francisco.

Margulis, L. and Guerrero, R. (1991) Kingdoms in turmoil, *New Scientist* 23 March, 46-50.

Maynard Smith, J. (1993) *The theory of evolution,* Canto Edition, Cambridge University Press, London, p. 122.

Maynard Smith, J. and Szathmary, E. (1993) The origin of chromosomes I. Selection for linkage, *J. Theor. Biol.* **164**, 437-446.

McKay, C.P. (1996) Oxygen and the rapid evolution of life on Mars, in Chela-Flores, J. and Raulin, F. (eds.), (1996) *Chemical Evolution: Physics of the Origin and Evolution of Life,* Kluwer Academic Publishers, Dordrecht, pp. 177-184.

Mojzsis, S. J., Harrison, T.M. and Pidgeon, R.T. (2001) Oxygen-isotope evidence from ancient zircons for liquid water at the Earth's surface 4,3000 Myr ago, *Nature* **409,** 178-181

Mossman, D. J., Gauthier-Lafaye, F. and Jackson, S. (2001) Carbonaceous substances associated with the Paleoproterozoic natural nuclear fission reactors of Oklo, Gabon: paragenesis, thermal maturation and carbon isotopic and trace element composition. *Precambr. Res.* **106**, 135–148.

Noll, K. S., Johnson, R. E., Lane, A. L., Domingue, D. and Weaver, H. A. (1996) Detection of ozone on Ganymede, *Science* **273**, 341-343.

Noll, K. S., Roush, T. L., Cruikshank, D. P., Johnson, R. E. and Pendleton, Y. J. (1997) Detection of ozone on Saturn's satellites Rhea and Dione, *Nature* **388**, 45-47.

Olsen, G. J. and Woese, C. R. (1993) Ribosomal RNA: a key to phylogeny, *The FASEB Journal* **7**, 113-123.

Oren, A. (1988) The microbial ecology of the Dead Sea, in: K. C. Marshall (ed.) *Advances in microbial ecology* (Plenum Publishing Company, New York, **10,** 193-229.

Prieur, D. Ercuso, G. and Jeanthon, C. (1995) Hyperthermophilic life at deep-sea hydrothermal vents, *Planet. Space Sci.* **43**, 115-122.

Rizzotti, M. (2000) *Early Evolution,* Birkhauser Verlag, Basel, Chapter 3, pp. 24-52.

Romero, A. B. and Thiemens, M. (2002) Mass-independent sulfur isotopic compositions in sulfate aerosols and surface sulfates derived from atmospheric deposition: Possible sources of the MI anomaly and implications for atmospheric chemistry. *Eos* **83** (Fall Meet. Suppl.), B71A–0731.

Runnegar, B. (1992) Origin and Diversification of the Metazoa, in Schopf, J. W. and Klein, C. (1992) *The Proterozoic Biosphere,* Cambridge University Press, New York, p. 485.

Schopf, J. W. (1993) Microfossils of the Early Archean Apex Chert: New Evidence of the Antiquity of Life, *Science* **260**, 640-646.

Seckbach, J. (1972) On the fine structure of the acidophilic hot-spring alga *Cyanidium caldarium*: a taxonomic approach, *Microbios* **5,** 133-142.

Seckbach, J. and Chela-Flores, J. (2011) Astrobiology: From extremophiles in the Solar System to extraterrestrial Civilizations. In: Tymieniecka, A.-T., Grandpierre, A. (eds.) *Astronomy and Civilization in the New Enlightenment.* Series: Analecta Husserliana, Vol. **107**; 1st Edition, 2011, Springer Science and Business Media, pp. 237-246.

Woese, C. R. (1983) The primary lines of descent, in D. S. Bendall (ed.), *Evolution form molecules to man,* CUP, London, pp. 209-233.

Woese, C. R. (1998) The universal ancestor, *Proc. Natl. Acad. Sci. USA* **95**, 6854-6859.

Zukerkandl, E. and Pauling, L. (1965) Molecules as documents of evolutionary history, *J. Theor. Biol.* **8**, 357-366.

6

The evolution of intelligent behavior

In evolutionary terms, we choose not to emphasize complex multicellular organisms. Instead, we have shifted our attention to the single-celled nucleated organism (eukaryote), whose evolution is known to have led to the evolution of intelligent behavior, at least on planet Earth.

The reader is advised to refer especially to the following entries in the Illustrated glossary: *Archaea, archaebacteria, chaos, De Duve, domain, eukaryotes, eukaryogenesis, hominid, hominoids phylogenetic tree, taxonomy, Woese.*

6.1 Modern taxonomy emphasizes single-celled organisms

The microscopic point of view on taxonomy is forced upon us by the present taxonomic classification of organisms into domains, which stresses single-celled organisms. This work is due to Carl Woese, an American microbiologist who defined the Archaea as a new domain in the taxonomic tree (cf., Fig. 6.1). In other words, a domain refers to the highest grouping of organisms, including kingdoms and lower taxons, such as phyla or divisions, orders, families, genera and species. The term "domain" stands for three new branches of life: Bacteria, Archaea, and Eucarya. Woese and co-workers shortened the name Archaebacteria to Archaea (Woese *et al.*, 1990).

A three-domain system is based upon genetic relationships rather than morphological similarities. An older taxonomic classification in terms of kingdoms as the highest taxa was replaced by the domain taxonomy. One evident advantage is that multicellular organisms are not highlighted. As we stated in the Introduction the older approach was due to biologists lacking an understanding of molecular biology that only made its first appearance in the early 1950s. Biology has been able to provide us with sufficient insights into the cell constituents to encourage wide acceptance of a comprehensive taxonomy. We can paraphrase Sir Julian Huxley's comments in the introduction of *"The Phenomenon of Man"* (Teilhard de Chardin, 1954) by remarking that there is an evident inexorable increase towards greater complexity in the transition from Bacteria to Eucarya.

J. Chela-Flores, *The Science of Astrobiology: A Personal View on Learning to Read the Book of Life*, Cellular Origin, Life in Extreme Habitats and Astrobiology 20, DOI 10.1007/978-94-007-1627-8_6,

To borrow the phrase of Christian De Duve, we may say that the laws of Physics and Chemistry imply an 'imperative' appearance of life during cosmological evolution (De Duve, 1995). This is a view that is not in contradiction with the relevant critical remarks of Sir Peter Medawar in his comment on Père Teilhard's work (Medawar, 1996).

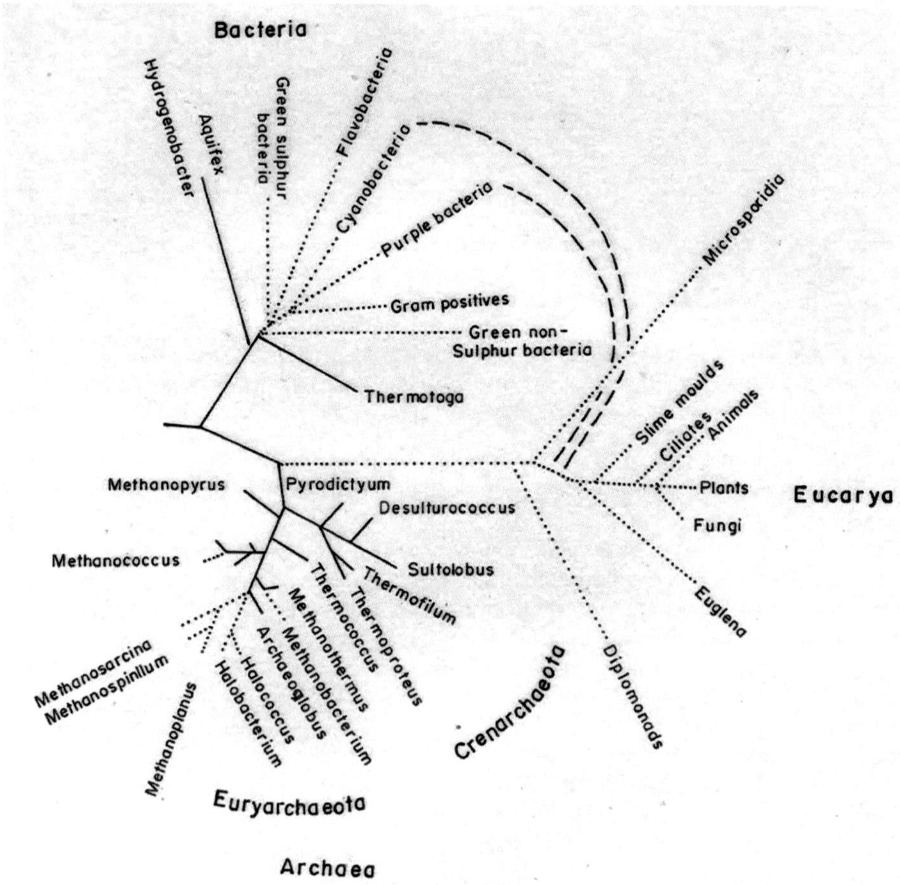

Fig. 6.1 The relationship between the three domains of life (Woese *et al.*, 1990). We have followed the original proposal and terminology used by Carl Richard Woese and co-workers, where the nucleated cells are grouped in the domain Eucarya, while the microorganisms themselves are called eukaryotes

6.2 The phenomenon of the eukaryotic cell

We may argue that not only life is a natural consequence of the laws of physics and chemistry, but once the living process has started, then the cellular plans, or blueprints, are also of universal validity: the lowest cellular blueprint (prokaryotic) will lead to the more complex cellular blueprint (eukaryotic). This is a testable hypothesis. Within a decade or two, a new generation of space missions may be operational. Some are currently in their planning stages, which are aiming to reach the Jovian satellite Europa in the second decade of this century by means of an international collaboration where we expect all the major space agencies will participate by joining NASA and ESA that are the two major partners. We shall present the rationalization behind this effort in the present chapter, and discuss some details in Chapter 8. First of all there is the question of the different positions regarding extraterrestrial life: *Is it reasonable to search for Earth-like organisms or should we be looking for something else?*

We will discuss in turn some of the arguments involved. Firstly, the more widely accepted belief on the nature of the origin of life is that life evolved according to the principles of deterministic chaos (Kauffman, 1993). Evolutionary developments of this type never run again through the same path of events. Secondly, the possibility for similar evolutionary pathways on different planets of the solar system has been defended in the past (Davies, 1998).

Indeed, even if some authors may consider this to be a remote possibility, there is an increasing acceptance that catastrophic impacts may have played an important role in shaping the history of terrestrial life. There may be some common evolutionary pathways between the microorganisms on Earth and those that may have developed on Mars during its ‘clement period’ (roughly equivalent to the early Archean in Earth stratigraphy). The means of transport may have been the displacement of substantial quantities of planetary surface, due to large asteroid impacts on Mars.

Finally, even in spite of the second possibility raised above, many researchers still see no reason to assume that the development of extraterrestrial life forms followed the same evolutionary pathway to eukaryotic cells, as it is known to have occurred on Earth. Moreover, it would seem reasonable to assume that our ignorance concerning the origin of terrestrial life does not justify the assumption that any extraterrestrial life form has to be based on just the same genetic principles that are known to us.

In sharp contrast to the position denying that of common genetic principles may underlie the outcome of the origin of life elsewhere, there is a fourth point of view in the discussion of the various possible ways of approaching the question of the nature of extraterrestrial life. We may conclude that we all agree that the final outcome of life evolving in a different environment would not be the same as the Earth biota. New ground is reached raising the question: *How different would be the outcome of the origin of life elsewhere?* (De Duve, 1995, pp. 294-296). This has led to a clear distinction that there is no reason for the details of our phylogenetic tree to be reproduced elsewhere (except for the possibility of biogenic exchange in the solar system discussed above). The evolutionary tree of life constituted by the Earth biota may be unique to planet Earth. On the other hand, there is plenty of room for the development of differently shaped evolutionary trees in an extraterrestrial environment, where life may have taken hold.

6.3 The phenomenon of multicellularity

A point worth emphasizing is that higher organisms are not only characterized by eukaryoticity, but also by multicellularity. This second feature evolved gradually form unicellular microorganisms. An advantage of cellular aggregation is its adaptive value.

An evident example of such advantage regards predator-prey interactions (Farmer, 1992). This follows from direct observation of contemporary colonies of the bacterium *Myxococcus xanthus*. This bacterium forms spherical colonies that increase their success in predation. The unicellular aggregation lets the prey enter the sphere through gaps, which are able to retain digestive enzymes that are produced by the bacteria. We have seen in previous chapters that the prokaryotic blueprint is a consequence of chemical evolution. (In a geologic time scale prokaryogenesis is almost instantaneous.) It follows from these considerations that multicellularity is also a consequence of chemical and biological evolution in a terrestrial-like environment.

The detailed current view on terrestrial evolution spanning from the first appearance of the eukaryotic cell to the evolution of multicellular organisms is as follows: In the Proterozoic Eon, a geologic period which extends from 2.5 to 0.545 Gyr BP, we come to the end of the Proterozoic, an era which had seen the first appearance of the eukaryotic cell. The Proterozoic was followed by the most recent eon that has seen the spread of multicellular organisms throughout the Earth on its oceans and continents, the Phanerozoic, which extends from the end of the Proterozoic to the present. We should comment on some aspects of the current Phanerozoic:

- The Paleozoic era saw the first appearance of fish (510 Myr BP) and some other vertebrates, a landmark in the evolution of life on Earth.
- The Mesozoic era, initiated after the massive extinction some 250 Myr BP. This era saw the first appearance of mammals (205 Myr BP).

Current evolutionary thinking sees the first appearance of Man, a question of central interest both to theology and philosophy, as a natural consequence of the series of events which began in the Mesozoic and culminated in the Cenozoic Era, particularly some 5 Myr BP, when according to anthropological research, the human ancestors first appeared.

6.4 Evolution of the hominoids

The last era of the Phanerozoic Eon is called the Cenozoic (cf., Fig. 6.2). This was initiated after the great extinction of the Upper Mesozoic. The Cenozoic saw the first appearance of primates some 60 Myr BP. These essentially small animals took up residence in trees. Natural selection eventually provided these small mammals with long arms, fingers and frontal vision. Some of these species remained unaltered to the present. They are represented by the lemurs of Madagascar, which include the indri, a prosimian that hardly ever comes to the ground (Attenborough, 1979).

The Rift Valley in Eastern Africa may hold the key to the divergence of hominids from the great apes (represented, for example by chimpanzees and gorillas). In fact, tectonic forces created the Rift Valley by about 8 Myr BP. The new mountain boundaries divided the hominoid primates, which are a group that includes the lesser apes (gibbons, the small arboreal anthropoid apes of Southeastern Asia), the great apes, and humans. Such a division formed two groups. Firstly, a western group bound to the forests, which were the ancestors of the modern great African apes; secondly, an eastern group was forced to live in the savanna, evolving directly into our ancestors, the first humans. Intelligence, in its highest form, had to await the emergence of the genus *Homo*, more than 2 Myr BP, when its first traces arose in our ancestors.

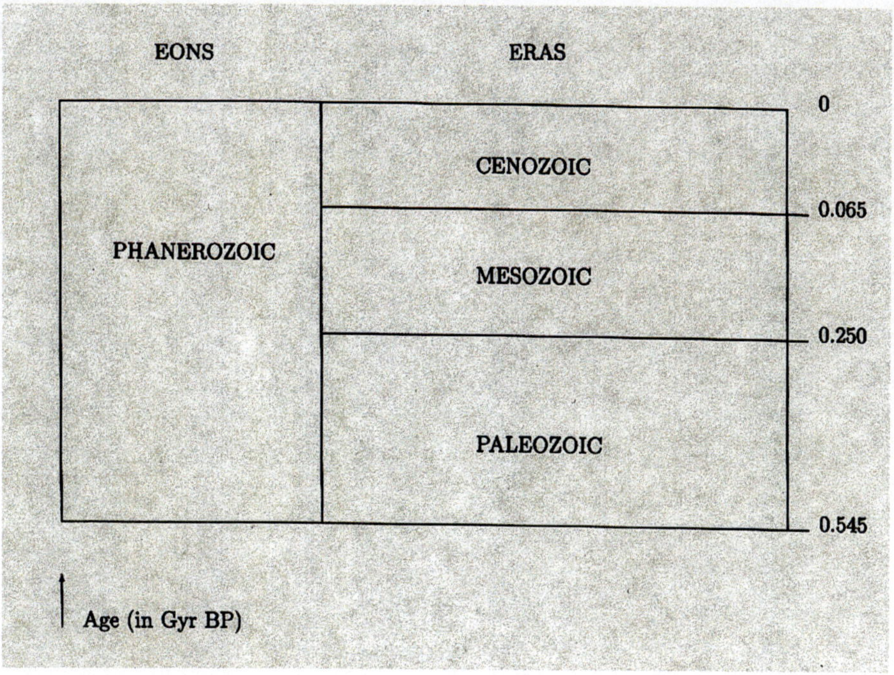

Fig. 6.2 Eras and suberas of the Phanerozoic Eon, the most recent eon of Earth history

A more evident demonstration of the appearance of intelligent life on Earth than the habilines' tools, or the ceremonial burials had to wait till the Magdalenian "culture" (cf., Table 6.1). This group of human beings was one of the later cultures of the Upper Paleolithic in Western Europe, dating from around 17,000 BP to 9,000 BP. It is named after the site of La Madeleine, a rock shelter located in the Vézère valley, commune of Tursac, in the Dordogne department of France. The Magdalenian is associated towards the end of the last ice age with sites that contain evidence for the hunting of large European mammals. The culture was geographically widespread, and later Magdalenian sites have been found from Portugal to Poland.

The Magdalenians and even earlier cultures—the Aurignatians—left some fine works of primitive art as, for instance, the 30,000-year-old paintings on the walls of the cave discovered in December 1994 by Jean-Marie Chauvet in South-Eastern France. Indeed, the birth of art in these caves is one of the most striking additions to the output of humans that entitle us to refer to the groups that produced these fine works, as *cultures,* rather than *industries*, a term that is reserved to the group of humans that produced characteristic tools, rather than works of art.

The purpose of this sketch has been to bring into focus different branches of science. These include the social sciences. In terms of them some answers have been provided to the questions humans have recorded since the most ancient times. Pioneers in these queries were the Israelites over three thousand years ago, when the events of the Torah (Old Testament of the Bible) were taking place. Those disciplines include anthropology, paleontology, prehistoric research, geochronology and geochemistry, amongst others.

Table 6.1 Archaeological classification with a reference to the European cultures. Some details are given of events, which have been relevant to the genus *Homo*. The continental classification for glaciations and interglaciations is highlighted

Time ***(10^3 yrs)***	***Stratigraphical reference***	***Glaciations***	***Archaeologic Classification (General)***	***Archaeologic Classification (SW Europe & North Africa)***
0			Iron Age	
5			Bronze Age	
10	Holocene	Post Glacial	Neolithic	
			Mesolithic	
10				Magdalenian
20				
30	Late			Aurignacian
40	Pleistocene	Würm	P	
50				Mousterian
			A	
60				
70			L	
80				
			E	
			O	
		Interglacial	L	
120		Riss	I	
250				Acheulian
		Interglacial	T	
300	Middle			
500	Pleistocene	Mindel	H	
		Interglacial		
800	Early	Günz	I	
900				
		Interglacial	C	
1400	Pleistocene	Danube		
1700	Proterozoic			

(van Eysinga, 1975; Chela-Flores, 1994)

6.5 Evolution of intelligent behavior in the hominids

Amongst the earliest humans, or hominids, the habilines (*Homo habilis*) were a group from East Africa that first appeared from about 3 to 2 Myr BP, just before the onset of the Quaternary Period about 2.6 Myr BP. The brain of these early hominids was large, but still about 60% of the size of our own brain (1360 cc, on the average, over the last 100,000 years). In spite of this difference in cranial capacity, there is evidence that the habilines used a wide variety of tools. One of the differences between *Homo* and *Australopithecus* is the smaller size of *Homo*'s jaw muscles. Some australopithecines needed powerful muscles for processing nuts and other hard foods (some were gracile and some were robust) that became necessary after the environmental change in Africa during the Pliocene.

On the other hand, muscle size can influence bone growth. Reduction of the jaw muscle attenuates the stress on skull bones. The larger and thinner skull of humans could be the effect of smaller jaw muscles. In conclusion there could be a strong connection between size of jaw muscles and size of the brain. In addition, a less massive jaw muscle could have allowed a better coordination of the mandible function, improving speech capabilities. At the time of the emergence of the habilines (cf., above), the climate was becoming drier and cooler, inducing dramatic changes of vegetation and fauna. This is a challenging moment for hominins in Africa, who needed to change diet and way of life in order to survive.

The group of tools produced by the habilines belongs to the so called "Oldowan industry", as they have been found in the Olduwai Gorge, where the East African Rift cuts through about 100 meters of sediments laid down in a former lake basin. More advanced tools are those of the Acheulian industry, which gradually replaced the earlier tools, but the new tradition in tool-making is characteristic of a more advanced hominid, *Homo erectus,* who lived from at least 1.7 to 0.5 Myr BP, and may have been the first hominids to use fire for cooking. The Acheulian industry, in turn, was gradually replaced by the Mousterian, which emerged during the last interglacial (cf., Table 6.1). Glaciations and interglaciations were major challenges to the development of the genus *Homo*, from the habilines to modern humans. The most recent humans besides ourselves are the Neanderthals (*Homo neanderthalens*), who lived in Europe and western Asia, approximately from the Riss glaciation (250,000 - 120,000 yr BP) to the Würm glaciation (80,000 - 10,000 yr BP), most of them living from 100,000 to 40,000 yr BP (cf., Table 6.1).

Instead of considering the concept of intelligence itself, the evolutionary background of this work induces us to stress cultural evolution with results that became more prominent since the last glaciation (the Holocene Epoch). These included the ability of the early humans to learn, communicate and teach their offspring skills, such as writing (some 5,000 yr BP), and even abstract and practical concepts such as religion, philosophy and rudimentary science. We have seen in Chapter 1 that these aspects of humanity were already established in Ancient Greece some 2,500 years ago. Such abilities of *H. sapiens* gave, to those that possessed them, increased advantage over the already selective advantage they had on other primates, which had been provided by genetics.

Cultural evolution may be considered as a consequence of the presence of larger brains. Not only humans, but also other mammals and birds, possess an additional selective advantage proportional to brain size. This advantage goes beyond favorable random mutations, as the capacity to learn is an additional pressure that will encourage, not only the adaptability to the environment, but also some special attributes, such as learning.

6.6 Did external events contribute to the evolution of intelligence on Earth?

The role of cataclysmic events in the evolution of life on Earth has been discussed in recent times (Chela-Flores *et al.*, 2009, 2010, on which some of this section is based). It has been suggested that there might be a possible causal connection between the evolution of hominids and large impacts, for instance, the Vredefot Dome, an impact structure in the Free State (Tobias, 2005). It is Precambrian in age with a diameter of over 140 km. More precisely it is an "astrobleme", namely the remains of an ancient meteorite-impact structure. It was made by a projectile estimated at 10-15 km in diameter, which collided with the Earth at 2.1 Ga.

This impact coincided with two significant events in the evolution of life on Earth, namely the oxygenation of the atmosphere and eukaryogenesis (Chapter 5 and Chela-Flores, 1998). The Vredefot impact illustrates the major implications of extraterrestrial events, not only related to Space Weather, but even planetary sterilizations going back into the Phanerozoic and Proterozoic, extending back to the Hadean.

This event may have been relevant in the evolution that led to *Homo sapiens* (Tobias, 2005). For instance, about 2.6-2.5 Ma BP marked climatic changes in Africa that were associated with uplift of its southern and eastern parts. The ensuing cooler and dryer weather was accompanied with significant changes in the paleontological record:

- Extinction of the small-brained hominids *Astralopithecus africanus.*
- The earliest appearance of *Homo* of the species *Homo habilis.*
- The first signs of the enlargement of the hominid brain, as compared with the smaller brains of the australopithecines.

The possibility of a supernova explosion near the Solar System has been discussed for a long time (Ruderman, 1974; Reid *et al.,* 1978; Ellis and Schramm, 1995). Such a nearby supernova explosion can be confirmed by the detection of radioisotopes on Earth that were produced and ejected by the supernova. A measurement of a well-resolved time profile of the ^{60}Fe concentration in a deep-sea ferromanganese crust showed a significant increase 2.8 Ma (Knie *et al.,* 2004). The amount of ^{60}Fe is compatible with the deposition of ejecta from a supernova at a distance of a few 10 pc. The well-defined time of the supernova explosion makes it possible to search for plausible correlations with other events in Earth's history.

The profile of the ^{60}Fe concentration in the deep-sea ferromanganese crust has been considered in terms of the environmental changes that were relevant for *Homo* evolution. According to the authors (Knie *et al.*, 2004), at the time of the supernova explosion there was an increase of the cosmic radiation of a few percent that lasted for some thousand years. They claim this might have triggered climate change in Africa, causing significant developments on hominid evolution. This effect would in any case be superimposed on other phenomena causing climate change, such as tectonic activities (like those that gave rise to the Great Rift Valley in Africa), as well as other global phenomena.

Discrepancies in the production rate of stable cosmogenic radionuclides, such as ^{21}Ne and radioisotopes with different half-lives such as ^{10}Be, ^{26}Al and ^{53}Mn might indicate time variation in the galactic cosmic-ray flux within the Solar System. The existing data do not support major variations in cosmic ray intensity within the past 5 million years, crucial period for the evolution of the *Homo* species (Moniot *et al.,* 1983). More recently it has been suggested that a comet or asteroid exploded over

North America 13,000 years ago, but these factors still need to be widely accepted. This event may have caused a major shift in the climate, the well-known Youger Dryas cooling event. The cooling produced may also have affected humans in Europe and Asia (Firestone *et al.,* 2007). A detailed analysis of the sediments corresponding to 13 ka ago reveal a high concentration of extraterrestrial markers such as glass-like beads, soot and fullerenes, materials that are absent in other layers of the stratigraphy. The glassy beads could only be produced by melting carbon at 4000 degrees C. Electron microscope analyses show the glassy spherules are reach in micro-diamonds. Diamonds are produced in the interior of the Earth by compressing carbon at the pressure of several gigapascals. These conditions could be produced on the surface of the planet only by the impact of a massive extraterrestrial body.

6.7 The pathway towards communication

We would like to stress that cultural evolution is a stage in human development that can be traced back to that crucial transition from prokaryotes to eukaryotes, which is estimated to have occurred on Earth some 2 Gyr BP (cf., Sec. 5.5). We are now aware of the existence of extrasolar planets. This suggests the question whether in those environments eukaryogenesis may have already occurred. Data available from a large number of earth and life science disciplines, mentioned above, lead us to the conjecture:

Provided the planets of a given solar system have the appropriate volatiles (particularly water and oxygen), not only prokaryotic life will appear, but also eukaryogenesis will emerge.

This conjecture has the merit that it is subject to observational confirmation, as we shall see in Chapter 10. The conjecture is not idle for it addresses a latent problem that underlies a large research effort, namely that of the search for extraterrestrial intelligent behavior (cf., Chapter 11). In fact, a defining attribute of humans is our cognitive ability. We seem to be the only species that has evolved on Earth, which is capable of wondering, as we are doing in this book, about the relation life-universe. It remains unknown whether the several other human species that may have co-existed with *Homo sapiens* in the last hundred thousand years, may have also been capable of wondering on the position of human beings in the universe. In other words, we are not the only species where some degree of intelligent behavior has evolved. (In this context we define intelligence in the usual manner—the faculty of understanding, the capacity to know or comprehend the environment. A lesser degree of intelligence in this sense is also present in dolphins and other cetaceans in the aquatic environment (cf., Chapter 12).

The problem of communication amongst dolphins and between dolphins and humans (Deacon, 1997, Impey, 2010) can be considered as a model study of a larger issue that we will have eventually to face, namely communication between our civilization and those potential products of natural selection on extrasolar planets, or their satellites. We shall return to the significant analogies between delphinids and anthropoids in Chapter 12. In the wider scenario of the present work, the problem of communication amongst intelligent beings from different stellar environments has eventually to be faced in truly scientific terms (cf., Chapter 11). We consider the two evolutionary pressures, terrestrial and aquatic that may have given rise to the largely different brains of dolphins and humans. But some general features are common to both of them, such as the modular arrangement of neurons (Manger *et al.*, 1998). The intelligent behavior of humans and dolphins are products of an evolutionary pathway that began from the general prokaryotic blueprint appearing on Earth almost instantaneously (in a geologic time scale, cf., Chapter 5), and continued through the sequence of eukaryogenesis, neuron, multicellularity and, finally, brains.

The present discussion of universal aspects of communication amongst humans at some point has to be extended to all life in the universe. In other words, our discussions should not be simply confined to our phylogenetic tree, in spite of the fact that ours is the first species we are aware of that has evolved cognitive ability.

6.8 The emergence of language and music in modern humans

This question is of fundamental importance to astrobiology. In particular, it is relevant to the search of other civilizations. Language is a natural target for astrobiological research. It is a difficult subject, one of the reasons being that brains do not fossilize. However, the existence of complex musical instruments that are known to be preserved in the archeological record can be taken as indication of fully modern behavior and advanced symbolic communication (Conrad *et al.*, 2009). The available evidence comes from southwestern Germany, where bone and ivory flutes have been retrieved from the early Aurignacian culture, namely a toolmaking industry of the Upper Paleolithic Europe. These findings demonstrate the presence of a well-established musical tradition at the time when modern humans colonized Europe, more than 35 kyr BP. This period corresponds to a tool making industry and artistic tradition of Upper Paleolithic Europe that followed the Mousterian industry over 100kyr BP (cf., Table 6.1).

In our planet natural selection seems to have produced universal characteristics in the first species that has reached cognitive ability. Indeed, Steven Pinker in *The Language Instinct* argues (Pinker, 1994) that language is compatible with gradual evolution due to natural selection. Pinker argues his case with a large amount of evidence in favor of a genetic basis for spoken language.

This possibility should be seen against the background of the development of a science of language. Noam Chomsky argued that language has underlying structural similarities. Since the evolutionary process has not led to our species having the faculty of expressing itself in one particular language, Chomsky defends the thesis that powerful constraints must be operative restricting the variety of languages (Chomsky, 1975). Hence, according to this view, one can conclude that we must be born with knowledge of an innate universal (or transformational) grammar. He supplied linguistics with a philosophical foundation (rationalism rather than empiricism). According to Chomsky the structure of language is fixed in the form of innately specified rules: during the process of development all a child has to do is to turn on a few 'biological switches' to become a fluent speaker of a given language. In other words, a child after being exposed to the speech of his parents is in a position to infer grammatical rules and use them to generate his own sentences. In Chomsky's view children are not learning at all, anymore than birds learn their feathers (Dennett, 1995). Yet this point of view has turned out to be controversial.

Pinker argues convincingly in favor of the adaptationist interpretation of the origin of language. In this view, language can be seen as the "imperative" by-product of natural selection, rather than a universal grammar. The outline of his argument is as follows: the abrupt increase in brain size may have been due to the evolution of language, instead of language arising only as brain size goes beyond a certain threshold. This faculty of the brain is to be searched as a consequence of natural selection, as any other attribute that characterizes a human being, such as size or skin color. Language, in this evolutionary context, evolved through social interaction in our hominid ancestors, as they adopted collective hunting habits (cf., Chapter 6). Better communication by means of rudimentary language would have been favored by natural selection.

Underlying all our efforts in the science of astrobiology, we find the assumption that the scientific laws are universal. Independent of the particular form given to an alien being by the evolutionary process on an extrasolar planet, or satellite (cf., Chapter 10), it seems reasonable to conjecture that Darwin's seminal theory of evolution through natural selection is probably the only theory that can adequately account for the phenomenon that we associate with the phenomenon of life anywhere in the universe. Some biologists have already defended this view for decades, beginning with Richard Dawkins (Dawkins, 1983).

According to the ideas expressed in Chapter 12, brain is bound to evolve, once the evolutionary process is set in motion anywhere in the cosmos. Provided there is sufficient geologic time available to a given solar system (cf., comments in Chapter 1 regarding the differential ages of stars that follow from the information summarized in the HR diagram, Fig. 1.6), natural selection is bound to play analogous roles on brain evolution across species, within a given phylogenetic tree.

Yet, we know form life on Earth that there are clear examples of evolution prizing one specialized faculty; this remark can be illustrated with the enormous development of the corresponding part of the brain. For example, 7/8ths of a bat's brain is devoted to hearing. The thesis of co-evolution of language and brain has already been developed in considerable detail (Medina-Callarotti, 2000).

Without entering into too much detail, it is sufficient to point out that recent human evolution demonstrates that certain adaptations are useful to other animals. For instance, fur and early sexual maturity were, in fact, lost in our own species in favor of language, speech and consciousness. The presence of such faculties in humans that have evolved on our own solar system, approximately half way through the lifetime of its star (i.e., the Sun), would argue in favor of *evolution of intelligent behavior in the plurality of new worlds* that were brought to our attention for the first time at the closing of last century, since 1995 (cf., Chapter 10).

Supplementary Reading

Bendall, GS, ed. (1983) *Evolution from molecules to men,* Cambridge University Press, London.

Gould, S.J. (1991) *Wonderful Life The Burgess Shale and the Nature of History,* Penguin Books, London.

Maynard Smith J. (ed.), (1982) *Evolution now,* Nature, London.

Maynard Smith, J. (1993) *The theory of evolution.* Cambridge University Press, London.
Patterson, C. (1978) *Evolution,* British Museum (Natural History), London.

References

Attenborough, D. (1979) *Life on Earth,* Fontana, London, p. 271.

Chela-Flores, J. (1994) La vita nell'universo: verso una comprensione delle sue origini. In: Origini: l'universo, la vita, l'intelligenza, in F. Bertola, M. Calvani and U. Curi (eds.), Padova, Il Poligrafo, pp. 33-50.

Chela-Flores, J. Jerse, G., Messerotti, M. and Tuniz, C. (2009) Astronomical and astrobiological imprints on the fossil records. A review. In: *From Fossils to Astrobiology*, J. Seckbach (ed.). Cellular Origins, Life in Extreme Habitats and Astrobiology, Springer, Dordrecht, The Netherlands, pp. 389-408. http://www.ictp.it/~chelaf/FOASfinal.pdf

Chela-Flores, J., Montenegro, M.E., Pugliese, N. Tewari, V.C. and Tuniz, C. (2010) Evolution of plant-animal interactions. In: *All flesh is grass: Plant-Animal Interactions, a love-hate affair.* J. Seckbach and Z. Dubinsky and (eds.). Cellular Origin and Life in Extreme Habitats and Astrobiology, Springer: Dordrecht, The Netherlands. http://www.ictp.it/~chelaf/PLAN.pdf

Chomsky, N. (1975) *Reflections on Language,* New York: Pantheon Books, pp. 10-11.

Conard, N. J., Malina, M. and Münzel, S. C. (2009) New flutes document the earliest musical tradition in southwestern Germany. *Nature* **460**, 737-740

Davies, P.C.W. (1998) Did Earthlife Come from Mars?). In J. Chela-Flores and F. Raulin, (eds.), *Chemical Evolution: Exobiology. Matter, Energy, and Information in the Origin and Evolution of Life in the Universe,* Kluwer Academic Publishers, Dordrecht, pp. 241-244.

Dawkins, R. (1983) *Universal Darwinism,* in Evolution form molecules to men, Bandell, D.S. (ed.), Cambridge University Press: London, pp. 403-425.

De Duve, C. (1995) *Vital dust: Life as a Cosmic Imperative,* Basic Books, N. Y.

Deacon, T. W. (1997) *The Symbolic Species. The Co-evolution of Language and Brain,* W.W. Norton & Company, New York.

Dennell, R. and Roerbroeks, W. (2005) An Asian perspective on early human dispersal from Africa. *Nature* **438**,1099–1104.

Dennett, D. (1995) *Darwin's Dangerous Idea Evolution and the Meanings of Life,* Penguin Books, London, pp. 384-393.

Ellis, J. and Schramm, D. N. (1995) *Could a nearby supernova explosion have caused a mass extinction. Proceedings of the National Academy of Sciences USA* **92**, 235-238.

Farmer, J. (1992) Origins of multicellular individuality, in J. W. Schopf and C. Klein (eds.), *The Proterozoic Biosphere. A Multidisciplinary Study,* Cambridge University Press, New York, pp. 429-431.

Firestone, R., Alan West, James P Kennett, Luann Becker (2007) New Insights into Younger Dryas Climatic Instability, Mass Extinction, the Clovis People, and Extraterrestrial Impacts. Contribution PP05, 2007 Joint Assembly: Acapulco, Mexico - 22-25 May 2007.

Impey, C. (2010) Chapter 16: Lori Marino. In: Talking about life. Conversations about Astrobiology. Cambridge University Press, UK, pp. 154-164.

Kauffman, S.A. (1993) *The origins of order: Self-Organization and Selection in Evolution,* Oxford University Press, London.

Knie, K., Korschinek, G., Faestermann, T., Dorfi, E.A., Rugel, G., Wallner, A. (2004) ^{60}Fe anomaly in a deep-sea manganese crust and implications for a nearby supernova source. *Physical Review Letters* **93**, 171103-1-171103-4.

Manger, P. Sum, M, Szymanski, M. Ridgway, S and Krubitzer, L. (1998) Modular Subdivisions of Dolphin Insular Cortex: Does Evolutionary History Repeat Itself?, *Journal of Cognitive Neuroscience* **10**, 153-166.

Medawar, P. (1996) *The strange case of the spotted mice and other classic essays on science,* Oxford University Press, London.

Medina-Callarotti, M. E. (2000) *Origins of Langiuage. The evolution of human speech* in J. Chela-Flores, G. A. Lemarchand and J. Oro (eds.) *Astrobiology from the big bang to civilization.* Kluwer Academic Publishers, Dordrecht, The Netherlands, pp. 225-232.

Moniot, R. K., Kruse, T. H., Tuniz, C., Savin, W., Hall, G. S., Milazzo, T., Pal, D., and Herzog, G.F. (1983) The Ne-21 production rate in stony meteorites estimated from Be-10 and other radionuclides. *Geochim. Cosmochim. Acta* **47**, 1887.

Pinker, S. (1994) *The Language Instinct. How the Mind Creates Language,* William Morrow, New York.

Reid, G. C., McAfee, J. R. and Crutzen, P. J. (1978) Effects of intense stratospheric ionization events. *Nature* **275**, 489-492.

Ruderman, M. A. (1974) *Possible consequences of nearby supernova explosions for atmospheric ozone and terrestrial life. Science* **184**, 1079-1081.

Sponheimer, M., Passey, B. H., de Ruiter, D. J., Guatelli-Steinberg, D., Cerling, T. E. and Lee-Thorp, J. A. (2006) Isotopic Evidence for Dietary Variability in the Early Hominin Paranthropus robustus. *Science* **314,** 980 – 982.

Teilhard de Chardin, P. (1954) *The Phenomenon of Man,* Collins, New York.

Tobias, P. V. (2005) *Catastrophism and the history of life,* Quest **2** (1), 24-25.

van Eysinga, F. W. B. (1975) *Geological Time Table,* (3rd ed.), Elsevier Scientific Publishing Company, Amsterdam.

Woese C, Kandler O, Wheelis M (1990) *Towards a natural system of organisms: proposal for the domains Archaea, Bacteria, and Eucarya.* Proc. Natl. Acad. Sci. USA **87** (12), 4576-4579.

THE BOOK OF LIFE

PART 3: THE DISTRIBUTION OF LIFE IN THE UNIVERSE

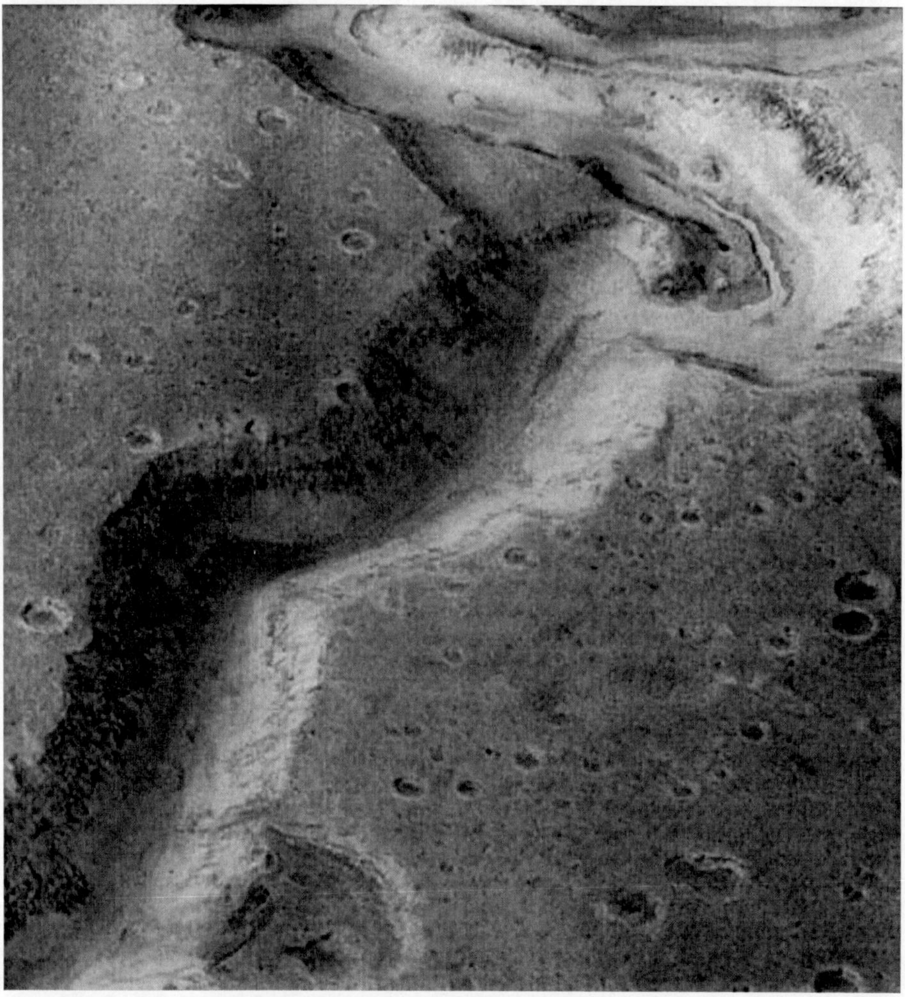

In a search-for-water strategy, NASA and ESA have produced some evidence that water is ubiquitous in the Solar System, enhancing our insights as to what is possible in the main objective of astrobiology- evidence that life beyond the Earth is not rare- on the contrary- life is a consequence of cosmic evolution.

We will go beyond Mars to enquire what is possible in the outer solar system, in the satellites of Jupiter and Saturn. In the case of the Red Planet a 1998 Mars Orbital Camera image of the Nanedi Vallis canyon illustrates the question that some evidence of an Eden-like era may have taken place on Mars (cf., Section 7.2). Channels suggest that liquid water flowed on the red planet sometime in the past.

7

On the possibility of biological evolution on Mars

The main interest in Mars from the point of view of astrobiology is centered on the fact that in the past this planet may have been hospitable to life. The details of the search for life on Mars shall be discussed briefly in this chapter.

The reader is advised to refer especially to the following entries in the Illustrated glossary: *Gas chromatography, mass spectrometer, plate tectonics, pyrolysis, Viking.*

7.1 What have we learnt from previous missions to Mars?

Unlike the Outer Solar System giant planets, Jupiter and Saturn, there has been fleet of missions sent to Mars (Table 7.1 shows some of the significant missions that were launched successfully last century, and many others are now currently being planned).

Table 7.1 A selection of some of the early missions sent to Mars

Mission	*Launch date*	*Results*
Mariner 4	28/11/64	Fly by on 15/7/65 (First close-up images)
Vikings 1,2	20/8/75; 9/9/75	Landed on 20/7/76, 3/9/76 (search for life)
Pathfinder	4/12/96	Landed on 4/7/97

(Carr, 1999)

The reasons for these multiple and frequent missions to the Red Planet are not difficult to find. Mars is a near neighbor and in many senses the most similar to our own world (cf., Table 7.2).

J. Chela-Flores, *The Science of Astrobiology: A Personal View on Learning to Read the Book of Life*, Cellular Origin, Life in Extreme Habitats and Astrobiology 20, DOI 10.1007/978-94-007-1627-8_7,

Table 7.2 Physical parameters of Mars

Parameters	***Values***
Mass (Earth mass=1)	0.1
Radius (Earth radius =1)	0.53
Orbital eccentricity	0.0934
Eccentricity (Earth eccentricity=1)	5.471
Mean surface temperature	-60 ° C
Mean surface pressure	6 mb
Mean surface gravity	372 cm/sec^2
Mean orbital velocity	24.1 km/sec
Mean density	3.933 gm/cc
Visual albedo or reflectance	0.159
Tilt of axis	25°
Sidereal day	24 hr 37 min 22.663 sec
Sidereal period	687 mean solar days
Mean distance from the Sun	227,941,000 km

To these reasons we should add a much stronger argument. If we look attentively into the Martian geologic past (Barlow, 2008, Carr, 2006), we discover what has been called an "Eden-like era" where it is likely that liquid water may have been abundant on the Martian surface.

7.2 An Eden-like early Martian environment

Surface conditions on Mars are generally regarded not favorable for life. The reason is mainly due to the extreme conditions compared to those enjoyed by the Earth biota. This withdraws the possible biotopes to deep underground; some estimates allow the possible Mars biota several kilometers beneath the surface, where there may be some liquid water. Life may have evolved during an early 'clement period' that may have occurred in the Noachian Epoch, or Early Hesperian in Mars stratigraphy, according to the standard terminology (Sleep, 1994). Possible candidates for sites in which life may have evolved are located in the Tharsis region in the northern hemisphere, where volcanic activity has taken place since, by analogy with the Earth, the heat from underground magma may have produced hot springs, which are known to be possible sources of hyperthermophilic microorganisms. Knowledge of these possible locations raises the question whether life may have survived till the present confined to regions where pockets of liquid water may occur.

There are several reasons why Mars may have experienced a more rapid environmental evolution towards an atmosphere rich in oxygen that may have favored for a restricted period a rapid pace of evolution of the background population of prokaryotic cells towards eukaryogenesis. As we have seen in Chapter 4 such an environment is favorable to the first appearance of the eukaryotic blueprint. On Earth eukaryogenesis occurred as far back as 2 Gyr BP, according to the micropalentological data. Clearly such an "Eden" may not have lasted for long on a geologic time scale. At most we should expect some interesting work of "exopalentology" (Farmer, 1997). In Table 7.3 we have enumerated some favorable factors. We have to wait for missions to Mars planned for the near future, particularly for the sample-return mission.

7.3 The early Viking Missions

All terrestrial planets have a crust similar to the silicate crust of the Earth. In Table 7.3 we show the elemental composition of Mars soil. In the case of Mars, for example, Viking missions analyzed samples, in this case are also called 'fines' (cf., Figure 7.1).

Fig. 7.1 The 4000-km long Vallis Marineris is shown near the equator in an image taken by the Viking spacecraft

The Viking landing sites of the NASA mission of 1976 were two, the *Chryse Plains* that were located 20° N, 40-50° W and the *Utopia Plains* at 47.89° N, 225.86° W.

(The usual convention is that the prime meridian is defined by a small crater named *Airey-O*, i.e., this location defines longitude).

Table 7.3 Elemental composition of Mars soil

Element	***% of total composition (average over the landing sites)***
Silicon	21
Iron	13
Magnesium	5
Calcium	4
Sulfur	3
Aluminum	3
Chlorine	0.5
Titanium	0.5
All others (probably mostly oxygen)	50

(Soffen, 1976)

The Gas Chromatograph Mass Spectrometer (GCMS) test for organic molecules was the specific experiment responsible for the consensus that the Viking Mission found no persuasive evidence for life at Martian surface in the two locations chosen for the landing. A chemical interpretation was provided for a set of remarkable results reported by the Viking mission to Mars. The Viking lander performed a series of experiments, including one designed by John Oro, involving a small gas chromatograph and mass spectrometer.

In one of these experiments, where a set of nutrients was mixed with Martian soil samples, a sudden production of carbon dioxide was reported, initially suggesting the presence of Martian microbes, which would have shown some kind of metabolic processing of nutrients. Oro showed that a simpler, abiotic interpretation was more likely to be the correct one: the catalytic chemical oxidation of test nutrients.

GCMS results indicated no organic matter on the Martian surface within the sensitivity of the instrument winch was a few parts per billion. In fact, in the whole extensive area denominated the *Oxa Palus* region, extending from the equator to a latitude of 25° north, the *Chryse Plains* were chosen out of all the possible locations in this area only for safety considerations. (The Earth analogy would be to safely land the spacecraft in the Sahara desert to look for life, rather than in the more favorable environment for living organisms located at the basin of the Amazon River.)

7.4 Mars Pathfinder and the Sojourner Rover

This was a NASA mission that intended to demonstrate an alternative way for a lander with the additional novelty of possessing and an independent rover. Pathfinder returned significant images from the Red Planet. The main results gave additional support to the Eden-like era in the Martian past. Mars Pathfinder landed on Mars on the 4th July 1997. Like one of the two Vikings it landed in the *Chryse Planitia* (19° N, 33° W), about 850 km southeast of the location of the Viking 1 lander. The spacecraft consisted of two small elements, a 370-kg (816-pound) lander and a 10.6-kg (23-pound) rover. Once on the surface, the lander was formally named the Carl Sagan Memorial Station. The six-wheeled rover was named Sojourner in honor of the 19th-century African American civil rights advocate Sojourner Truth (cf., Fig. 7.2). For almost two years it returned 550 images to Earth through the lander. The lander itself returned over 16,000 images. The images portrayed Sojourner at work and provided a vivid view of the Martian surface. Pathfinder ended its mission in September 1997.

Fig. 7.2 Sojourner on the surface of Mars. The visible tracks in the terrain are due to the displacement of the robot from the lander ramp at the lower left to the rock at the upper right

7.5 The Mars Global Surveyor

The Global Mars Surveyor (MGS) has added significant information to our knowledge of the geologic evolution of the planet (Table 7.4). In the year 2000, MGS sent images of gullies suggesting to many that water has recently flowed on the Martian surface.

The water inventory on Mars has been estimated from morphological evidence. An evident source of water ice is the north polar cap, which also contains dust ('dirty water') and carbon dioxide. The MGS altimeter of (cf., Fig.7.3) has made a map of the northern polar ice cap (cf., Fig. 7.2).

On the basis of these measurements it has been estimated that the diameter of this polar cap is 1,200 km and its maximum depth is 3 km. It is sufficient to cover the Martian surface to a depth of 10-30 m (or, equivalently about 4% of the total amount of water locked up in the ice of the Antarctic). An uncertainty is the unknown dust-to-water ratio. However, it is possible that liquid water existed in ancient times on the Martian surface, rather than simply water ice.

Fig. 7.3 A composite image of the orbital probe Mars Global Surveyor over the extinct volcano Olympus Mons, the largest volcano in the Solar System, 27 km high

Indeed, the 'outflow channels' (surface structures measuring 10 km or more in width and hundred of kilometers in length), are known form aerial photographs; their slopes can be measured in terms of a few kilometers along their length. The origin of these features has been traced back to the catastrophic release of liquid water from the interior of Mars.

We can estimate when liquid water flowed from a fairly accurate 'chronometer': the counting of craters on its surface. In the most remarkable outflow channels, which are in the northern hemisphere and drain into the *Chryse* basin, the crater count points towards an age of 3.5 Gyr BP.

The total amount of water estimated to have flowed along these channels is equivalent to covering the entire surface of Mars to a depth of 35 m. Less dramatic morphologic features that also were produced by water flowing on the Martian surface are the so-called *run-off channels.*

Table 7.4 Factors favoring rapid evolution of the Martian environment

Martian process compared to Earth	***Comments***
The Global Mars Surveyor has discovered some signs of ancient magnetic stripes. On Earth analogous magnetic banding is interpreted as evidence for plate tectonics.	One possibility is that the magnetic anomaly patterns detected may be some evidence for ancient plate tectonics.
Faster rate of hydrogen escape due to lower gravity.	Gravity on Mars = 0.38 gravity on Earth
Smaller oceans.	Mars being a smaller planet would be less likely to intercept the trajectory of comets, hence it should have less volatiles.

(Kerr, 1999)

These structures are valley networks with dendritic drainage systems. They were produced by gradual erosion and slowly moving water. They are found in the heavily cratered terrain (hence very ancient) of the southern hemisphere. Their age is compatible with the period that followed the end of the heavy bombardment period some 3.8 Gyr BP. What fed these valley networks may have been rainwater, or even glacial melt. This water activity is an indicator of warmer climate and thicker atmosphere, conditions that have led to calling this period the Martian 'Eden', mentioned above. During this period, life may have originated on Mars at a time when the Sun was not as luminous as it is now. There must have been a green house effect due to more abundant atmospheric carbon dioxide.

A special comment must be reserved for the images of the MGS studied during the year 2000 which led the scientists responsible for the MGS camera, to infer that many of Mars's meteoritic craters were the sites of lakes during part of their history. According to their analysis the craters that have been considered contain accumulations of sedimentary rock that are several kilometers thick. The rock is divided into strata similar in color and thickness throughout Mars. The general distribution suggests that the sedimentation process was a global phenomenon rather than the result of local events.

7.6 Mars Express

The ESA Mars Express mission was a European orbiter (cf., Fig 7.4) and lander mission, whose main objective was to study the interior, surface and atmosphere of Mars. The mission included various experiments.

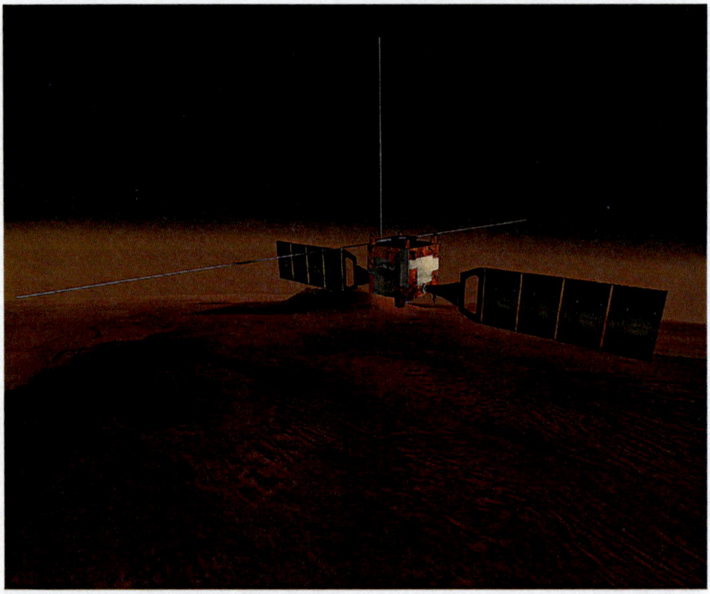

Fig. 7.4 Artist rendering of the Mars Express Orbiter. Although it was launched in 2003 in October 2009 ESA's Science Programme Committee approved the extension of mission operations for Mars Express until 31 December 2012

7.7 The Phoenix Mission

Phoenix was a NASA mission, which landed inside the Arctic Circle of Mars in May 2008 (the late spring). Unlike the Chryse and Utopia landing sites of the preceding Vikings landers, Phoenix landed on a valley surrounding the volcano (Alba Patera). Phoenix was designed to verify the presence of subsurface H_2O ice (Smith *et al.*, 2009) and was mapped at low resolution (~500 km) within 1 m of the surface by using Odyssey's Gamma-Ray Spectrometer (GRS) instrument. The Phoenix mission investigated patterned ground and weather in the northern arctic region of Mars for 5 months starting in May 2008 (cf., Fig. 7.4). The robotic arm uncovered water ice at depths of 5 to 18 centimeters.

This was NASA's sixth successful landing out of seven trials, as well as the first successful landing in a Martian polar region (cf., Fig. 7.5). The lander completed its mission in August 2008, and made a last communication on November 2 as available solar power dropped with the Martian winter. The mission was declared concluded on November 10, 2008, after engineers were unable to re-contact the craft.

The mission had two goals: (a) to study the geologic history of water. (b) To evaluate past or potential planetary habitability in the ice-soil boundary. In late summer, snowfall and frost blanketed the surface at night; H2O ice and vapor constantly interacted with the soil. The soil was alkaline (pH = 7.7) and contained calcium carbonate ($CaCO_3$), aqueous minerals, and salts up to several percent in the in surface soil samples that were available.

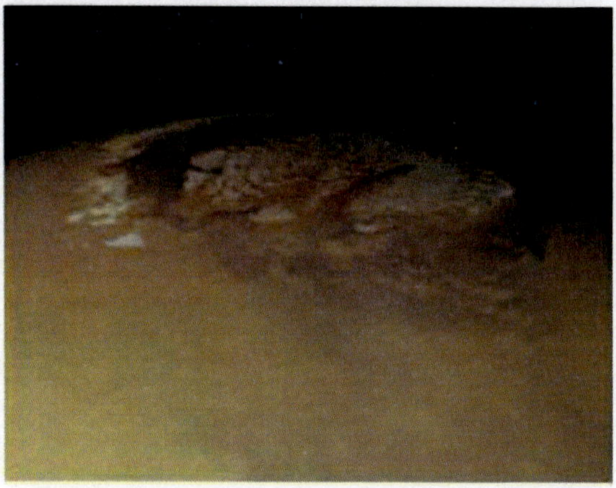

Fig. 7.5 The Martian northern polar cap

Their formation likely required the presence of water. Indeed, the amount of calcium carbonate is most consistent with formation in the past by the interaction of atmospheric carbon dioxide with liquid water films on particle surfaces (Boynton *et al.*, 2009).

Table 7.5 Minerals in Martian samples

Mineral	***% of total composition of a sample***
Quartz (SiO_2)	27 *(Chryse)*
Feldspar ($KAlSi_3O_8/CaAl_2SiO_3$)	19 *(Chryse)*
Pyroxene ($MgSiO_3CaSiO_3$)	30 *(Chryse)*
Hematite (Fe_2O_3)	22 *(Chryse)*
Calcium carbonate ($CaCO_3$)	3-5 (Northern arctic region of Mars)

(Soffen, 1976, Smith *et al.*, 2009)

It has been remarked that the pressure at the landing site is always higher than the triple point pressure. As we have mentioned above, there is some evidence supporting our belief that liquid water was present on Mars in the past: $CaCO_3$ was identified by TEGA (cf., Acronyms and Abbreviations). This compound is known to form in the presence of water. This indirect evidence for liquid water in an alkaline environment with the presence of various salts and a source of energy (Hecht *et al.*, 2009) implies that this region of Mars could have been a candidate for a habitable ecosystem. The presence of perchlorate has been discussed: Since Phoenix's discovery of carbonate requires liquid water in the past to produce the mineral, the high concentration of perchlorate (salts of perchloric acid $HClO_4$) found by Phoenix offers a means of forming brines that could remain liquid even at today's temperatures (Stoker *et al.*, 2010).

7.8 The question of the origin, evolution and distribution of Martian life

The question of the possible existence of microorganisms in the solar system has been raised in the past, even independently of the Viking results. Our nearest-neighbor planet is a candidate for having supported life in the past especially for its similarity with the Earth (cf., Table 7.1).

We cannot exclude its presence in some isolated environments. The possibility of extending the biosphere deep into the silicate crust in another terrestrial planet (Mars) deserves special attention; this will be considered below. The present status of the search for life on Mars consists essentially of three Viking results (cf., Table 7.2):

• The nature of organic molecules on the surface.
• The presence of objects that may suggest living organisms, or fossils, through their motion or appearance, respectively.
• The presence in the soil of factors that may suggest metabolic activity.

The first aspect of the research was addressed by means of a very careful chemical analysis at the two landing sites. The second question was investigated by means of the camera borne by the lander spacecraft. Finally, the third question was addressed by means of three experiments (cf., Table 7.6).

Table 7.6 The question of life on Mars

Experiment	***Results***
Test for signs of photosynthesis or chemosynthesis induced by samples from the soil (the 'Pyrolytic Release experiment').	Small incorporation of CO/CO_2 into organic compounds.
Measurement of any gaseous products from a soil sample (the 'Gas-Exchange experiment').	Initial rapid release of O_2; slow release of CO_2 and N_2.
Search for the release of radioactive gas when the soil sample was exposed to a radioactive organic nutrient solution (the 'Labeled Release experiment').	Initial rapid release of labeled gas, followed up by slow release.

(Soffen, 1976)

However, UV radiation in the absence of a large fraction of oxygen in the Martian atmosphere (i.e., an ozone layer) prevents the possibility of life on the surface of that planet since, at least on the planetary surface, there are no evident UV defense mechanisms. Not only have the biology experiments been repeatedly questioned, particularly in Japan, but also the possible existence of carbon compounds, including amino acids as well (Kobayashi *et al.,* 1996).

The Japanese group addressed the question of the possible existence of organic molecules in the polar caps and in permafrost elsewhere as a result of the interaction of cosmic rays and CO_2 in the Martian atmosphere. One possibility for living microorganisms surviving in Mars is represented by life underground, as a layer of permafrost could serve as the necessary UV defense mechanism (Farmer and Doms, 1979). This question seems pertinent to astrobiology, since we cannot exclude at present that the organisms that have been found to inhabit deep in the silicate crust of the Earth may have been deposited with the original sediment.

The question of water on Mars is of major importance for ascertaining the habitability of the Red Planet is some of the evidence to the present time for what we have called above the Martian 'Eden", when the Red Planet was covered to a large extent by liquid water, as we have emphasized in the previous missions, notably, MGS and Phoenix. This, in turn, adds some support to the possibility that life originated on Mars, or at least, that as a consequence of some large impact on the surface of the Earth, terrestrial life traveled to Mars during the Archean, closer to the period of heavy bombardment (cf., Chapter 3 for an account of the SNC meteorites that may provide examples of this possibility. Unfortunately, we must leave the reader at the very frontier of current knowledge. It will be left for him to follow up ongoing discussions that concern the presence of water on the past, or present (cf., Fig 7.6).

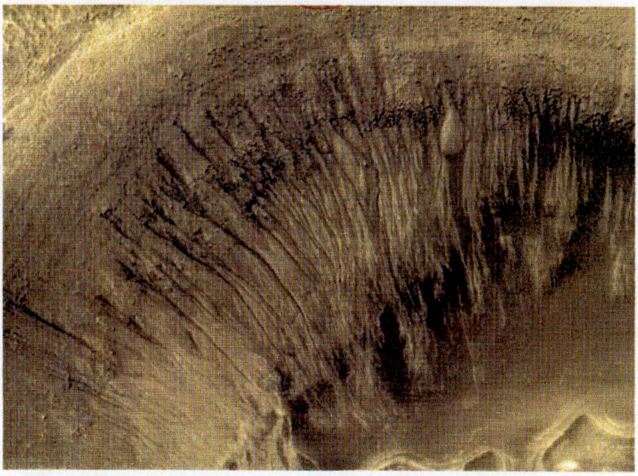

Fig. 7.6 A high-resolution image from the Mars Global Surveyor of a detail of a small 7-km crater inside the Newton Crater. It has been hypothesized that the channels that seem to flow down-slope could be the consequence the outflow of water from an underground layer

One of them is the idea that Mars may at one time have had an ocean (Head *et al.*, 1999). This follows form data that has been gathered with instruments on MGS by James Head and his collaborators.

The idea behind their work is that if an ocean existed at one time in the past, there should remain traces of a shoreline with an elevation that remains relatively constant.

At the same time the plains below should be smooth, due to a normal sedimentary process that oceanographers already have observed on our own planet. Other scientists have argued against this possibility. Another idea that will have to be evaluated is whether the liquid that modified some of the Martian surface was indeed water, or some other fluid.

7.9 Mars Odyssey and the Exploration Rover Opportunity

In May 2002, the gamma ray spectrometer on 2001 Mars Odyssey found a huge deposit of hydrogen in shallow polar soil, which is a reliable sign of water ice (cf., Fig. 7.7). The Mars Odyssey spacecraft made the discovery. Many lines of evidence had suggested that the Red Planet had an Eden-like era, and suggests that water lies in the regolith (a layer of rock and dust on the surface). The spacecraft contained an instrument called a gamma-ray spectrometer that looks for gamma-rays with a specific signature showing that they come from hydrogen less than one meter beneath the Martian surface.

Fig. 7.7 2001 Mars Odyssey was a tribute Arthur C. Clarke's famous Science Fiction books "2001"

The main instruments on this mission were:

• THEMIS (Thermal Emission Imaging System), for determining the distribution of minerals, particularly those that can only form in the presence of water.

• GRS (Gamma Ray Spectrometer), for determining the presence of 20 chemical elements on the surface of Mars, including hydrogen in the shallow subsurface (which acts as a proxy for determining the amount and distribution of possible water ice on the planet), and,

• MARIE (Mars Radiation Environment Experiment), for studying the radiation environment.

In December 2004, researchers using NASA's Mars Exploration Rover Opportunity discovered rocks may support the idea that Mars was warm and wet billions of years ago: the Eden-like early Mars environmental conditions. In fact, the evidence regards a certain rock on the southwestern slopes of "Endurance Crater." The rock's fractures, which divide the surface into polygons, may have been formed by one of several processes: They may have been caused by the impact that created Endurance Crater, or they might have arisen when water leftover from the rock's formation dried up. A third possibility is that much later, after the rock was formed, and after the crater was created, the rock became wet once again, then dried up and developed cracks.

7.10 Is methane as a possible biomarker in the Martian atmosphere?

In recent times multiple efforts have been made to reach the Red Planet with instrumentation that has been adequate for getting deeper insights into its geology. But some of the discoveries that have been made also concern the possibility of testing for a Martian biology. There is a large body of discussion regarding an Eden-like era in the evolution on the Martian surface during the geological period from c. 4,500 to 3,500 million years (Ma) before the present (BP). Multiple visual evidence of surficial morphological features argue in favor of the emergence of life on Mars in the geological era that was contemporary with the terrestrial Lower Archean (Grady and Wright, 2006). The relevant observations include, amongst others, remote sensing from orbit (Mars Global Surveyor, Mars Odyssey and Mars Express) and robotic landers (Pathfinder, Spirit and Opportunity).

The Martian atmosphere has been tested for the presence of methane, a gas that is closely related with terrestrial microbes: the methanogens. These are microbes that produce methane in oxygen-free ('anoxic') conditions. Methanogens are classified in the Domain Archaea. We have a clear idea of their presence in our past geologic eras. How these microorganisms have conquered different environments can be traced right back to the Archean, especially in microfossils from reliable sources in Western Australia and South Africa. Indeed, two reliable and easily accessible windows on the nature of the early habitable ecosystems on Earth are available: Firstly, the Pilbara Craton is an old and stable part of the continental crust and the uppermost mantle, which constitutes the hard and rigid outer layer of the Earth. The Pilbara Craton includes its microfossil-rich reservoir, the Dresser Formation in Western Australia. Secondly, another easily accessible window on the early Earth is the South African Kaapvaal Craton with its Barbeton Greenstone Belt containing some of the oldest exposed rocks on Earth.

We can safely infer relevant lessons from these two windows—real archives of the early steps of biological evolution around the Archean hydrothermal vents. For instance, both cratons have preserved details of ancient hydrothermal vents. In these sites we have learnt about the presence of a complex set of both sulfur-reducing bacteria in ~3.47 Gyr barite deposit in the Pilbara Craton, North Pole Dome (Shen and Buick, 2004), and methanogens of ~3.46 Gyr from close to the same location (Ueno *et al.,* 2006). It is remarkable that all of these microorganisms were already in existence a mere 1 Gyr after the formation of the Earth itself, at a time when speculations on an Eden-like Martian era abound. Extraordinary fractionation of the isotopes of sulfur and methane has been recorded in these ancient sites.

Are microbes making the methane that's been found on Mars, or does the hydrocarbon gas come from geological processes?

Organisms make almost all the methane on Earth, directly, or indirectly. A small proportion comes from buried, decomposing plants, whose insoluble parts become a material called kerogen. When kerogen breaks down through thermal "cracking," the result is methane, as well as longer-chain hydrocarbons like ethane, propane, and butane. Methane, the simplest hydrocarbon, has one carbon and four hydrogens (CH_4). Ethane has two carbons and six hydrogens (C_2H_6). The formula for propane is C_3H_8, and butane is C_4H_{10}. Much more methane comes from anaerobic microbes called methanogens. Some methanogens are even extremophiles because they can prosper under extreme acidity, alkalinity, or saltiness. Such conditions once were thought to be intolerable to life. Methanogens can also tolerate extreme temperatures. *Methanopyrus kandleri*, for example, lives in the 80 to 100 degrees C water around black smokers in the Gulf of California. Other methanogens live below 0 degrees C in Antarctica.

7.11 Testing for life on Mars on terrestrial analogs

In 1984 the National Science Foundation mission retrieved a meteorite from Antarctica. It was found in a field of ice called the Allan Hills (cf., Fig. 7.8). Even though it did not settle the question of whether there is life on Mars, it nevertheless reinvigorated NASA's search for extraterrestrial life in the 1990s. At present the discussion of how to search for life on the surface and subsurface of Mars is not generally agreed upon. There is a cautious shift from the search for water to identify likely ecosystems, to focusing on traces of carbon as a biomarker of ancient life (cf., Section 7.8).

Fig. 7.8 The Allan Hills meteorite ALH84001,0 The small black box on the left represents a scale of 1 cm

In fact, from Mars Global Surveyor (Fig. 7.6), it seems reasonable to assume that the channels that seem to flow down-slope could be the consequence the outflow of water from an underground layer in geologically recent times.

We have also seen in Sec. 7.10 that methane could be a possible biomarker in the Martian atmosphere. But additional hints are suggested by a terrestrial version of the Viking search for organic matter on Mars (cf., Sec. 7.3). In the Viking experiments, a Martian soil sample was heated to 500°C and the resulting gases analyzed. Instead of organics from the soil, Viking detected only carbon dioxide and two chlorinated methane compounds. The latter were considered to be contaminants brought from Earth, even though they did not appear when the experiment was run without soil while Viking was in space, the Viking experiments in the lab using the most Mars-like soil available were repeated. One such experiment was in the Atacama Desert of Chile, where a trace of organic matter was identified. When they added a bit of perchlorate before the heating, in line with the Phoenix discovery, they duplicated the Viking results: no volatilized organics, some carbon dioxide, and the same two chlorinated methane compounds. These findings underline the relevance for searching molecular traces of life in Martian organic matter (cf., Kerr, 2010 for additional references to this interpretation on the search for biomarkers, and Sec. 7.7 for the discussion of perchlorate from the Phoenix mission point of view.)

7.12 Mars in the second and third decades of the 21st century

In the second decade of the present century a fleet of spacecrafts are either on Mars, on their way or in the planning stages. In Table 7.7 we give a brief description of some of these exciting missions.

Table 7.7 Some of the missions for Mars in the first and second decade of the 21st century

Mission	*Expected date of launch*
Mars Express	2003
Phobos-Grunt (Roscosmos)	2011
Mars Science Laboratory, MSL (NASA)	2012
Exo-Mars (ESA)	2016
Mars sample return mission, MSR (NASA, ESA)	2018

Ever since the end of the Apollo missions and a smaller sample retrieval by the then Soviet Union (cf., Chapter 3), no sample return missions has been attempted by any of the space agencies. A Mars sample return mission (MSR) is expected to be a spaceflight mission for retrieving rock and dust samples from Mars and to return them to Earth for geological, micropaleontological and biological enquiries. A joint project of NASA and ESA is expected to launch in 2018. The actual sample return itself is expected in the time frame of 2020-2022.

Supplementary Reading

Barlow, N. (2008) *Mars: an introduction to its interior, surface and atmosphere*. Cambridge planetary science. Cambridge University Press, 264 pp.

Carr, M. H. (2006) *The surface of Mars*, Cambridge planetary science series, Cambridge University Press, 307 p.

Sullivan, W. T. and Baross, J. A. (eds.) (2007) *Planets and life: the emerging science of astrobiology*. Cambridge, Cambridge University Press, 604 p.

References

Barlow, N. (2008) *Mars: an introduction to its interior, surface and atmosphere*. Cambridge Planetary Science; [new ser.], 8, Cambridge University Press.

Boynton, W. V., D. W. Ming, S. P. Kounaves, S. M. M. Young, R. E. Arvidson, M. H. Hecht, J. Hoffman, P. B. Niles, D. K. Hamara, R. C. Quinn, P. H. Smith, B. Sutter, D. C. Catling, R. V. Morris (2009) Evidence for Calcium Carbonate at the Mars Phoenix Landing Site, *Science* **325**, 61.

Carr, M. H. (1999) *Mars: surface and interior*, in P. R. Weissman, L.-A. McFadden and T. V. Johnson (eds.), *Encyclopedia of the Solar System,* Academic Press, San Diego, pp. 291-308.

Carr, M. H. (2006) *The surface of Mars*, Cambridge planetary science series, Cambridge University Press, 307 p.

Farmer, J. D. (1997) Implementing a strategy for Mars Exopaleontology, in R. B. Hoover, (ed.), *Instruments, Methods and Missions for Investigation of Extraterrestrial Microorganisms,* The International Society for Optical Engineering, Washington, Proc. SPIE **3111,** pp. 200-212.

Farmer, C. B. and Doms, P. E. (1979) Global seasonal variation of water vapor on Mars, *J. Geophys. Res.* 84, 2881-2888.

Head III, J.W., Hiesinger, H., Ivanov, M.A., Kreslavsky, M.A., Pratt, S.,and Thomson, B.J. (1999) Possible ancient oceans on Mars: Evidence from Mars Orbiter Laser Altimeter Data, *Science **286,*** 2134-2137

Hecht, M. H. et al. (2009) Detection of Perchlorate and the Soluble Chemistry of Martian Soil at the Phoenix Lander Site. *Science* **325**, 64-67.

Kerr, R. A. (1999) Signs of plate tectonics on an infant Mars, *Science* **284**, 719-720;

Kerr, R. A. (2010) Growing Prospects For Life on Mars Divide Astrobiologists. *Science* **330,** 26.

Kobayashi, K., Sato, T., Kaneko, T., Ishikawa, Y. and Saito, T. (1996) Possible formation of carbon compounds on Mars, in 11th International Conference on the Origin of Life, Orleans. Book of Abstracts, p. 61.

Shen, Y. and Buick, R. (2004). The antiquity of microbial sulfate reduction. *Earth-Science Reviews* **64**, 243-272.

Sleep, N. H. (1994) *J. Geophys. Res.* **99,** 5639-5655. The reader is referred to Table 1 in this paper for references to the original literature.

Smith, P. H., L. K. Tamppari, R. E. Arvidson, D. Bass, D. Blaney, W. V. Boynton, A. Carswell, D. C. Catling, B. C. Clark, T. Duck, E. DeJong, D. Fisher, W. Goetz, H. P. Gunnlaugsson, M. H. Hecht, V. Hipkin, J. Hoffman, S. F. Hviid, H. U. Keller, S. P. Kounaves, C. F. Lange, M. T. Lemmon, M. B. Madsen, W. J. Markiewicz, J. Marshall, C. P. McKay, M. T. Mellon, D. W. Ming, R. V. Morris, W. T. Pike, N. Renno, U. Staufer, C. Stoker, P. Taylor, J. A. Whiteway, A. P. Zent (2009) H2O at the Phoenix Landing Site, *Science* **325**, 58-61.

Soffen, G. A. (1976) Scientific results from the Viking Mission, *Science* **194**, 1274-1276.

Stoker, C. R., Zent, A. Catling, D. C., Douglas, S., Marshall, J. R., Archer Jr., D., Clark, B., Kounaves, S. P., Lemmon, M. T., Quinn, R., Renno, N., Smith, P. H., Young, S. M.M. (2010) Habitability of the Phoenix landing site. *J. Geophys. Res.* **115**, E00E20, □doi:10.1029/2009JE003421.

Ueno, Y., Yamada, K., Yoshida, N., Maruyamaand, S., Isozaki, Y. (2006) Evidence from fluid inclusions for microbial methanogenesis in the early Archaean era. *Nature* **440**, 516-519.

8

On the possibility of biological evolution on the moons of Jupiter

The satellites of the planet Jupiter were discovered early in the 17th century. The most intriguing possibilities for detecting extraterrestrial microbial life lie within the Jovian system, particularly in Europa, the second of the Galilean satellites.

The reader is advised to refer especially to the following entries in the Illustrated Glossary: *Carbonaceus chondrites, Galileo Mission, hydrothermal vents, Jovian planets, Voyager missions.*

8.1 The discovery of Europa

On January 10, 1610 while pointing his telescope at Jupiter Galileo Galilei, a lecturer of Mathematics at the University of Padua (Rosino, 1992) observed three objects that he interpreted to be stars. He further noticed that two of these stars were to the east of the planet, while the third one was to the west. During the next evening he returned to his telescope and remarked that all three stars were at the west of Jupiter. Finally, during the night of January 13 he saw a fourth object in the vicinity of Jupiter. These evenings were the crucial period in which there were the first intimations from observations that there was a center of motion other than the Earth. This put an end to one of the ideas about the immovable heavens maintained in ancient Greek philosophy that had lasted into medieval times.

The German astronomer Simon Marius subsequently (1614) gave these four satellites names from Greek mythology, corresponding, in part, to the maidens that Jupiter fell in love with. Marius called the second innermost satellite 'Europa', a choice deriving from Greek mythology, rather than Roman mythology. Zeus, the Greek equivalent of the Roman god Jupiter, was attracted by the charms of several mortal women. Amongst them was Europa, the beautiful daughter of Agenor, King of Phoenicia, from whose union was born Minos King of Crete. Io was another mortal woman with whom Jupiter fell in love. Io later married Telegonus, a mythical King of Egypt. Io was later confused with Isis, the Egyptian divinity. On the other hand Callisto was a minor divinity represented by a beautiful maiden (a nymph) who, after being

J. Chela-Flores, *The Science of Astrobiology: A Personal View on Learning to Read the Book of Life*, Cellular Origin, Life in Extreme Habitats and Astrobiology 20, DOI 10.1007/978-94-007-1627-8_8,

seduced by Zeus, was killed by the virgin huntress Artemis, daughter of Zeus himself. Finally, the name of the fourth of the Galilean satellites, Ganymede, unlike the other Galilean satellites, corresponds to that of a youth taken to Olympus to become the gods' cup The satellite Europa is a second candidate for life in the solar system, after the more prominent planet Mars. This follows from the discovery by Carl Plicher that a large proportion of the spectroscopically detectable material on its surface is water (cf., Horneck, 1995 for references). Europa is covered by an icy surface, according to the results obtained by the Voyager 2 mission and subsequently confirmed by the Galileo Mission. Europa has a diameter of 3,130 km, comparable with the Moon, as shown in Table 8.1.

Table 8.1 Europa, some information and statistics compared with the Moon

Statistics	***Moon***	***Europa***
Mass (Earth = 1)	0.0123	0.0083021
Diameter	3,476 km	3,138 km
Density	3.34 gm/cc	3.01 gm/cc
Distance from its planet	Apogee: 406,740 km Perigee: 356,410 km	670,900 km 9.5 Rj (Rj=Jupiter's radius)
Surface composition	Solid; the lunar basalt consists of *silicon dioxide.*	Water Ice
Surface gravity (Earth = 1)	0.165	0.135
Escape velocity	2.37 km/sec	2.02 km/sec
Orbital eccentricity	0.0549	0.009
Sidereal period of revolution	27.322 days	3.551181 days
Sidereal period of rotation	27.322 days	3.551181 days

Europa orbits Jupiter at a distance of about $6R_J$, which is equivalent to some 671,000 km. Its density of 3.0 grams per cubic cm suggests mainly a rocky interior. Numerical models plus gravity data for the interior suggest that an iron-rich core about 1,250 km in diameter would be present surrounded by a rocky mantle. There is a water layer about 150 km thick most of which is liquid (cf., the detailed discussion of the most significant discovery of the Galileo Mission described in the following Section 8.2). The forthcoming missions of the 2020s will explore Europa focusing on this fundamental Solar System interior ocean.

The great excitement that Europa has raised is due to the possibility of finding traces of life on that satellite. The surface is crisscrossed by an intricate array of grooves and ridges. The interpretation of these intriguing Galileo images has led to much discussion whether the surficial Europan ice bears some “smoking-gun” uncontroversial evidence for a liquid interior (cf., Fig. 8.1).

Fig. 8.1 A Galileo image of the intricate morphology of the Europan ice

However, in Fig. 8.2 we show prominent features much wider than grooves and ridges, which are typically only a couple of kilometers wide. Indeed, such prominent features can have up to several tens of kilometers in width and measure up to thousands of kilometers. Their origin has been a challenge taken up by many researchers, but a leading explanation is that they may be fractures caused by the stretching of Europa’s crust due to the influence of the tides caused by its giant Jovian neighbor.

It is useful to list, in Table 8.2, some of the largest features on the surface of Europa that we have just mentioned.

Table 8.2 Some prominent features on the surface of Jupiter's moon Europa

Feature	*Longitude/latitude*	*Size (km)*
Asterius Linea	17.7N/265.6W	2753
Belus Linea	11.8N/228.3W	2580
Minos Linea	45.3N/195.7W	2134
Phineus Linea	33.0S/269.2W	1984

(Greely and Batson, 1997)

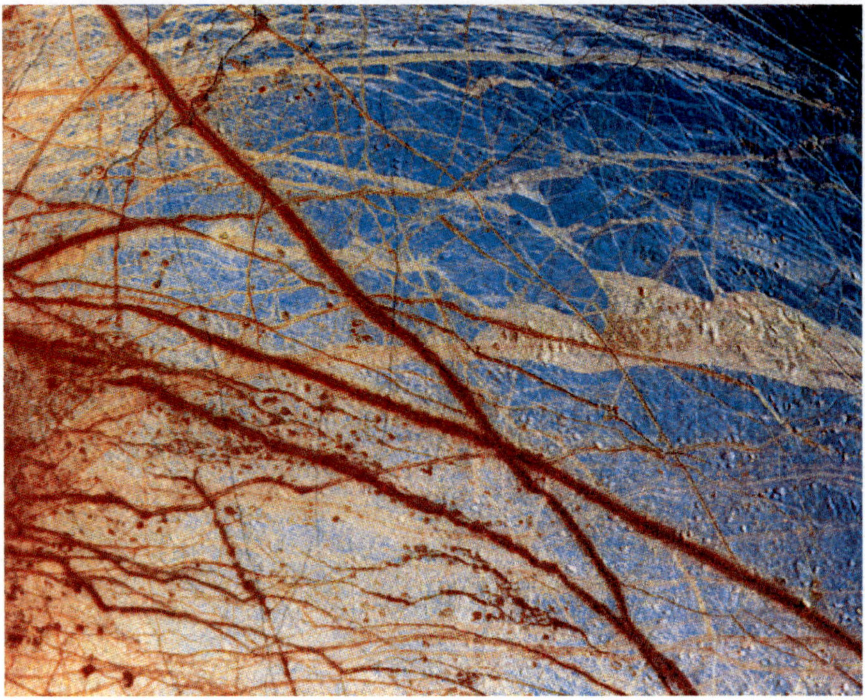

Fig. 8.2 A composite image from the Galileo spacecraft highlighting some of the prominent features on the surface of Jupiter's moon Europe, including the Minus Linea region (cf., Table 8.2)

8.2 The Galileo Mission

The Galileo Mission began to be planned in the 1970s, as the Voyager missions were being launched. James Van Allen had proposed the mission to NASA with the intention that some 12 orbits to the Jovian system be undertaken. Eventually, after some initial delays due to the Challenger accident, the mission took off and arrived at the Jovian system in December 1995. A year later (on 19 December 1996), the fourth orbit of the satellite Europe passed about 700 km from its surface and the data were processed the following month (cf., Fig. 8.3).

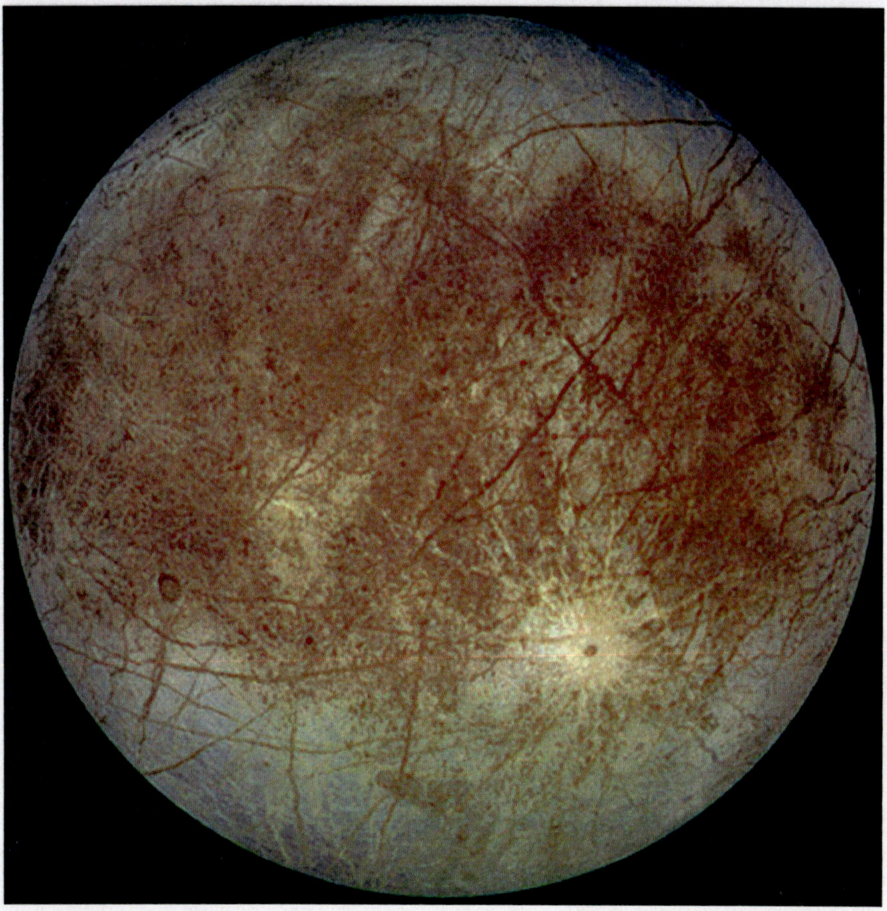

Fig. 8.3 During the Galileo orbit G2 a full hemisphere of Europe was imaged in great detail capturing most of the 'trailing hemisphere', so called since the satellite motion is in the visual direction penetrating the page. The remarkable detail includes the famous Will crater lower than the centre slightly below

Long before the Galileo Mission succeeded in reaching the Jovian System, in 1976 the Voyager missions provided low-resolution images of the surface of Europe. These images showed a series of intersecting ridges and line, namely, cracks on the surface. The Galileo Mission has shown us that the central parts of some line are of lower Aledo than the surrounding terrain. Some planetary scientists believe that these bright surface features may represent fresh ice that has come from below. The darker parts of the line may represent silicate contamination from below, or alternatively ice that may have been darkened by other external or internal factors.

Besides, we learnt that craters were not abundant, suggesting that Europe has been geologically active until a relatively recent date (or, alternatively, there may have been 'resurfacing' by the emergence of liquid water from below). To sum up, Voyager supported the intuition that planetary scientists already had. Such confidence was based on two facts. Firstly, from Earth-bound spectroscopy we knew that Europe was covered with water ice. Secondly, its density is not radically different from the Moon; from a combination of these remarks it follows that Europe has a silicate core. The Galileo mission has added much to the early insights we already had. One example is some form of 'ice tectonics'. The Jet Propulsion Laboratory, which handled the mission for NASA, has released some images that suggest that part of the surface is understood in terms of shifting plates of ice. From all the information gathered from Voyager and Galileo reasonable guesses have been put forward regarding possibly a substantial amount of liquid water between the silicate crust and the icy surface (cf. Fig. 8.4).

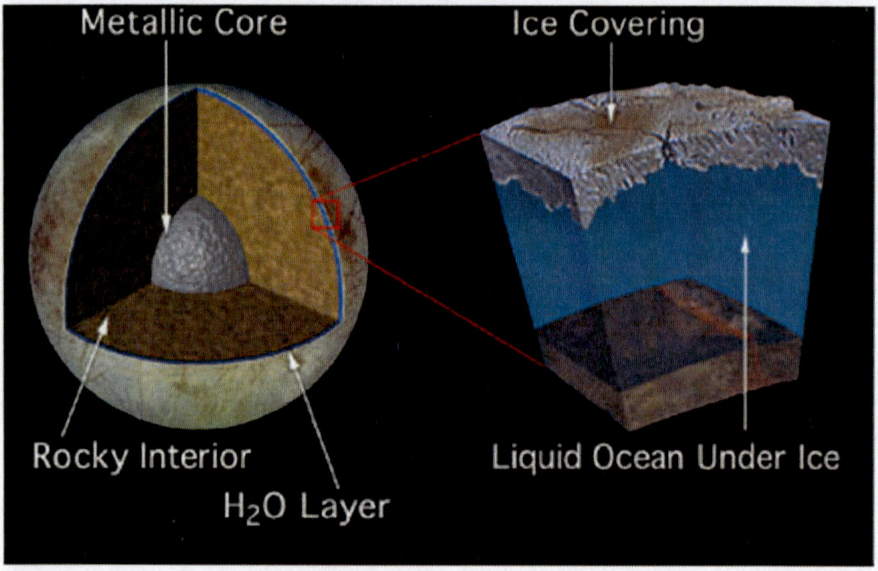

Fig. 8.4 These artist's drawings depict a proposed model of the subsurface structure of the Jovian moon, Europe. Geologic features on the surface, imaged by the Galileo spacecraft might be explained by the existence of a layer of liquid water with a possible depth of more than 100 kilometers

The onset for the melting of the ice that we 'see' spectroscopically from the Earth could be tidal heating. We recall that in the vicinity of a satellite to a giant planet heating may be produced due to repeated stressing arising from orbital motion in the planetary field. (This source of heating may be enhanced due to the additional radioactive dacay of nuclei, a phenomenon known as radiogenic heating.) It may be useful to summarize the reasons for the excitement raised by Europa's intriguing surface and mysterious interior. Liquid water is rare in the Solar System. Water is universally required for the biochemistry of life. For instance, water is responsible for the structure assumed by lipids in the cell membrane bilayer, a characteristic of both eukaryotes and bacteria. Some of the evidence for liquid water may be inferred from the linear features of the surface of this satellite (cf., Table 8.2). Triple bands provide one particular case. Each of these structures consists of two dark bands separated by a bright intermediate ridge. The dimensions involved are colossal: the width of the triple bands is about 18 km, while their length can extend for over a thousand kilometers. These fractures of the icy surface are assumed to be the result of tidal deformations that are filled in with water and silicates from its interior. After water and minerals are separated on the surface, freezing would result by water expanding to form the central ridge.

We may conclude that the possible presence of liquid water in Europa forces upon us the possible presence of autochthonous life, as we were encouraged to do throughout this chapter. Both Galileo images and measurements, have reinforced, but not yet proved the existence of an Europan ocean (Carr *et al,* 1998). One of the more significant measurements by the Galileo mission was by Margaret Kivelson and collaborators. Their method was based on remarking that the presence of a magnetic field that varies with time implies that electric currents are induced in a conductor (in the present case it is water). Jupiter's axis of rotation is not lined up with its magnetic axis. This configuration induces the Jovian moons to feel a fluctuating magnetic force. In the case of Europa, an electric current is generated in the ocean. As a consequence, the corresponding magnetic field is oriented in the opposite direction of the Jovian magnetic field. The instruments on board of Galileo were able to measure both of these magnetic fields, giving additional support to the presence of a sub-surface ocean. Actually, the first observational evidence for the liquid water was the work of Richard Greenberg (Hoppa *et al.,* 1999).

8.3 Tentative inventory of organic elements in the Europan ocean

Under the hypothesis that the Galilean satellites were of matter similar to the carbonaceous chondrites, it is possible to reach some estimates of their total 'volatile content' (substances with a relatively low boiling temperature). For this purpose consider Table 8.3, where the approximate correlation of the densities of the chondrites and satellites is given.

Table 8.3 Comparison between densities (g/cm^3) of the Jovian satellites and carbonaceous chondrites

Jovian satellite name (density)	***Carbonaceous chondrite type (density)***
Io (3.57)	III (3.5)
Europa (2.97)	II (2.5-2.9)
Ganymede (1.93)	I (2)

(Adapted from Oro *et al.*, 1992)

On the other hand, Table 8.4 suggests that CII carbonaceous chondrites have a sufficiently high fraction of water (13.35 %) to make up the volume required to fill up the presumed Europan ocean, provided that planetesimals of this composition made up the main part of the proto-satellite during the process of accretion (cf., Chapter 2, "Origin of the Satellites of the Jovian Planets").

Table 8.4 Chemical composition of meteorites made primarily of silicates (carbonaceous chondrites), but including some volatile elements

	SiO_2 (%)	MgO (%)	C (%)	H_2O (%)	S (%)
Type I	22.56	15.21	3.54	20.08	6.20
Type II	***27.57***	***19.18***	***2.46***	***13.35***	***3.25***
Type III	33.58	23.75	0.46	0.99	2.27

(Adapted from Oro *et al,* 1992)

In the formation of Europa from the protonebula, thermal heating of the initial carbonaceous chondrites would have led to the solution of biogenically important elements in the ocean (cf., Table 8.5: Mg, C, S). Table 8.5 suggests that there is a substantial amount of carbon in Europa. Besides the evidence for liquid water in the Europan ocean, the likelihood of substantial supplies of carbon provide two of the three basic ingredients that we have learnt to appreciate from our studies of the origin of life on Earth. The third major ingredient for life is also very likely to be present on Europa, a source of energy, since Europa is very close to Jupiter what we see happening on Io—volcanic activity—is also likely, but underneath Europa's icy surface (cf., Fig. 8.5).

Fig. 8.5 A 1997 image of Io taken by the Galileo Mission. The almost circular and dark effect of a volcano (Pele) on the satellite surface can be appreciated near the center

8.4 The habitability of Europa

In the present search for life in the solar system there are two prominent candidates: The first one is Mars already considered in Chapter 7, but the Jovian satellite Europa is a second environment in the Solar System where we have insisted for some time that life may have emerged (Chela-Flores, 1996). Some of the reasons for our expectation are listed in Table 8.5.

Table 8.5 The question of life on Europa

Voyager 2 Mission and observations from Earth.	***Conclusions from the common origin of the Solar System.***
Europa is slightly less dense than the Moon (3.04g/cm^3).	From typical Earth- crust rocks (3 g/ cm^3), Europa may be inferred to be made mainly of rocky material.
Covered by a smooth layer of ice criss-crossed by a pattern of long cracks.	The absence of mountains and craters suggests it to be covered by water.
Under the surface there may be an ocean of water, whose temperature is 4°C.	Hot springs at the bottom of the ocean may imply the presence of microorganisms.

One of the leading factors that may have triggered the living process on Europa is he existence of an internal ocean. For the Jovian satellites this possibility was already conjectured as early as 1980, when Gerald Feinberg and Robert Shapiro speculated on this possibility in the specific case of the satellite Ganymede. More recently, Shapiro has further refined the Feinberg-Shapiro hypothesis by suggesting the presence of hydrothermal vents at the bottom of the ocean in Europa, in analogy with the same phenomenon on Earth (Shapiro, 1994) an assumption that is widely held at present by most researchers. In fact, for these reasons Europa is a candidate for a program for the search for extraterrestrial microbes.

8.5 What are the constraints on the putative Europan biotope?

A biotope is understood as a uniform region in environmental conditions together with a population of organisms for which it is the habitat. Speculating on the putative Europan biotope is not an idle exercise, for technology has reached a point in which funding seems to be the only barrier preventing us from a concentrated campaign of exploration of the biotope.

On the basis of the Galileo images we have argued that many hints already suggest that part of the water surrounding the silicate core is liquid water. The biotope that may have been formed may be analogous to those existing in Antarctica, especially in the dry valley lakes and Lake Vostok (cf., Chapter 3). As we have shown in Chapter 3 the Vostok lake may be considered to be an analog of the Europan environment that should be taken with a degree of caution, since we expect the Europan ocean to be linked to its surface by cracks. The Galileo images allow us to speculate further. There has been major resurfacing events, since the craters on the icy surface are few in number, but from the few that are present: Pwyll, Cilix and Tyre (cf., Table 8.2), we draw additional support for the assumption that beneath there is an ocean of liquid water. Besides, resurfacing also hints at submerged geologic activity.

Europan and terrestrial volcanism may resemble each other. This follows from the analogies between the formation of the terrestrial planets and the Jovian satellites (cf., Chapter 2). Geologic activity on Earth has suggested a candidate source for the origin of life, as we know it namely, hydrothermal vents. It is clear that the chemical disequilibrium arising from thermal gradients present in the water that circulates through the hot volcanic rock could drive chemical reactions that may lead to life. One of the remaining questions is whether an ample supply of organics could be found on the seafloor of the ocean.

In Chapter 2 we sketched some aspects of the formation of satellites such as Europa. In fact, the carbon abundance on Europa should be correlated with solar abundance (cf., Chapter 1, Table 1.2). This is further supported by evidence supplied by the carbonaceous chondrites (cf., Table 8.5). Since the above arguments suggest that the sea floor could be a suitable site for life, the microbial community of extremophilic microorganisms may have raised the prokaryotic blueprint up to the more complex eukaryotic blueprint.

8.6 Technological challenges for reading further into the Book of Life

A major scientific discipline for astrobiology is oceanography that is concerned with oceans and seas, including their physical and chemical properties, their origin and geology, and the microbial and multicellular organisms that inhabit the marine environment. If there is an ocean on Europa, our knowledge of oceanography will offer significant hints as to what to expect on the Jovian satellite. With the experience on Earth this scientific subjet has traditionally been concerned with studying the oceans and seas, taking into account their physical and chemical properties, but a major effort is devoted to their origin, geology and life in the aquatic medium that dominates the major part of the surface of our planet. Having discovered other oceans and seas elsewhere in the Solar System since the closing years of the last century, oceanography has been forced to extend its scope to situations that are most intriguing. In order to cope with this larger field of research, theoretical and experimental basic sciences are not sufficient for the new challenges we face on Europa and Titan. For this reason engineering and technological innovation are fundamental

THE ORIGINAL HYDROBOT-CRYOBOT

A lander is a possibility, which could look for evidence of life while sampling the surface composition and possibly performing seismic studies of the subsurface structure. Subsequent missions of the type of hydrobot-cryobot mission may try to melt down to the ocean or even perform a sample return (Horvath *et al.*, 1997). A vehicle dubbed a "cryobot" was suggested to carry a small deployable submersible (a "hydrobot") equipped with a complement of instruments (cf., Fig. 8.6).

A cryobot is an adaptation of previous melter probes designed for terrestrial ice exploration; the hydrobot would draw on previous submersible experience from oceanic research. The design of an instrument package to search for life across the wide range of thermal and pressure environments expected on Europa raises some challenging issues in sample handling, and long-term reliability. Finally, this mission in the distant future may present opportunities for performing high-value science on Earth, particularly in Antarctica, in order to test these instruments.

Fig. 8.6 There are several objections to our forward-looking, but unrealistic preliminary thoughts that were conceived at the closing of the 20th century and excited the author and his co-workers. Some of the difficulties have been summarized in this image of Horvath *et al*: The submersible (hydrobot) is seen in the foreground and the ice-melting penetrator is depicted in the top left hand side of the figure. This original hydrobot painting has an unrealistic seafloor with respect to its distance to the bottom of the icy crust, as explained in the text. The image, however, has the undeniable merit of suggesting a conceivable technology for the search for microorganisms in Europa's ocean that may be feasible with more modern technologies. At the present time it is not possible to develop a coupled hydrobot-cryobot as described in our paper, of 1997, neither can a combined effort of the main space agencies afford such technology, but in the future the ingenuity of astrobiologists should not be underestimated

Thus the case for life's origins, either through chemosynthesis at hydrothermal vents or deep underground is still an open question. Such forward-looking plans for exploration in the distant future are aiming to allow us to read some of the unknown passages of the Book of life: testing the nature of biology— within the Solar System. This is gradually becoming more feasible with technology that is available modern design for the old concept of a hydrobot and penetrators, a new technology that is currently being developed beyond the early efforts of the American Space Agency NASA, the Japanese Space Agency JAXA and the 1996 efforts of the Russian Space Agency Roscosmos. These two technologies will be the topic of the next two sections.

ADVANCED DESIGN FOR HYDROBOTS

A hydrobot concept has been used by ENDURANCE, a potentially valuable tool for the eventual exploration of Europa's ocean (cf., Fig. 8.7 and Bortman, 2010).

Fig. 8.7 The hydrobot concept has been updated by ENDURANCE, potentially a valuable tool for the eventual exploration of Europa's ocean

The possibilities of primary deep-sea or, alternatively, deep-underground evolution, are at present open questions. What remains to be shown in microbiology is that some barophilic and thermophilic microorganism has a metabolism that can proceed in completely anoxic conditions, deprived from carbon and organic nitrogen derived from surface photosynthesis.

PENETRATORS: SURFICIAL LANDERS FOR EUROPA

A most appropriate technology for the exploration of the Solar System planets and satellites—the penetrator—is being further developed by the UK Penetrator Consortium. These instruments consist of small projectiles that can be delivered at high velocity to reach just beneath the surface of planets or their satellites for probing samples of surficial chemical elements, amongst other investigations (cf., Weiss, 2010 for a review and additional references). If budgetary constrains force a choice between penetrators and landers, some advantages of the penetrator approach are evident: the low mass of these instruments combined with their agility in deployment, make them worthy complements to otherwise orbiter missions launched without landers. Penetrators have a long history of feasible technological development by several space agencies. For a detailed description of the previous attempts we refer the reader to a previous review (Chela-Flores, 2010).

The question of separating signals form life and non-life requires some of the geochemistry knowledge that was referred to in the Introduction: sulfate-reducing bacteria use sulfate as an oxidizing agent, reducing it to sulfide. Most sulfate-reducing bacteria can also use other oxidized sulfur compounds such as sulfite and thiosulfate, or elemental sulfur. This type of metabolism is called dissimilatory, which means that sulfur is not assimilated into any organic compounds. Of the possible fractionation mechanisms responsible for sulfur isotope variations that may occur in nature, dissimilatory sulfate reduction produces the largest fractionations in the sulfur stable isotopes (Brunner and Bernasconi, 2005). This is a ubiquitous process on Earth, since a large number of species of prokaryotes are known to possess this capability (Detmers *et al.,* 2001).

These microbes are especially exciting for the potential identification of an Europan habitable ecosystem. They are widely distributed in anoxic environments. Their extremophilicity is characterized by toleration of temperatures in the range from just below freezing point to their highest recorded ambient temperatures, over the boiling point of water. They can also survive in salinities from fresh water to brines.

A great deal (but not exclusively) of what is known about sulfate reducers comes from pure cultures, where the outcome of the depleted sulfide ranges between -4 to -46‰ (McCready, 1975). In natural environments some significant insights on the relation between sulfur and life has been retrieved from the Black Sea called 'Pontus Euxinus' by the Romans (its waters may be referred to as euxinic). The Black Sea derives its name from its dark sulfur-laden waters: unlike the Mediterranean Sea, where visibility extends down to a depth of about 30 meters, visibility reaches only as far as about 5 meters in the Black Sea. At the Black Sea's oxic/anoxic boundary, anaerobic organisms, such as sulfate-reducers consume the rain of organic material falling from above, producing H_2S, while anaerobic methanotrophs reduce the concentration of methane (Grice *et al.,* 2005). More relevant for the main theme of this work is to underline that naturally occurring sulfides in sediments, and in euxinic waters, can be depleted in ^{34}S by as much as –70‰.

However, through a repeated cycle of sulfide oxidation to elemental sulfur, followed by a reaction in which a single compound is simultaneously oxidized and reduced (a process known as "disproportionation"), these microbes can generate large fractionations that go beyond the upper bound of -46‰ (suggested by Rees, cf., Canfield and Thamdrup, 1994). We must underline that if sulfur fractionation concerns us as a possible trigger for the Europan patches on its icy surface, then detecting large fractionations would be a fingerprint of life. Measuring this effect is within reach of the penetrator technology coupled with the current capabilities of miniaturized mass spectrometry.

Especially relevant for the putative Europan habitable ecosystem is the hydrothermal reduction seawater sulfate, which would take place, according to equilibrium fractionation processes, varying with temperature. The corresponding heavy sulfates are easy to distinguish from marine sulfate in the geologic record. Such depletions can also take place during the magmatic reduction of gaseous SO_2 to H_2S, but the expected range of $\delta^{34}S$ depletions lie constrained in the range 15 - 20‰ (Rollinson, 2007).

Thus, abiotic sulfur isotope fractionation tends to be the result of hydrothermal reactions and is quite constrained with respect to the large effects that sulfur bacteria are capable of achieving. Bacterial sulfate reduction is characterized by large δS^{34} (reaching over -70‰), whereas thermogenic sulfate reduction generally leads to smaller S^{34} depletions (Hoefs, 2009). We should keep in mind that we have discussed the possibility of anaerobic life in the Europan ocean, by other forms of life are also possible. As will be discussed in Chapter 12 in more detail, exogenous oxygen may also play a role in maintaining an ecosystem in the Europan ocean, even allowing more advanced life (Hand *et al.*, 2009; Greenberg, 2010).

8.7 Returning to the Jovian System

Such forward-looking new technologies, as anticipated in the previous section, can indeed find its use in the forthcoming missions that will return us to the Jovian System. Preliminary plans for returning to Europa and the Jupiter System were developed under the name of LAPLACE Mission (cf., Fig. I.13, left, and Blanc *et al.*, 2009). As we have seen above, one of the most likely places in the solar system for the existence of extraterrestrial life forms is the Jovian moon Europa. We have discussed above the possibility that a tidally heated ocean exists underneath Europa's icy surface. After a first reconnaissance is made of Europa and if it is determined that an ocean does exist under the ice, *in-situ* measurements will be needed to directly explore the Europan ocean and the ice that lies above it. We have maintained in our research papers that for exploring the Europan biota, penetrating right into the ocean is unnecessary (as we hoped in our early work, cf., Fig. 8.5) for the major problem of astrobiology of identifying an unambiguous biomarker (Chela-Flores, 2006, 2008, 2010, Gowen *et al.*, 2009, 2010) a point of view shared by others (Greenberg, 2005).

Given the above-mentioned possibility of designing an advanced lander mission that may melt through the ice layer above the putative Europan ocean, so as to deploy a tethered submersible, the question arises as to which suitable complement of instruments may be developed for testing for the presence of biogenic activity. Feasible experiments can be made, as the technological capability is consolidated for delivering equipment underneath or on the surface of satellites with iced surfaces, such as Europa and Ganymede (cf., Section 8.8 for details of the missions that will be in a position to deliver payloads with sufficient experimental innovation.

8.8 Are there biosignatures revealing a second biology on Europa?

In view of considerable technological progress, the central problem of astrobiology (the existence of life elsewhere in the universe) is no longer the exclusive domain of organic chemistry. As we have seen in Book I, Part 1, this field has been extensively reviewed over the last decade (as mentioned in the Preface of this book, especially in the Trieste Series, cf., Appendix). We expect that radio astronomy and space exploration will be ever-increasing partners with a significant relevant role to play (cf., Chapter 12).

There are significant strategies for identifying those places where future landers could search for the biomarkers. The Galileo Near-Infrared Mapping Spectrometer (NIMS) provided some evidence for the presence of sulfur compounds has been discussed in elsewhere (Chela-Flores, 2006). The most likely sites would be where the salt deposits, or organics, are concentrated, as suggested by the NIMS data. For instance, the search for biomarkers on Europa could focus on the area north of the equatorial region, between 0 and 30 N and between the longitudes 240 and 270 (McCord *et al.*, 1998). But a more intriguing and smaller patch would be the narrow band with high-concentration of non-ice elements that lies east of the Conamara Chaos, between the Belus and Asterius lineae, namely, between 18-20 N, and longitudes 198-202. Definite answers can be searched in situ on the icy surface.

The exploration of the Jovian System is one of the main priorities to be considered in future exploration of the Solar System. Amongst the scientific objectives a major issue is the question: *Is Europa habitable?* From the earliest stages of planning for the post-Galileo era this question has been framed more tightly in two formulations as follows: Does Europa represent a "habitable zone" of the Jupiter System? Does Europa actually harbor life? The Jovian System indeed displays many facets that are fundamental for astrobiology: It is virtually a small planetary system in its own right, built-up out of the mixture of gas and icy material that was present in the external region of the solar nebula. Unique among Jupiter's satellites, Europa is believed to shelter an ocean between its active icy crust and its silicate mantle (Anderson *et al.,* 1997; Kivelson *et al.,* 1997). In this liquid environment the main conditions for habitability may be fulfilled. The likelihood of habitability rests on the fulfillment of some conditions: the presence of liquid water, an adequate energy source to sustain the necessary metabolic reactions and a source of the biogenic elements (C, N, H, O, P, S), which can be used as chemical, rather than as photosynthetic means for the synthesis of biomolecules. Appropriate pressure and temperature are also required. Europa is not unique among the three outer Galilean satellites, but according to current models, Europa represents the only case in the Jovian System in which liquid water is in contact with a silicate core (Bland *et al.,* 2009). The presence of hydrothermal activity at the interface of silicate core-ocean could provide a variety of chemicals that could play a role in sustaining putative life forms at the ocean floor and, especially, in its subsurface, since autotrophy at the lowest branches of the phylogenetic tree does not require sunlight (Jannasch and Mottl, 1985). Hence, emerging life in the early Solar System is more likely to have lived in subsurface, rather than surface environments (Shock, 2001).

The question of habitability on Europa by the identification of reliable biomarkers is a major aim in a truly worldwide collaboration for exploring the Jovian System beyond 2020 as was briefly mentioned in the Introduction, Sec. I.7. This project was originally called the LAPLACE Mission (Blanc *et al.,* 2008). At present the mission is being planned under the name of the "Europa Jupiter System Mission" (EJSM), in which the main partners are the American space agency NASA and the European space agency ESA. The mission consists of two flight elements operating in the Jovian System: the NASA-led Jupiter Europa Orbiter (JEO), and the ESA-led Jupiter Ganymede Orbiter (JGO). JEO and JGO will explore Europa and Ganymede, respectively. But significant contributions for the eventual mission are expected to come from the Japanese JAXA space agency, whose main interest is centered on the Jovian magnetosphere. The Russian Federal Space Agency Roscosmos intends to contribute a lander for probing the Europan surface, but the JAXA and Roscosmos proposals are not yet part of the current baseline of the mission (Grassett *et al.*, 2009).

Rather than facing the complex problem of the origin of life on Earth, by means of the time-honored reductionist methods of chemical evolution, we can instead focus on the origin of habitable ecosystems on Earth and elsewhere. Fortunately, two reliable and easily accessible windows on the nature of the early habitable ecosystems on Earth are available: Firstly, the Pilbara Craton is an old and stable part of the continental crust and the uppermost mantle, which constitutes the hard and rigid outer layer of the Earth. The Pilbara Craton includes its microfossil-rich reservoir, the Dresser Formation in Western Australia.

Secondly, another easily accessible window on the early Earth is the South African Kaapvaal Craton with its Barbeton Greenstone Belt. Extraordinary fractionation of the isotopes of sulfur and methane has been recorded in these ancient sites (>3.6 Gyr BP). In this respect we recall that significant contemporary sulfur reduction is now known to be possible and to take place in pore waters retrieved by the Ocean Drilling Project (ODP), which is now called the Integrated Ocean Drilling Program (IODP): This activity studies the nature of the Earth's seafloor. In 110 expeditions IODP has collected about 2000 deep-sea cores from major geological features located in the ocean basins of the world. These searches have revealed natural populations that are able to fractionate efficiently S-isotopes up to $\delta^{34}S$ of -70‰ (Wortmann *et al.*, 2001; the definitions of the delta parameters are given in the Chapter 1). These extremely high delta values are present, in spite of the Rees upper bound of -46‰ that had been suggested earlier (Rees, 1973).

We have learnt a great deal of relevant information concerning the geologic and evolutionary history of how biogeochemical interactions in hydrothermal vents are able to turn rocks and hot water into habitable ecosystems (Reysenbach and Shock, 2002). On the Archean Earth we have gathered sufficient insights to extrapolate these ecosystems to other environments of the early Solar System, when evolving terrestrial planets, or satellites, may have had an analogous geologic structure. Another aspect that emerges from these ancient settings is that discoveries in the Dresser Formation point towards an active subsurface biosphere (Canfield, 2006). A prolific current deep terrestrial biosphere can be documented with comparable magnitude to that corresponding to the biota on the Earth surface (Whitman *et al.*, 1998). These remarks raise the question that Europa's seafloor, including its presumably prolific subsurface (in as the terrestrial case), could be a significant ecosystem. The instrumentation for reaching such biota is not beyond present technology's grasp, since there are mechanisms capable of raising the processed (microbially fractionated) nutrients from the seafloor and its subsurface to the icy cover of Europa, in analogy with the processes that take place in the Dry Valley Lakes (Antarctica, cf., Chapter 3).

8.9 Sulfur patches and space weather in the neighborhood of Europa

The Galileo Mission discovered the sulfur patches of non-ice elements on Europa's icy surface. These patches have been inferred and confirmed over a twenty-year period. They match the distribution of an ultraviolet absorber that was suggested all the way back by the Voyager data (McEwen, 1986). The New Horizons Mission on its way to Pluto has confirmed the patches of non-ice elements (Grundy *et al.*, 2007). Based on combined spectral reflectance data from the Solid State Imaging experiment (SSI), the Near Infrared Mass Spectrometer (NIMS) and the Ultraviolet Spectrometer (UVS), it has been argued that the non-water ice materials are endogenous in three diverse, but significant terrains (Fanale *et al.,* 1999). Effusive cryovolcanism is clearly one possible endogenous source of the non-water-ice constituents of the surface materials (Fagents, 2003).

The most striking feature of the non-water surficial elements is certainly their distribution in patches. Implantation would be expected to produce a more uniform surface distribution if the source were ions from the Jovian plasma; it may be argued that if the plasma from the magnetosphere were responsible for the sulfur distribution, some geologic process has to be invoked to allow for a non-uniform distribution (Carlson *et al.,* 1999). Alternatively, the sulfurous material on the surface may be endogenous. In other words, the cryovolcanism on Europa would be from the bottom of the global ocean, more like the "black smokers" that are found on the Earth seafloor. Europa's surface is an extremely hostile environment, as a result of space weather, namely the constant exposure to Jupiter's intense radiation belts. The topmost ice layer is subject to harsh radiation, but changes other than chemical reactions (photolysis, radiolysis) are not to be taken into account since, for instance, there is no high-energy (several GeV) deep-inelastic scattering that would change the nuclear identity of the atomic components of the surficial molecules participating in the above-mentioned S-cycles.

Meteoroid gardening might lead to the contamination of the upper surface 1-2 meters with regolith (Cooper *et al.,* 2001). This would require penetrating deeper than this depth in order to reach *pristine material*, which would be challenging, even for penetrators. However, since sulfur processing by bacteria may lead to such radical and extreme depletions of ^{34}S compared to exogenous sulfur, even if some of the contaminated surface by the gardening process contains non-biological sulfur, and penetrators do not reach the above-mentioned depths (1 – 2 meters), robotic mass-spectrometry analysis of the non-water ice elements would identify biologically processed sulfur (for example, δS^{34} reaching over –(50 to 60) ‰).

Indeed, if seafloor microbes were present, they would be able to fractionate substantially all the endogenous sulfur that would reach the icy surface. The biogeochemical signatures would be robust and could in principle be revealed by mass spectrometry. In fact, our original hope to get underneath the ice with a cryobot and a hydrobot (Fig. 8.5) has to be seen in the light of the advent of feasible technologies that can probe the Europan surface searching for biogenicity. Although, in the present discussion we argue in favor of detection of biomarkers imprinted on the icy surface itself with the help of penetrators, alternative instrumentation with new feasible technology is in the process of being developed for penetrating even deeper into the ice (Ulamec *et al.,* 2007; Weiss *et al.,* 2008). Such additional instrumentation would give us a better understanding of the icy crusts of Europa and Ganymede, providing additional tests of our hypothesis. Our main interest is the search for biosignatures on Europa on the sulfur patches discovered by the Galileo mission (Singer, 2003; Chela-Flores and Kumar, 2008; Chela-Flores *et al.,* 2009).

Supplementary Reading

Bertola, F. (1992) *Da Galileo alle Stelle.* Biblos Edizioni, Padua. [This book was published during the celebrations at the University of Padua to commemorate the fourth century of the nomination of Galileo Galilei as a lecturer of Mathematics in that university, where he spent 18 years of very productive work.]

Europa Ocean Conference at San Juan Capistrano Research Institute. 12-14 November, 1996. Book of summaries (pp. 90). San Juan Capistrano Research Institute. 31872 Camino Capistrano. San Juan Capistrano, CA 92675.

Feinberg, G. and Shapiro, R. (1980) *Life beyond Earth: The intelligent Earthling's Guide to Life in the Universe,* William Morrow and Co., New York, pp. 328-332.

Greenberg, R. (2005) *Europa: The Ocean Moon: Search for an Alien Biosphere*, Springer Praxis Books.

Greenberg, R. (2008) *Unmasking Europa: the search for life on Jupiter's ocean moon.* Copernicus Books, 277 pp.

Johnson, T. V. (1995) The Galileo Mission. *Scientific American.* December, p. 35.

Pappalardo, R.T., McKinnon, W.B. and Khurana, K.K. (eds.) (2009) *Europa*. Tucson, Univ. Arizona Press. 727 pp.

Schopf, J. W. (2001). *Cradle of Life: The Discovery of Earth's Earliest Fossils.* Princeton University Press, 336 pp.

References

Anderson, J. D., Lau, E. L., Sjogren, W. L., Schubert, G. and Moore, W. B. (1997) Europa's Differentiated Internal Structure: Inferences from Two Galileo Encounters. *Science* **276,** 1236 - 1239□.

Blanc, M. *et al* and LAPLACE Team Members (2009) LAPLACE: a mission to Europa and the Jupiter System for ESA's Cosmic Vision Programme, *Experimental Astronomy,* **23,** Issue 3, 849-892. The LAPLACE Team Members are in: http://www.ictp.it/~chelaf/ss164.html

Bland, M. T., Showman, A. P. and Tobie, G. (2009) The orbital–thermal evolution and global expansion of Ganymede. *Icarus* **200**, 207-221.

Bortman, H. (2010) Antarctic diving robot practices for Europa. http://www.ictp.it/~chelaf/Antarctic_robot_for_Europa.pdf

Brunner, B. and Bernasconi, S.M. (2005) A revised isotope fractionation model for dissimilatory sulfate reduction in sulfate reducing bacteria. *Geochim. Cosmochim. Acta* **69**, 4759–4771.

Canfield, D.E. (2006) Gas with an ancient history. *Nature* **440,** 426-427.

Canfield, D., Thamdrup, B. (1994) The production of ^{34}S-depleted sulfide during bacterial disproportionation of elemental sulfur. *Science* **266**, 1973-1975.

Carlson, R. W., Johnson, R. E. and Anderson, M. S. (1999) Sulfuric Acid on Europa and the Radiolytic Sulfur Cycle. *Science* **286**, 97-99.

Carr, M. H., Belton, M. J. S., Chapman, C. R., Davies, M. E., Geissler, P., Greenberg, R., McEwen, A S., Tufts, B. R., Greely, R., Sullivan, R., Head, J. W., Pappalardo, R. T., Klaasen, K. P., Johnson, T. V., Kaufman, J., Senske, D., Moore, J., Neukum, G., Schubert, G., Burns, J. A., Thomas, P. and Veverka, J. (1998) Evidence for a subsurface ocean in Europa, *Nature* **391**, 363-365.

Chela-Flores, J. (1996) Habitability of Europa: possible degree of evolution of Europan biota, *Europa Ocean Conference,* San Juan Capistrano Research Institute, San Juan Capistrano, California, USA, 12-14 November, 1996. p.21.

Chela-Flores, J. (2006) The sulphur dilemma: Are there biosignatures on Europa's icy and patchy surface? *Int. J. Astrobiol.* **5**, 17-22.

Chela-Flores, J. (2010) Instrumentation for the search of habitable ecosystems in the future exploration of Europa and Ganymede. *Int. J. Atrobiol.* **9**, 101-108. http://www.ictp.it/~chelaf/jcf_IJA_2010.pdf

Chela-Flores, J., Bhattacherjee, A. B., Dudeja, S., Kumar, N. and Seckbach, J. (2009) Can the biogenicity of Europa's surfical sulfur be tested simultaneously with penetrators and ion traps? *Geophysical Research Abstracts* **11**, EGU2009-0, 2009, EGU General Assembly 2009. Vienna, 22 April.

Chela-Flores, J. and Kumar, N. (2008) Returning to Europa: Can traces of surficial life be detected? *International Journal of Astrobiology* **7**, 263-269. http://www.ictp.it/~chelaf/JCFKumar.pdf

Cooper J. F., Johnson R. E., Mauk B. H., Garrett H. B., Gehrels N. (2001) Energetic Ion and Electron Irradiation of the Icy Galilean Satellites. *Icarus* **149**, 133-159.

Detmers, J., Brüchert, V., Habich, K. S and Kuever, J. (2001) Diversity of Sulfur Isotope Fractionations by Sulfate-Reducing Prokaryotes. *Applied and Environmental Microbiology* **67**, 888-894.

Fagents, S. A. (2003) Considerations for the Effusive Cryovolcanism on Europa: The Post-Galileo Perspective. *J. Geophys. Res.* **108,** No. E12, 5139.

Fanale, F. P., Granahan, J. C., McCord, T. B., Hansen, G.,.Hibbitts, C. A., Carlson, R., Matson, D., Ocampo, A., Kamp, L., Smythe, W., Leader, F., Mehlman, R., Greeley, R., Sullivan, R., Geissler, P., Barth, C., Hendrix, A., Clark, B., Helfenstein, P., Veverka, J., Belton, M., Becker, K., Becker, T., and the Galileo instrumentation teams NIMS, SSI, UVS (1999) *Galileo's* Multiinstrument Spectral View of Europa's Surface Composition. *Icarus* **139,** 179-188.

Gowen, R. A., Smith, A., Fortes, A.D., Barber, S., Brown, P., Church, P., Collinson, G., Coates, A. J., Collins, G., Crawford, I. A., Dehant, V., Chela-Flores, J., Griffiths, A. D., Grindrod, P. M., Gurvits, L.I., Hagermann, A, Hussmann, H., Jaumann, R., Jones, A.P., Joy. A. Sephton, K.H., Karatekin, O., Miljkovic, K., Palomba, E., Pike, W.T., Prieto-Ballesteros, O., Raulin, F., Sephton, M. A., Sheridan, M S., Sims, M., Storrie-Lombardi, M. C., Ambrosi, R., Fielding, J, Fraser, G., Gao, Y., Jones, G. H., Kargl, Karl, W. J., Macagnano, A., Mukherjee, A., Muller, J. P., Phipps, A., Pullan, D., Richter, L., Sohl, F., Snape, J., Sykes, J., Wells, N. (2010) Penetrators for *in situ* sub-surface investigations of Europa. *Advances in Space Research*, doi: 10.1016/j.asr.2010.06.026.

Grasset, O., Lebreton, J.-,P., Blanc, M., Dougherty, M., Erd, C, Greeley, R., Pappalardo, B. and the Joint Science Definition Team (2009) The Jupiter Ganymede Orbiter as part of the ESA/NASA Europa Jupiter System Mission (EJSM). *EPSC Abstracts* **4**, EPSC2009-784, European Planetary Science Congress.

Greely, R. and Batson, R. (1997) *The NASA Atlas of the Solar System,* Cambridge University Press, London.

Greenberg, R. (2005) *Europa, the Ocean Moon*. Springer, Berlin.

Greenberg, R. (2010) Transport rates of radiolytic substances into Europa's ocean: implications for the Potential Origin and Maintenance of Life. *Astrobiology* **10**, 275-83.

Grice, K., Cao, C., Love, G. D., Böttcher, M. E., Twitchett, R. J., Grosjean, E., Summons, R. E., Turgeon, S. C., Dunning, W. and Y. Jin (2005) Photic Zone Euxinia During the Permian-Triassic Superanoxic Event. *Science* **307**, 706 – 709.

Grundy, W. M., Buratti, B. J., Cheng, A. F., Emery, J. P., Lunsford, A., McKinnon, W. B., Moore, J. M., Newman, S. F., Olkin, C. B., Reuter, D. C., Schenk, P. M., Spencer, J. R.; Stern, S. A., Throop, H. B., Weaver, H. A. (2007) New Horizons Mapping of Europa and Ganymede. *Science* **318**, 234-236.

Hand, K. P., Chyba, C.F., J.C. Priscu, Carlson, R.W. and K.H. Nealson (2009) Astrobiology and the Potential for Life on Europa. In: *Europa*. R. Pappalardo, W. McKinnon and K. Khurana (eds.). Univ. of AZ Press, pp. 589-629.

Hoefs, J. (2009) *Stable Isotope Geochemistry*. 6th ed., Springer Berlin, Heidelberg, pp. 76-77.

Hoppa, G. V., Tufts, B. R., Greenberg, R. and Geissler. P. E. (1999) Formation of cycloidal features on Europa. *Science* **285**, 1899–1902.

Horneck, G. (1995) Exobiology, the study of the origin, evolution and distribution of life within the context of cosmic evolution: a review, *Planet. Space Sci.* **43**, 189-217.

Horvath, J., Carsey, F., Cutts, J. Jones, J. Johnson, E., Landry, B., Lane, L., Lynch, G., Chela-Flores, J., Jeng, T-W. and Bradley, A. (1997) Searching for ice and ocean biogenic activity on Europa and Earth. *Instruments, Methods and Missions for Investigation of Extraterrestrial Microorganisms,* (R.B. Hoover, ed.), *Proc. SPIE* **3111**, pp. 490-500. http://www.ictp.it/~chelaf/searching_for_ice.html

Jannasch, H. W. and Mottl, M. J. (1985) Geomicrobiology of Deep-Sea Hydrothermal Vents. *Science* **229**, 717 - 725□.

Kivelson, M. G., Khurana, K. K., Joy, S., Russell, C. T., Southwood, D. J., Walker, R. J. and Polanskey, C. (1997) Europa's Magnetic Signature: Report from Galileo's Pass on 19 December 1996. *Science* **276**, 1239 – 1241.

McCord, T. B., Hansen, G. B., Clark, R. N., Martin, P. D., Hibbitts, C. A., Fanale, F. P., Granahan, J. C., Segura, N. M., Matson, D. L., Johnson, T. V., Carlson, R. W., Smythe, W. D., Danielson, G. E. and the NIMS team (1998) Non-water-ice constituents in the surface material of the icy Galilean satellites from the Galileo near-infrared mapping spectrometer investigation, *Jour. Geophys. Res.* **103**(E4), pp. 8603-8626.

McCready, R. G. L. (1975) Sulphur isotope fractionation by *Desulvovibrio* and *Desulfotomaculum* species. *Geochim. Cosmochim. Acta* **39**,1395-1401.

McEwen, A. S. (1986) Exogenic and endogenic albedo and color patterns on Europa. *J. Geophys. Res.* **91**, 8077–8097.

Oró, J. Squyres, S. W., Reynolds, R. T., and Mills, T. M. (1992) Europa: Prospects for an ocean and exobiological implications, in G. C. Carle, D. E. Schwartz and J. L. Huntington (eds.), *Exobiology in Solar System Exploration,* NASA SP 512, pp. 103-125.

Rees, C.E. (1973) A steady-state model for sulphur isotope fractionation in bacterial reduction processes. *Geochim. Cosmochim. Acta* **37**,1141–1162.

Rollinson, H. (2007) *Early Earth Systems*. Blackwell, London. p. 225.

Shock, E. L. (2001) Geochemical habitats in hydrothermal systems, in *The First Steps of Life in the Universe*. Eds. J. Chela-Flores, T. Owen and F. Raulin. Kluwer Academic Publishers, Dordrecht, pp. 179-185

Singer, E. (2003) Vital clues from Europa. *New Scientist magazine* Issue N. 2414, 22-23. http://www.ictp.it/~chelaf/VitalClues.pdf

Ulamec, S., Biele, J., Funke, O., Engelhardt, M. (2007) Access to glacial and subglacial environments in the solar system by melting probe technology. *Rev. Environ. Sci. Biotechnology* **6**, 71-94.

Weiss, P. (2010) System Study and Design of a Multi-Probe Mission for Planetary In-Situ Analysis. PH. D. Thesis The Hong Kong Polytechnic University. 178 pp.

Weiss, P., Yung, K. L., Ng., T. C., Komle, N., Kargl., G. and Kaufmann, E. (2008) Study of a melting drill head for the exploration of subsurface planetary ice layers. *Planetary and Space Science* **56**, 1280-1292.

Whitman, W.B., Coleman, D.C., and Wiebe, W. J. (1998) Prokaryotes: The unseen majority. *Proceedings of the National Academy of Sciences USA* **95**, 6598-6583.

Wortmann, U.G., Bernasconi, S.M. and Bottcher, M.E. (2001) Hypersulfidic deep biosphere indicates extreme sulfur isotope fractionation during single-step microbial sulfate reduction. *Geology* **29**, 647–650.

9

On the possibility of biological evolution on the moons of Saturn

The current interest of astrobiology in Saturn's satellite Titan, which is the second largest satellite of the solar system, is mainly due to the Cassini-Huygens Mission, an ESA-NASA collaboration that in 2004 dropped a probe through the thick atmosphere of Titan and in 2005 discovered plumes of water ice and dust emanating form the tiny moon Enceladus.

The reader is advised to refer especially to the following entries in the Illustrated Glossary: *Archaea (archaebacteria), Cassini-Huygens Mission, gas chromatography, mass spectrometer, Voyagers.*

9.1 The discovery of Titan

In 1655 the Dutch astronomer Christiaan Huygens discovered a moon rotating around the planet Saturn. The Italian-born astronomer of the Paris Observatory, Jean Dominique Cassini observed the gap in Saturn's rings (Cassini's Division) in 1675. The joint mission ESA/NASA took his name. The 'Cassini' probe explored the Saturn system and the probe Huygens was dropped into the Titan atmosphere (Coustenis and Lorenz, 1999). Following the tradition that we have discussed in the previous chapter in relation with the Jovian system, names for the Saturn system were chosen from Greek mythology. In the case of the largest satellite, it referred to the Titans, who formed the first divine race. (The name probably comes from the Cretan word for 'king'.)

In the beginning of last century (1908), the Spanish astronomer Josep Comas Solá from the Fabra Observatory in Barcelona discovered that Titan was the first satellite in the Solar System to have an atmosphere. The distinguished English physicist Sir James Jeans studied the problem of atmospheric physics of the bodies in the solar system. In the case of Titan he was able to determine the factors that might prevent atmospheric gases from escaping even a body as small as a satellite. To counteract the weak gravitational hold on the gases, Jeans established a limit for retaining the atmosphere, provided that the temperature was sufficiently low. The Jeans limit allows for gases whose molecular weight was larger than or equal to that corresponding to oxygen to remain in the atmosphere over geologic time. The gases that could be candidates for the

J. Chela-Flores, *The Science of Astrobiology: A Personal View on Learning to Read the Book of Life*, Cellular Origin, Life in Extreme Habitats and Astrobiology 20, DOI 10.1007/978-94-007-1627-8_9,

Titan atmosphere included methane (CH_4, since at the low temperatures in the Jeans calculation methane still remains in its gaseous state.)

For the discovery of methane in Titan's atmosphere, we had to wait till 1944, when the Dutch astronomer Gerard Kuiper added a second clue to the singular features of Titan. He discovered that its atmosphere contained the gas methane. It was soon realized that the reason for the sizeable satellite to hang on to an atmosphere was more due to its extremely low temperatures, rather than by the force of gravity (confirming observationally the theoretical calculations of Jeans). Indeed, Kuiper identified the spectral signatures of methane at wavelengths longer than 6000 Å (there are two such lines). Before proceeding to a more careful discussion of Titan, and the other bodies of the Saturn System, the reader is advised to consider some of the satellite statistics in Table 9.1:

Table 9.1 Physical characteristics of Titan

Parameters	***Values***
Surface radius	2,575 km
Mass (Earth mass = 1)	0.022
Surface gravity (Earth gravity = 1)	0.14
Mean density	1.88 gm/cc
Distance from Saturn	1.226×10^6 km (20 R_S)
Orbit period around Saturn	15.95 days
Obliquity	26.7 °
Surface temperature	93.7 K
Surface pressure	1470 mbar

(Owen et al, 1992; Raulin , 2010, Fulchignoni et al., 2005)

9.2 The Voyagers: The first missions to Saturn and Titan

The Voyagers consisted of two space probes sent by NASA in 1977 to explore the solar system, but the results for Saturn in particular are worthy of some comment. Voyager 1 reached Titan on 12 November 1980. Voyager 2 was also able to explore Saturn, but the distance of closest approach of Voyager 1 was a hundred times nearer. Two features of Titan were remarkable in the Voyager 1 images (cf., Fig. 9.1):

• There is a difference in brightness between the two hemispheres. It has been referred to as the north-south asymmetry. It is probably due to circulation of its thick atmosphere. Subsequent observations form the Earth's orbit by means of the Hubble Space Telescope have shown that the asymmetry can change with time. Hubble was able to focus on Titan in 1990; but the observation of Hubble and Voyager were at different wavelengths.

• The other significant feature in the Voyager 1 images is that a 'cap' can be seen close to the north pole of Titan (strictly, the feature resembles a dark ring and is referred to as a polar hood).

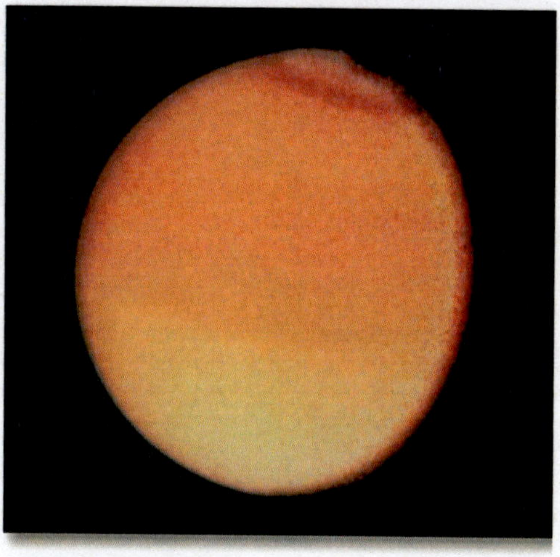

Fig. 9.1 This early Voyager 2 image of Titan was taken on August 23, 1981 from a range of 2.3 million kilometers. It shows some detail in the cloud systems discussed in the text. The southern hemisphere appears lighter in contrast. A well-defined band is seen near the equator, and a dark collar evident at the North Pole. Cassini's instruments were able to penetrate the thick atmosphere

It covers a large part of the northern hemisphere from 70º—90º north latitude. Planetary scientists assumed that this feature may be caused by either the lack of illumination in the winter or alternatively it may be caused by the dynamics of the atmosphere.

9.3 The Cassini-Huygens Mission

After the very successful Voyager encounters with Saturn and Titan (cf., Fig. 9.1) The initial plan to make a specific mission for revisiting the Outer Solar System, was in 1982 due to Europeans and Americans, including Daniel Gautier, Wing Ip and Tobias Owen (Raulin, 2010). From the very beginning, in the joint NASA–ESA Mission, approved in 1989, NASA was responsible for the spacecraft that was eventually called Cassini and ESA was in charge of a probe that was to descend on Titan (Huygens). This was one of the most advanced and successful space exploration missions undertaken in the later part of last century.

In addition to what was said above, we should mention the many insights that have been gained into the nature of Titan. Outstanding amongst these is a scientific investigation of its atmosphere and surface; but Cassini-Huygens has also investigated the rings of Saturn, the magnetosphere of the giant planet; it also studied the surface of the icy satellites of Saturn, especially Enceladus that was hiding some unexpected secrets. The Cassini Mission has brought to our attention another small satellite, Enceladus, which although not as large as Titan is nevertheless extraordinary.

Table 9.2 Some components of Titan's atmosphere (a few references to the original reports are given for a fuller account of all the groups involves we advice the reader to refer to (Raulin, 2010)

	Gas	*Mixing ratio*
Hydrocarbons		
	Ethane, C_2H_6	1×10^{-5} (Vinatier et al., 2007)
	Acetylene, C_2H_2	2×10^{-6} (Vinatier et al., 2007)
	Propane, C_3H_8	5.0×10^{-7} (Vinatier et al., 2007)
	Ethylene, C_2H_4	4.0×10^{-7} (Vinatier et al., 2007)
	Diacetylene, C_4H_2	1.0×10^{-9} (Vinatier et al., 2007)
Nitriles		
	Hydrogen cyanide, HCN	1.4×10^{-7} (Teanby *et al.*, 2006)
	Cyanoacetylene, HC_3N	10^{-8}-10^{-7} (Teanby *et al.*, 2006)
	Cyanogen, C_2N_2	10^{-8}-10^{-7} (Teanby *et al.*, 2006)
Oxygen compounds		
	Carbon dioxide, CO_2	1.6×10^{-9} (de Kok *et al.*, 2007)
	Carbon monoxide, CO	4.5×10^{-5} (de Kok *et al.*, 2007)
	Water	4×10^{-10} (Coustenis *et al.*, 1998)

(Owen et al., 1992, Israël *et al.*, 2005, Raulin, 2010)

The spacecraft Cassini began the exploration of Saturn's satellite system in 2004. The probe Huygens was wrapped in a heat shield that protected its equipment during the descent in the Titan atmosphere (cf., Fig. 9.2).

Fig. 9.2 An artist's conception of the landing of the Huygens probe, a true landmark and utmost achievement of the early 21st century achievement in space exploration

Not only was Titan a major puzzle for chemical evolution, but also as mentioned above, the satellite Enceladus, with intriguing presence of water remains a challenge to planetary science that will be discussed in the next and subsequent sections. The Saturn System contains rings encircling the planet itself, many icy satellites, of which Enceladus is probably one of the most mysterious and fascinating with evidence of some geologic activity in spite of its small dimension (cf., below). In addition the Saturn magnetosphere is also of considerable interest with the myriad particles.

The careful preparation for the successful Cassini mission can be appreciated by the successful landing in 2005, when a trial was performed in northern Sweden. From a balloon carrying a full-scale model of Huygens in a gondola, the replica of the probe was dropped from a height of almost 40 km. Once the process of shedding its shields and deploying the parachutes was completed, the mission proceeded satisfactorily with all the work allowed by the payload components and planned objectives (cf., Table 9.3).

Table 9.3 The payloads and main research objectives of the Cassini Huygens mission

Cassini's payload and investigations	**Huygens payload and investigations**
Optical remote-sensing instruments	*Scientific instruments*
Composite infrared spectrometer	
Imaging science subsystem	Gas chromatograph–mass spectrometer
Ultraviolet imaging spectrograph	
Visual and IR mapping spectrometer	Aerosol collector and pyrolyser
Field particle and wave instruments	Huygens atmospheric structure
Cassini plasma spectrometer	
Cosmic dust analysis	Instrument
Ion and neutral mass spectrometer	
Magnetometer	Descent imager and
Magnetospheric imaging instrument	spectral radiometer
Radio and plasma wave spectrometer	
	Doppler wind experiment
Microwave remote sensing	
Cassini radar	Surface science package
Radio science subsystem	
Interdisciplinary investigations	
Magnetosphere and plasma	*Interdisciplinary investigations*
Rings and dust	
Magnetosphere and plasma	Aeronomy
Atmospheres	
Origin and evolution	Atmosphere–surface interactions
Satellites and asteroids	
Aeronomy and solar-wind interaction	Chemistry and exobiology

(Adapted from Raulin, 2010)

9.4 Titan and Enceladus

To appreciate Titan's features it is important to compare it with another satellite, of the outer solar system, the Neptune's moon Triton, particularly because they both have atmospheres, mainly of molecular hydrogen (cf., Fig. 9.3). In Table 9.4 we summarize the main physical parameters of both satellites. Triton is probably a captured object.

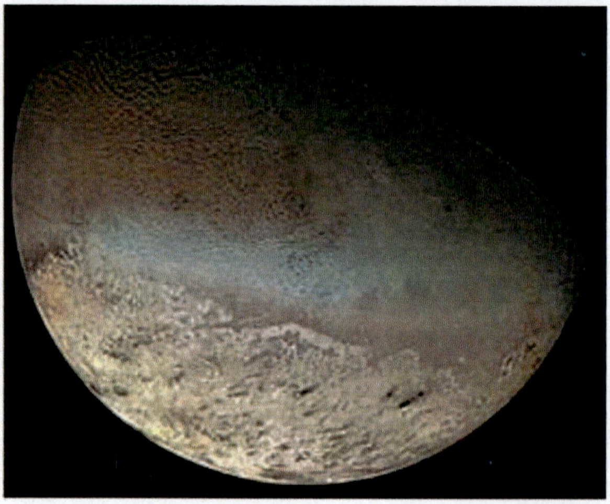

Fig. 9.3 Triton the largest moon of the planet Neptune

The formation of Triton would have been in the solar nebula, unlike the case of Titan, which may have been formed in the disk of material that surrounded Saturn. Its density is the same as that of the two other giant moons of the outer Solar System, the Galilean satellites Ganymede and Callisto.

Table 9.4 Comparison of the main physical parameters of Triton and Titan

Parameters	***Triton***	***Titan***
Radius (km)	1,352	2,575
Mass (10^{22} kg) $Mass_{Earth} = 597x10^{22}$ kg	2.14	13.5
Average density (g/cm^3)	2.1	1.88
Mean distance from the planet	14.33 R_N	20 R_S
Distance from the Sun (AU)	30	9.6
Period in orbit (days)	5.88	15.95
Composition of its surface	Ices of N_2, CH_4, CO, H_2 O, CO_2.	Hydrocarbons

(Coustenis and Lorenz, 1999; Morrrison and Owen, 1996)

Triton's rotation around Neptune is in a direction opposite to the planet's rotation ('retrograde motion'), an unusual case for the Solar System, adding further support to the hypothesis of the captured-object.

Surprisingly Enceladus retains an atmosphere that suggests this moon is sufficiently active geologically to reintroduce the water vapor as some of it may be escaping into interplanetary space due to its low gravity (cf., Fig. 9.4 and section 9.8 below).

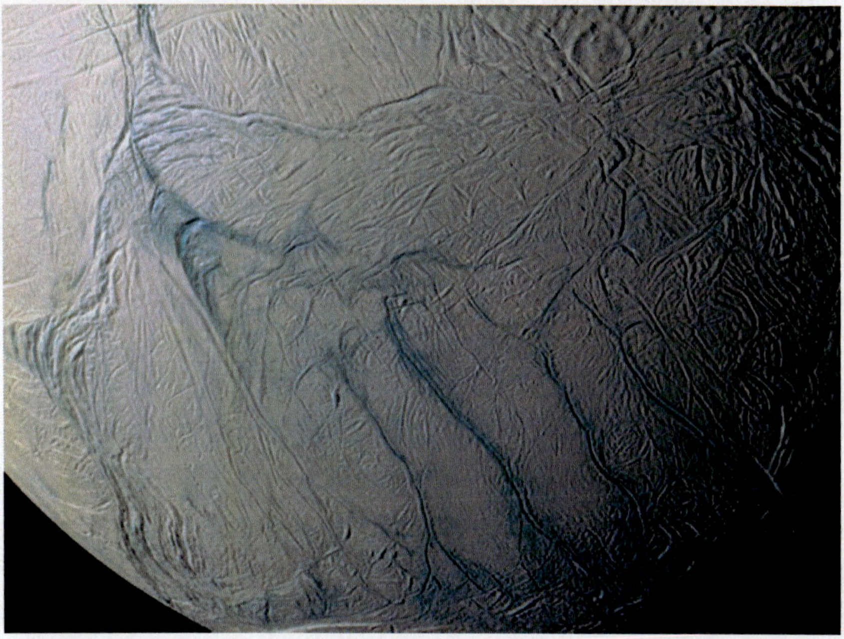

Fig. 9.4 Enceladus. Could microbial life exist inside Enceladus, where no sunlight reaches, photosynthesis is impossible and no oxygen is available?

Could microbial life exist inside Enceladus, where no sunlight reaches, photosynthesis is impossible and no oxygen is available? To answer that question, we need look no farther than our own planet to find examples of the types of exotic ecosystems that could make life possible on Saturn's geyser moon (McKay and Matson, 2008). The answer appears to be, yes, it could be possible. It is this tantalizing potential that brings us back to Enceladus for further study. In recent years, life forms have been found on Earth that thrive in places where the Sun doesn't shine and oxygen is not present because no photosynthesis takes place. Microbes have been discovered that survive on the energy from the chemical interaction between different kinds of minerals, and others that live off the energy from the radioactive decay in rocks.

9.5 The atmosphere and hydrosphere of Titan

The atmosphere of Titan is one of its most interesting features. As we mentioned above, Kuiper's discovery of spectral signatures in the atmosphere correspond to methane.

The atmosphere of Titan is made mainly of nitrogen and methane with small amounts of hydrogen cyanide, cyanoacetylene, and other organic compounds. As we have seen in the previous section, from the Earth Titan has the appearance of an orange haze, whose major components have been studied with various means, going back to the old data from the Voyager 1 flyby. In Table 9.5 we have listed some components of the atmosphere.

With the Cassini Mission (cf., Sec. 9.2), we are beginning to grasp the nature of a possible geophysical characteristics of Titan's diverse landscapes (Lunine *et al.*, 2008) and especially its hydrosphere; in fact, a dark, lake-like feature was observed in a series of images showing an outline of a surface darker than anything else in its surroundings: it has shore-like boundaries (cf., Fig. 9.5).

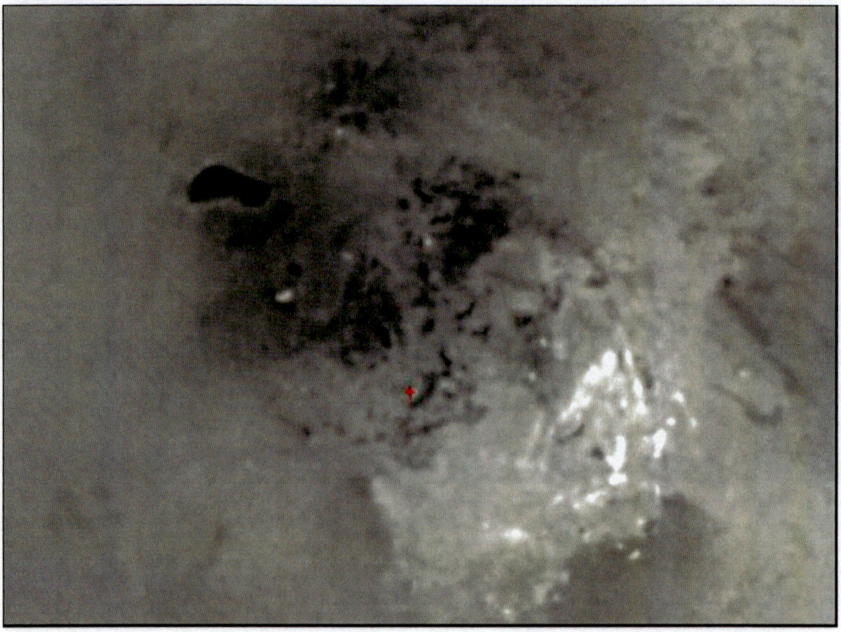

Fig. 9.5 This view of Titan's south polar region reveals an intriguing dark feature that may be the site of a past or present lake of liquid hydrocarbon

Carl Sagan and Bishun Khare in the early 1970s did a set of chemical evolution experiments in which they irradiated several methane-rich atmospheres with ultraviolet light or electrons. This led them to solids that were colored in a manner reminiscent to the color of Titan's atmosphere (reddish/brownish).

After a seven-year voyage, a probe, the Huygens separated from Cassini (cf., Fig. 9.2); it then be parachuted through the atmosphere (cf., Fig. 9.5), initially at a speed of some 20,000 km/hr. When Huygens reached the surface, as shown in Table 9.1, it was working at temperatures of -200 °C (94 K).

The mission was equipped with a radar altimeter and imaging devices, in order to probe, not only the atmosphere, but also the surface of Titan.

Table 9.5 Some components of Titan's atmosphere

Gas	*Mixing ratio*
Nitrogen, N_2	0.98 (Niemann *et al.*, 2005)
Argon, Ar^{40}	~3 x 10^{-5} (Niemann *et al.*, 2005)
Argon, Ar^{36}	~2 x10^{-7} (Niemann *et al.*, 2005)
Methane, CH_4	0.014 (Niemann *et al.*, 2005)
Hydrogen, H_2	~0.001 (Courtin *et al.*, 1995)

(Owen *et al.*, 1992; Raulin, 2010)

One of the Casssini Mission was to identify *in situ* the atmosphere constituents. We have tabulated the main components in Table 9.2.

9.6 Titan's surface

If we return to Chapter 1, Table 1.2, we can verify that if Titan's atmosphere had been captured from the Saturn protonebula, the abundance of the element nitrogen (N) would have been similar to that of neon (Ne). The abundance of both of these elements is almost identical in the Sun. By implication, both N and Ne should have been equally abundant in the solar nebula. We can verify from Table 9.5 some information on the elements in Titan's atmosphere. This data has been gathered since the time of the Voyager missions. We remark that since neon is depleted with respect to nitrogen, it follows that Titan's atmosphere was not formed from the Saturn protonebula itself.

The most likely hypothesis for the formation of the atmosphere is outgassing from the materials that made up this giant satellite. The primitive atmosphere received subsequently contributions from comets coming form outside the Saturn system. From this work educated guesses were formulated as to the nature of its surface that is invisible from either the giant Hubble Space Telescope, or from the previous close encounter of the Voyager mission at the beginning of the 1980s. The Voyager mission was able to detect the exact measurement of the satellite radius. Before this early mission what was known was the size of the satellite and its thick atmosphere. An outstanding step forward was taken by Cassini, through the Huygens lander provided us at an early stage of the mission with a clear image of the surface (cf., Fig. 9.6).

The relevant experiment was named a radio occultation (Jakosky, 1998): as the spacecraft went behind Titan the radio signal transmitted was interrupted exactly as its solid surface interrupted the transmission. From careful observation from the tracking station, the radius was obtained. (The atmosphere is not dense enough to stop the signal.)

Coupled with some calculations, the radio occultation measurements of Voyager demonstrate that there is a depth of some 200 km from the surface to the visible limb. Once the radius is known, only one more parameter is needed to infer the density of Titan, namely its mass.

Fig. 9.6 First color view of Titan's surface

The Voyager mission made this measurement simply by detecting the deflection of the radio signal sent by the spacecraft due to the gravitational interaction. The result shows that the density is much lower than the corresponding value for the Earth (cf., Table 9.1). This is the basis for our conjectures regarding the nature of the surface of Titan. Indeed, there must be materials on the satellite that will account for the much lower density of Titan.

A structure similar to the iron core of the Earth cannot be present. But if we consult the corresponding value for the Moon in Chapter 3, Table 3.1 and recall that its constituents are predominantly silicates, we still can infer that to reach such a low value of density, as observed on Titan, lighter materials must be present.

In fact, we may conclude that a large quantity of lighter materials such as liquid or solid hydrocarbons must be present on the Titan surface. But we may ask, what might be the distribution of the liquid and solid hydrocarbons on the surface? Although at visible wavelengths the atmosphere is opaque, infrared observations carried out with the help of the Hubble Space Telescope show variations in surface albedo that are consistent with the presence of continents and oceans (Dermott and Sagan, 1995). At such low temperatures the oceans, if they exist, are most likely of liquid hydrocarbons, not of liquid water. We may formulate conjectures as to the nature of the solid surface. For we assumed that the number of craters on the surface of Titan could be interpolated from the crater density observed elsewhere in the Saturn system, in which so many satellites lacking an atmosphere have been photographed by Voyager and now by Cassini. Typically there should be about 200 craters larger than 20 km diameter per million square kilometers.

Evidence of craters had to wait for Cassini's results. Very exotic landscapes in fact were discovered, in which large craters filled by liquid hydrocarbons appear to be present.

9.7 Is there a liquid ocean on Titan?

We have seen in Chapter 8 that one of the key questions to study in the future is the possibility of searching for extraterrestrial microorganisms. From what we have discussed in this chapter, prokaryogenesis on Titan—the first stage of the evolution of life, as we know it on Earth—seems a very unlikely possibility. Nevertheless, Titan still presents itself as the world most similar to the early Earth during its stage of prebiotic evolution before the dawn of life. In Table 9.3 we have illustrated the very rich organic chemistry that we already know to exist on our sister world. For this reason the question of the presence of liquid water on Titan remains as a high priority in the questions to be answered in the future, even in the post-Cassini-Huygens era.

The question of the presence of liquid water on Titan is not irrelevant and was raised even before the descent of the Huygens probe (Lunine, 1993; Fortes, 2000). From Table 9.1 the reader can infer that for such a low value of its mean density (0.56 of the Moon's mean density, cf., Chapter 3, Table 3.2), a large proportion of its mass must consist of volatiles. Yet, independently, we do know that water is one of the abundant compounds in the universe. Hence we expect a fraction of its mass to consist of water. On the other hand, we also know from Table 9.1 that the surface temperature of Titan is 94 K. Most of its water content is sequestered in the form of water ice. Hence, a leading question for the future research to decide is if hidden underneath its thick clouds, Titan may have some volcanic activity that could lead to some permanent liquid water.

Nevertheless, independent of the presence of volcanic activity, liquid water may be present. The above comparison of Titan's mean density with that of a silicate body like the Moon suggests that perhaps 30% of its mass should consist of an ice mantle over the silicate core (Taylor and Coustenis, 1998. A few hundred kilometers beneath the surface, the likely prevalent physical conditions are: low enough pressure coupled with low temperature; yet, these conditions would be sufficient to allow a subsurface ocean of liquid water. This would be a second example, much in analogy with the Europan ocean, which we have already discussed in Chapter 8. It should be noted that all the ice would probably not be all pure water, but a substantial fraction of ammonia is also possible. Episodes of liquid water can clearly happen due to the collision of meteorites and comets. It should be underlined that even though Io is relatively much closer to Jupiter ($6R_J$ in units of planetary radii) than Titan is to Saturn ($20R_S$), our inability to predict the effect of tidal heating of a giant planet on its satellite was recently demonstrated by the unexpected high temperature of silicate volcanism, as seen by the Galileo mission (McEwen *et al.*, 1998).

9.8 The plumes and habitability of Enceladus

Saturn's icy moon Enceladus displays evidence for active ice volcanism (cf., Figs. 9.6 and 9.7). In the year 2005 the Cassini spacecraft found a cloud of water vapor over the moon's south pole. Warm fractures were detected where evaporating ice is probably supplying the observed vapor. This is remarkable as Enceladus is the smallest body known to us with active cryovolcanism. The geologically young surface of this satellite is reminiscent of Jupiter's moon Europa discussed in Chapter 8.

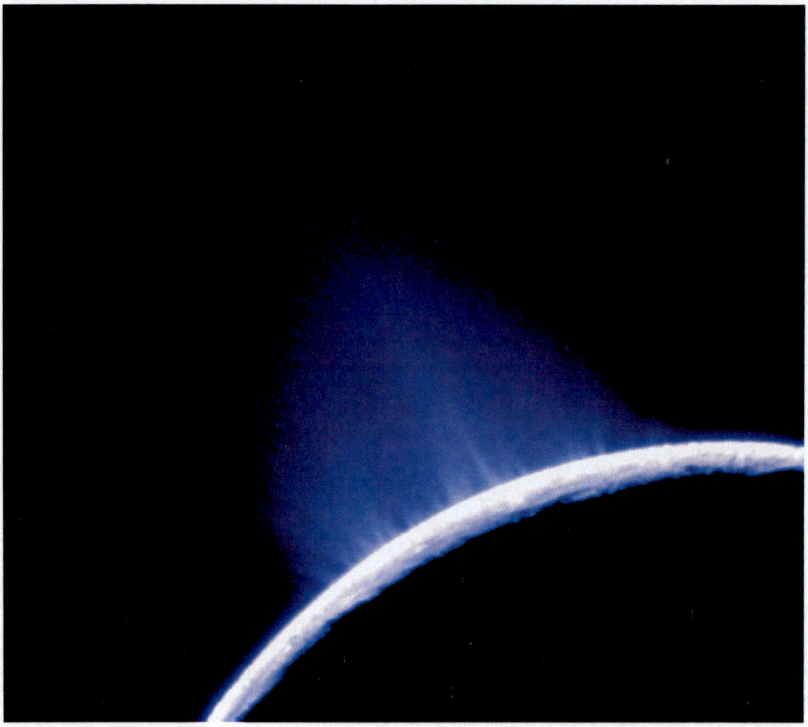

Fig. 9.7 The understanding of the plumes in the South Polar Terrain of Enceladus is a remaining challenge

The Enceladus atmosphere was first detected as Cassini flew within 175 kilometers of Enceladus. Data collected during that flyby confirm an extended and dynamic atmosphere. Water vapor was detected by the sophisticated instrumentation onboard of Cassini: the ion and neutral mass spectrometer and the ultraviolet imaging spectrograph.

On the other hand, the mass spectrometer found the water vapor to be about 65 % of the atmosphere, while the fraction of molecular hydrogen is approximately 20 %. The remaining 15 % is mostly CO_2 and a combination of molecular nitrogen and CO. The variation of water vapor density varies with altitude that may imply that the source may resemble a geothermal hot spot, in analogy with terrestrial geothermal hotspots. A narrow stream of hot mantle convecting up from the Earth's core-mantle boundary may cause these geophysical phenomena. Alternatively the source could be upper-mantle convection.

The ecosystems are completely independent of oxygen or organic material produced by photosynthesis at Earth's surface. These extraordinary microbial ecosystems are models for life that might be present inside Enceladus today. There are three such ecosystems found on Earth that would conceivably be a basis for life on Enceladus. Two are based on methanogens, which belong to an ancient group related to bacteria, called the Archaea, the extremophilic bacteria that thrive in harsh environments without oxygen. Deep volcanic rocks along the Columbia River and in Idaho Falls host two of these ecosystems, which pull their energy from the chemical interaction of different rocks. The third ecosystem is powered by the energy produced in the radioactive decay in rocks, and was found deep below the surface in a mine in South Africa. So the evidence points to the feasibility of life in Enceladus. But how would it get its start? A major problem in answering that question is that we don't know how life originated on Earth, nor have we been able to reproduce Earth's first spark of life in the laboratory. But here's the good news: there are a lot of theories for how life originated on Earth. Now the question is -- do they apply to Enceladus? Two of the theories for the origin of life on Earth do seem to apply to Enceladus--the "primordial soup" theory and the deep-sea vent theory.

9.9 An eventual return to Titan and Enceladus

A mission originally called the Titan Saturn System Mission (TSSM) is of great interest for the scientific community (Coustenis *et al.*, 2008). It would be possible to consider a nine-year cruise on an Earth-Venus-Earth-Earth gravity-assist trajectory to the Saturn system, augmented for the first five years by solar electric propulsion. Unfortunately, to the regret of the author we are unable to live through this wonderful possibility in the short term, given the unbelievable situation that one mission at a time can be envisaged by all the space agencies of the world collaborating; the equally fascinating possibility of returning to Europa will take precedence (cf., Chapter 8). On arrival, the spacecraft would perform an orbit insertion burn to capture into Saturn orbit. The montgolfière, targeted for Titan, would be dropped off just prior to the first Titan flyby, following Saturn orbit insertion.

Data relay from the montgolfière would continue through its six-month mission via the orbiter telecommunications system. The lander element, targeted for Kraken Mare (a northern lake) would be dropped off at the second Titan flyby and the orbiter would be dedicated to science data capture and to relay for the nine-hour length of the lander's mission. During a two-year Saturn tour phase, the orbiter would perform seven close flybys of Enceladus as well as 16 Titan flybys.

Finally, the Titan orbit phase would commence with a Titan orbit insertion burn, placing the orbiter in an elliptical orbit that would be used for concurrent aerobraking and aerosampling. The orbit would be circularized over two months, beginning a 20-month Titan orbit phase. Very challenging instrumentation can be considered for this eventual step of mankind into knowing our own cosmic backyard, such as the montgolfière (cf., Fig 9.8).

9.10 The habitability of Titan

The Cassini spacecraft have not found sign of acetylene on Titan. A separate study found evidence that hydrogen is disappearing near the moon's surface. The discoveries support a theory that Titanic microbes could survive by breathing hydrogen gas and eating acetylene, producing methane as a result.

Fig. 9.8 Diagram of the eventual mission to return to Saturn, Enceladus and Titan. The floating balloon is called a montgolfière and would be a challenge to space exploration later on during the present century

A non-biological explanation is difficult to conceive, since a strong catalyst would be needed to remove the hydrogen and acetylene falling from Titan's atmosphere under the very cold surface temperatures on Titan. Indeed, what is observed is exactly what we would expect if life on Titan is present that uses hydrogen and acetylene in its metabolic pathway and produces methane as a result.

A computer simulation indicates that hydrogen molecules flow downward from Titan's atmosphere, but are somehow missing from the surface. (Strobel, 2010). Hydrogen molecules are generated in Titan's atmosphere when ultraviolet light breaks down methane and acetylene molecules. Using data from two spectrometers on the Cassini spacecraft, it has been found that hydrogen molecules flow out of the atmosphere at an exceedingly high rate. But the analysis found that there is no corresponding buildup at the surface. It seems therefore unlikely that hydrogen is sequestered in a cave or beneath the surface. And because Titan is so cold, a catalyst would be required to convert hydrogen molecules and acetylene back into methane.

By using another Cassini spectrometer it was found that there was a lack of acetylene (Clark, 2010). Both acetylene and benzene are expected to be produced when sunlight strikes the methane gas in Titan's atmosphere, and then falls to the surface.

Abundant benzene was found on the surface, but no acetylene, even though it is predicted to be the more abundant of the two compounds in the atmosphere (cf., Table 9.2). The lack of acetylene, as well as a previously found deficit of ethane could have been due to a smaller than expected production of these compounds. But low acetylene and low ethane, plus low hydrogen, could make a good case for astrobiological studies. Several researchers have suggested that hydrocarbons on Titan could be the basis for life, playing the same role that liquid water does on Earth.

Just as organisms on Earth combine molecular oxygen with organic compounds to get energy, organisms on Titan might react molecular hydrogen with organic materials such as acetylene. In other words, the suggestion is that methane-based organisms called methanogens on Titan could consume hydrogen, acetylene, and ethane. Hence, results of the Huygens probe could indicate the presence of such life by anomalous depletions of acetylene and ethane as well as hydrogen at the surface (McKay and Smith, 2005). And because the combination of acetylene and molecular hydrogen produces methane, it would also solve a major mystery on Titan. Methane is easily broken down by sunlight and destroyed; yet, somehow the moon maintains a plentiful supply of it. Organisms that react hydrogen and acetylene might be the answer.

Each of the observations separately can be explained by abiotic processes, but the coupling of the two together is intriguing, and really calls for more detailed in situ measurements on a future mission that could test more sensitively for evidence of life, especially in the lakes and seas.

Supplementary Reading

Brown, R., Lebreton, J. -P. and Waite, H. (2009) *Titan from Cassini-Huygens.* Springer, 535 pp.

Gautier, D. (1997) Prebiotics on Titan: from available to expected measurements, in C.B. Cosmovici, S. Bowyer and D. Werthimer (eds.), *Astronomical and Biochemical Origins and the Search for Life in the Universe,* Editrice Compositore, Bologna, pp. 219-226.

Coustenis, A., Lellouch, E., Combes, M., McKay, C. P. and Maillard, J. -P. (1997) Titan's atmosphere and surface from infrared spectroscopy and imagery, in C.B. Cosmovici, S. Bowyer and D. Werthimer (eds.), *Astronomical and Biochemical Origins and the Search for Life in the Universe,* Editrice Compositore, Bologna. pp. 227-234.

Coustenis, A. and Taylor, F. W. (2008) Titan: exploring an earthlike world. World Scientific Publishers, 393 p.

Jakosky, B. (1998) *The Search for Life in Other Planets,* Cambridge University Press, London.

Lorenz, R. Mitton, J. (2008) *Titan Unveiled.* Cambridge, 243 p.

References

Clark, R. N., Curchin, J. M., Barnes, J. W., Jaumann, R., Soderblom, L., Cruikshank, D. P., Brown, R. H., Rodriguez, S., Lunine, J., Stephan, K., Hoefen, T. M., Le Mouelic, S., Sotin, C., Baines, K. H., Buratti, B. and Nicholson, P. (2010) Detection and Mapping of Hydrocarbon Deposits on Titan. *J. Geophys. Res.* **115**, E10005, doi:10.1029/2009JE003369.

Courtin, R., Gautier, D. and McKay, C. P. (1995) *Icarus* **114,** 144.

Coustenis, A. and Lorenz, R. D. (1999) Titan, in P. R. Weissman, L.-A. McFadden and T. V. Johnson (eds.), *Encyclopedia of the Solar System,* Academic Press, San Diego, pp. 377-404.

Coustenis, A., Salama, A., Lellouch, E., Encrenaz, T., Bjoraker, G. L., Samuelson, R. E., De Graauw, T., Feuchtgruber, H. and Kessler, M. F. (1998) *Astron. Astrophys.* **336**, L85.

Coustenis, A. and 155 co-authors (2008) TandEM: Titan and Enceladus mission. *Experimental Astronomy,* doi: 10.1007/s10686–008–9103-z.

de Kok, Irwin, P. G. J., N. A. Teanby, E. Lellouch, B. Bézard, S. Vinatier, C. A. Nixon, L. Fletcher, C. Howett, S. B. Calcutt, N. E. Bowles, F. M. Flasar, and F. W. Taylor, (2007) *Icarus* **186**, 354.

Dermott, S.F. and Sagan, C. (1995) Tidal effects of disconnected hydrocarbon seas on Titan, *Nature* **374**, 238-240.

Fortes, A. D. (2000) Exobiological implications of a possible ammonia–water ocean inside Titan. *Icarus* **146** , 444 –452.

Fulchignoni, M. and 42 co-authors (2005) Titan's physical characteristics measured by the Huygens atmospheric instrument (HASI). *Nature* **438**, 785 –791.

Israël G. and 21 co-authors (2005) Evidence for the presence of complex organic matter in Titan's aerosols by in situ analysis. *Nature* **438**, 796 –799.

Lunine, J. I. (1993) Does Titan have an ocean? A review of current understanding of Titan's surface. *Reviews of Geophysics*, **31**, 133 –149.

Lunine, J. I. and 43 co-authors (2008) Titan's diverse landscapes as evidenced by Cassini RADAR's third and fourth looks at Titan. *Icarus* **195**, 415 –433.

McEwen, A. S., Kezsthelyi, L., Spenser, J. R., Schubert, G., Matson, D. L., Lopes-Gautier, R., Klaasen, K. P., Johnson, T. V., Head, J. W., Geissler, P., Fagents, S., Davies, A. G., Carr, M. H., Breneman, H. H. and Belton, M. J. S. (1998). High-Temperature Silicate Volcanism on Jupiter's Moon Io, *Science* **281**, 87-90.

McKay, C.P., Smith, H. D. (2005) Possibilities for methanogenic life in liquid methane on the surface of Titan. *Icarus* **178,** 274-276.

McKay, C and Matson, D. (2008) The Extraordinary Ecosystems of Enceladus, The Astrobiology Magazine, April 2.

Morrrison, D. and Owen, T. (1996) *The Planetary System,* (2nd ed.), Addison-Wesley Publishing Co., Reading, Mass.

Niemann, H. B., Atreya, S. K., Bauer, S. J., Carignan, G. R., Demick, J. E., Frost, R. L., Gautier, D., Haberman, J. A., Harpold, D. N., Hunten, D. M., Israel, G., Lunine, J. I., Kasprzak, W. T., Owen, T. C., Paulkovich, M., Raulin, F., Raaen, E. and Way, S. H. (2005) *Nature* **438**, 779.

Owen, T., Gautier, D. and Raulin, F. (1992) Titan. In: Exobiology in Solar System Exploration, in G. C. Carle, D. E. Schwartz and J. L. Huntington (eds.), *Exobiology in Solar System Exploration,* NASA SP 512, pp. 126-143.

Raulin, F. (2010) Titan: New Insights from Cassini-Huygens. Chapter 8, in: *Astrobiology Emergence, Search and Detection of Life*. Basiuk, V. A. (ed.) American Science Publishers, Stevenson Ranch, California, USA, pp. 404-426.

Raulin, F., Bruston, P., Coll, P., Coscia, D., Gazeau, M.-C., Guez, L. and de Vanssay, E. (1995) *Exobiology on Titan,* in C. Ponnamperuma and J. Chela-Flores (eds.), *Chemical Evolution: The Structure and Model of the First Cell,* Kluwer Academic Publishers, Dordrecht, pp. 39-53.

Taylor, F. W. and Coustenis, A. (1998) Titan in the Solar System, *Planet. Space Sci.* **46**, 1085-1097.

Teanby, N. A., Irwin, P. G. J., de Kok, R, Nixon, A. C. Coustenis, A., Bézard, B., Calcutt, S. B., Bowles, N. E., Flasar, F. M., Fletcher, L, Howett, C. and Taylor, F. W. (2006) *Icarus* **181**, 243

Vinatier, S., Bézard, B., Fouchet, T, Teanby, N. A., de Kok, R., Irwin, P. G. J., Conrath, B. J, Nixon, C. A., Romani, P. N., Flasar, F. M., and Coustenis, A. (2007) *Icarus* **188**, 120.

10

How different would life be elsewhere?

Prior to discussing the possible environments for life and intelligent life in the universe, (a topic addressed in the next two chapters), it is appropriate to discuss the following question: How can we test both the habitability of planets around other stars whether life may have evolved on a habitable Earth-like planet outside our Solar System.

The reader is advised to refer especially to the following entries in the Illustrated glossary: *Astrometry, brown dwarf, chaos, Jovian planet, Kepler Mission, Kuiper belt, Orion Nebula.*

10.1 Possible degrees of evolution of life

We wish to discuss in the first place the different positions that are possible regarding the question of extraterrestrial life. It is against this background that we should test the validity of the main conjecture of this book. We have understood sufficient aspects of the question of the origin and evolution of life on Earth, so that we may begin the study of life in the universe with a more advanced approach. We will ask more than whether life is present in other environments, such as the Mars permafrost, in the Jovian satellite Europa or the Saturn moon Enceladus. We wish also to know the habitability of other planets outside our own Solar System.

These questions do not arise from simple curiosity. We are approaching a unique moment in the development astrobiology with new vigorous efforts are being followed by the main space agencies. Time and funding are crucial for this endeavor, so that if the general question of habitability of ecosystems can be faced from the start, we should make every effort to do so. Distinguishing between alternative possibilities require ingenious experiments either *in situ*, by sample-return-missions or by means of advanced missions with novel instrumentation, as for instance the Kepler mission that will be discussed below. We list a few alternative views on the nature of the origin of life:

• The more widely accepted belief on the nature of the origin of life is that life evolved according to the principles of deterministic chaos. Evolutionary developments of this type never run again through the same path of events (Kauffman, 1993).

• The possibility that similar evolutionary pathways may have been followed on different planets of the Solar System has been raised (Davies, 1998). This work is

J. Chela-Flores, *The Science of Astrobiology: A Personal View on Learning to Read the Book of Life*, Cellular Origin, Life in Extreme Habitats and Astrobiology 20, DOI 10.1007/978-94-007-1627-8_10,

based on the increasing acceptance that catastrophic impacts have played an important role in shaping the history of terrestrial life. Thus there may be some common evolutionary pathways between the microorganisms on earth and those that may have developed on Mars during its clement or Eden-like period (roughly equivalent to the early Archean in Earth stratigraphy). The means of transport may have been the displacement of substantial quantities of planetary surface due to large asteroid impacts on Mars.

• Regardless of the possibility raised above many researchers still see no reason to assume that the development of extraterrestrial life followed the same evolutionary pathway known to have occurred on earth. Moreover, a widespread point of view is that our ignorance concerning the origin of terrestrial life does not justify the assumption that any extraterrestrial life form has to be based on just the same genetic principles that are known to us.

• It has also been argued that lessons should be drawn from principles of genetics (De Duve, 1995). Amongst the various positions regarding the possible implications of our current understanding of life's origins, such an approach is original. This forward-looking vision may be said to add a fourth point of view to the discussion of the various possible ways to approach the question of expectations of the nature of extraterrestrial life.

• We all agree that the final outcome of life evolving in a different environment would not be the same as the Earth biota, but new ground has been broken with the question used as a title for the present Chapter, more precisely:

How different would such alternative extraterrestrial evolutionary pathways be?

This question has led to the conclusion that there is no reason for the details of our phylogenetic tree to be reproduced elsewhere, except if catastrophic events have led to some exchange of living organisms between the planets, as suggested by above. The tree of life constituted by all the living organisms may be unique to our planet. On the other hand, there is plenty of room for the development of differently shaped evolutionary trees in an extraterrestrial environment, where life may have taken hold. But, following the above De Duve line of reasoning, "certain directions may carry such decisive selective advantages as to have a high probability of occurring elsewhere as well".

10.2 The search for other solar systems

As we have seen earlier, the Sun and planets condensed from a large rotating disk of dust and gas, themselves the product of condensation of the interstellar medium. The bottom line of this hypothesis is that most of the angular momentum of the parent solar nebula is presently concentrated in the orbital motion of the planets. For this reason, it is only reasonable to assume that the leading candidates for possessing planetary systems are single stars. Besides, it is considerably difficult to infer their presence from the interaction of the planets with their parent star. For instance, the partial eclipse of a given star can occur if the ecliptic plane is in a favorable alignment with the plane of the planets that rotate around other solar systems.

It is instructive to consider a sample of stars, emphasizing some known to be somewhat similar to the Solar System, as they have Jovian-like planets, although in several known cases these giant planets are in very small orbits around their stars; for instance, in the case of the first extrasolar planet discovered by Michel Mayor and Didier Queloz, (Mayor and Queloz, 1995). In *Pegasi 51,* the Jupiter-like planet was only about seven million kilometers away. This raises the question as to the mechanisms that might either give birth to a giant planet close to the star or that would induce the orbit of a giant planet, such as our own Jupiter to decay into a Mercury-like orbit.

However, initially only Jupiter-like planets were discovered since the first report of the *Pegasi 51* system. But as we shall see below subsequently as the observational techniques improved smaller super-Earths began to be added to the ever-increasing number of exoplanets. Such discoveries favor life having been provided a variety of appropriate environments in our own galaxy, as illustrated in Table 10.1, were we listed just a few of the early discoveries out of the large number known at present (at the time of writing the lists contain over 500 exoplanets, a number that will inevitably increase into thousands, as the observational techniques improve, especially in the era of the Kepler mission).

In the 1970s Carl Sagan and co-workers used a computer program in order to assess what type of distribution one should expect in other solar systems (Sagan, 1980). However, the theoretical models did not anticipate the surprising results of the planets discovered since a planet was detected around *Pegasi 51.* Most of the early exoplanets were giant gas planets like Jupiter due to the limited instrumentation available. Some of these were in highly eccentric orbits and some excessively close to their parent star compared to our experience with the Solar System. Evidence for Earth-like planets remained just beyond the limits of detection in the pre-Kepler era. We expect that it will eventually be possible to detect Earth-like planets not excluding that feasible technologies might also detect similar environmental conditions that would be compatible with habitable ecosystems.

10.3 How are extrasolar planets found?

The first discoveries were made with the identification of periodic signals in the radial velocities discerned from time series of spectra obtained for nearby Sun-type stars. Other search methods include (Fridlund and Lammer, 2010):

(1) Analysis of a star's light curves,

(2) Identification of the signal that results from the diminishing of light when a planet passes between the observer and the star,

(3) Analysis of signatures in the caustics of gravitational lensing.

The key concept is that when a sufficiently large planet rotates around its star, say Jupiter rotating around the Sun, the gravitational interaction between planet and star is reciprocal and there is a mutual perturbation.

Given the development of astrometry to this date, the pair Jupiter-Sun produces a perturbation of the Sun that would be detectable within a radius of 100 light years. The star goes into a minuscule orbit, but not negligible. The stellar motion is referred to technically as the *reflex motion*. There are two ways to determine the reflex motion (Jakosky, 1998). We can first detect the change of velocity of the star with respect to the Earth. In physics the technique is known as the Doppler effect: as the star is moving towards or away from us, there will be a measurable difference in the wavelength of the light reaching us. This is a general phenomenon, which consists of the apparent change in frequency of a source of light (or sound, or indeed any wave motion) due to the motion of the source and observer. The second means of identifying the reflex motion of the star is by means of astrometry (the subject which concerns itself with the observation of the position of celestial objects and its variation over time).

10.4 Which are likely habitable zones?

The state of the art in detecting the reflex motion only allows one to detect the star motion due to the presence of a Jupiter-like planet. Indeed, the nine examples that we selected in Table 10.1 are all Jupiter-like planets (Fig. 10.1). It is remarkable that the planets themselves that have been detected up to the present are the least likely to be habitable. However, these Jupiter-like planets could have Europa-like satellites (cf., Chapter 8), which could be likely candidates for harboring life.

Fig. 10.1 Jupiter, the planet of our Solar System that serves as a reference for estimating the size of extrasolar planets mentioned in Table 10.1.Voyager 1 image

This opens the debate as to what are habitable zones, a subject of great interest that we shall not develop in detail here. A habitable zone was understood as one in which the amount of stellar energy reaching a given planet, or satellite, would be conducive to the process of photosynthesis. (Clearly, the amount of greenhouse gases, are relevant too.) Water contributes to the dynamic properties of a terrestrial planet, permitting convection within the planetary crust that might be essential to supporting Earth-like life by creating local chemical disequilibria that provide energy (Schneider *et al.*, 2009). Water absorbs electro-magnetic radiation over a broad wavelength range, covering part of the visible and most of the near-IR, and has a very distinctive spectral signature. The search for water at radio wavelengths (1.35 cm) started already 1999 (Cosmovici *et al.*, 2000; 2008). Water was also detected in absorption by the Spitzer IR Space Telescope. Abundant water may be likely on some planets since it was detected around the star HD 189733 b (Tinetti *et al.*, 2007; Swain *et al.*, 2008).

10.5 The multiplicity of exoplanets in our galactic neighborhood

We have seen in Chapter 1 the importance of the HR diagram. Once again it is useful to the astrobiologist: the mass of the companion to the star that wobbles-its reflex motion- can be calculated (Jakosky, 1998) as follows: the mass of the given star (for which we want to ascertain whether it has a companion planet) can be determined from its temperature and from the total amount of light that it emits. The HR diagram correlates the three parameters: mass M, temperature T and total amount of light (E) that it emits. Hence, with only two measurements (E, T) we can infer M. Once M is known and the amplitude of the reflex motion is carefully measured (by means of the variation of the star velocity with time), the mass of the companion can be determined.

Jupiter in our Solar System is the largest planet that is capable of introducing perturbations into the central star (the Sun) detectable from other worlds (cf., Figure 10.1). So far all the extrasolar planets have Jupiter-like and lower-mass planets called super Earths as a point of reference (cf., Table 10.1).

Some are the same size of the analogs of the Solar System, while others are substantially larger, so much so that even the question of the definition of a planet itself is debatable (cf., the next section). A complication worth noticing has already been made manifest in our Table 10.1, namely, if the velocity of the star is not precisely towards and away form us along the line of sight between us and the wobbling star (the Doppler shift to apply), the best we can do is to determine the component of the velocity that is along the line of sight-this is the significance of the sin (i) factor in the above-mentioned Table 10.1. The bottom line of this comment is that the mass of the companions that are cited in Table 10.1 are really lower bounds or equal to the real mass of the companion.

10.6 What is a planet?

Regarding the question as to what is a planet, in the outer solar system the dividing line between a large Kuiper Belt object and a small mass planet is rather blurred. The case in question concerns the discovery during the 1990s of over 60 mini-planets, whose diameters range between 100 and 500 km. By simple extrapolation, it is now believed that in the region between 30 to 50 AU on along the plane of the ecliptic (where the Kuiper Belt extends), there might be up to 40,000 such objects. At the other end of the spectrum, when we consider the extrasolar planets that we have been discussing in this chapter, the question of the brown dwarfs has to be faced. Indeed, the dividing line between a small brown dwarf and a large Jupiter-like planet is not easy to distinguish.

Table 10.1 Before the explosion in the rate of discovery of exoplanets (cf., Sec. 10.2), we have selected a few of the stars where exoplanets were first detected at the end of the 20th century. M_J is the mass of Jupiter; the results obtained are for the parameter M sin (i), where M is the new planet's mass, and i is the orbital inclination to our line of sight (this is an unknown parameter, and so the masses are maximum values)

Star	Distance from the Sun (pc)	Distance of its Jupiter-size planet: semi-major axis (AU); eccentricity	Size of its Jupiter-size planet (M_J)
51 Pegasi	15.4	0.051; 0.01	0.44
Upsilon Andromedae (*)	16.5	0.053; 0.03	0.63
Rho 1 55 Cancri (a double system)	13.4	0.12; 0.03 >4; -	0.85 >5
Rho Corona Borealis	16.7	0.23; 0.05	1.1
16 Cygnus B	~22	1.70; 0.68	1.74
47 Ursae Majoris	14.1	2.08; 0.09	2.42
Tau Bootis	~15	0.042; 0.00	3.64
70 Virginis	18.1	0.47; 0.40	6.84

(Mayor et al., 1997; Black, 1999; Marcy, 1998)

(*) *In the important case of Upsilon Andromedae, where a solar system was identified by two groups of astronomers, we have only inserted information on the first of the three Jupiter-like planets that were reported (the other two planets are of masses that double and quadruple the Jovian mass (Lissauer, 1999).*

In fact, a brown dwarf is an object formed in the same way as a star. However, the mass involved in the process of star formation is insufficient for triggering the thermonuclear reactions that are necessary for the birth of the star. The lower bound for star formation is about 80 Jupiter masses. However, the minimum mass is unknown for a brown dwarf.

Just like in many chapters in this book, the reader is invited not to be disappointed with the difficulty in getting definite answers to the question whether we are alone in the universe, but rather to wait enthusiastically for the technical developments of the future, when new missions beyond CoRoT and Kepler (discussed below) will bring us closer to detect earth-like planets, even if at the time of writing only super Earths have been identified. We have argued in Chapters 7 and 8 that our solar system may provide evidence of extraterrestrial life in the form of microorganisms. The current and new space missions may allow us a glimpse at a much wider sample of worlds.

10.7 The CoRoT Mission

The first space mission specifically designed to search for exoplanets similar to the Earth itself was launched in December 2006 (Friedlund *et al.*, 2010b). This is the French space agency (CNES) mission CoRoT, which was launched with a Soyuz-Fregat rocket and is a small (27 cm aperture) space telescope orbiting in a 900km high, Sun-synchronous polar orbit around Earth. The method used is the transit method, that is, observing a very large number of stars simultaneously where, assuming a random orientation in space, between 0.5% and tens of percent of the objects (depending on the orbital periods of the transiting planets) will experience an eclipse of a portion of the stellar surface, which will cause a temporary drop (timescale of hours) in the stellar flux. The mission is specifically designed to search for transiting super Earths (1–2 Earth radii) in short-period orbits (<15–20 days, though larger planetary radii are detectable up to periods of 50 days). During almost 900 days in space, CoRoT has found a total of seven exoplanets that have been confirmed.

10.8 The Kepler mission

CoRoT was followed almost immediately by the Kepler mission of NASA that was launched in March 2009 (Friedlund *et al.,* 2010b). Kepler has a larger telescope (95 cm, cf., p. 30). Kepler can remain pointed toward the same point in the sky for years and it is expected to make observations for a period of 3–4 years having the potential to detect small long-period planets. The star field that Kepler observes is in the constellations Cygnus and Lyra. With a significantly larger field of view, Kepler will follow about 100,000 stars. The Kepler Spacecraft has discovered six planets made of a mix of rock and gases orbiting a yellow dwarf star (2,000 light years away), known as Kepler-11 (Lissauer, *et al.*, 2011). This is an extraordinary solar system (cf., Fig. 10.2):

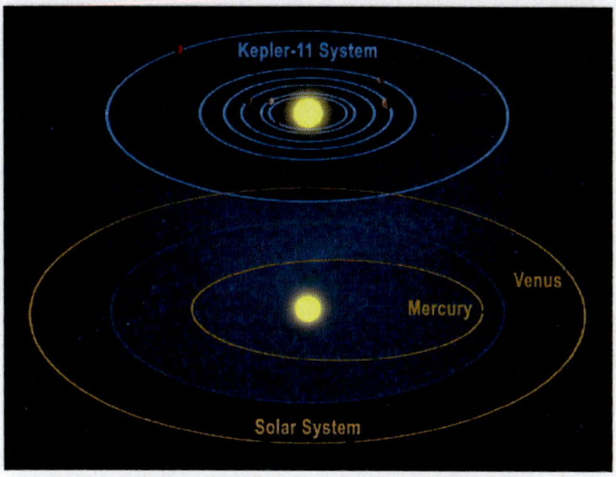

Fig. 10.2. An artistic view of the planetary system around the dwarf star Kepler 11 and the Solar System

Indeed, the orbits of the five inner planets have orbital periods less than 50 days around the dwarf star, while an additional one has a period of 118 days. They must be aligned edge-on to the Earth to have yielded this seminal result with the occultation method of detection ("eclipse of its star").

10.9 The discovery of the first super-Earths in habitable zones

The detection of 3 super-Earth planets orbiting a star located at only 13 pc from the Sun (HD 40307) has been reported (Mayor *et al.,* 2009). This discovery was achieved with the high precision radial velocities measured from data acquired with the HARPS spectrograph on the ESO 3.6-m telescope. The closest planet HD 40307 b is presently the lightest exoplanet detected around a main sequence star. The two other planets also belong to the category of super-Earths. All 3 planets are on circular orbits.

Available Spitzer IRS data of HD 40307 do not show any IR excess in the 10-40 micron region of the spectra. No warm dust disk is therefore detected in the inner regions of the system; this is unlike the earlier case of a star harboring a trio of Neptune planets (HD 69830) for which an observed IR excess indicated the presence of a debris disk, possibly in the form of an asteroid belt (Beichman *et al.,* 2005).

The characterization of multi-planetary systems with very low-mass planets requires a large number of measurements. With the HARPS program and with improved data reduction software, several dozens of planets of masses lower than 30 Earth-masses and period shorter than 50 days have been detected. In fact, at least 30% of main sequence stars have one or several super-Earth companions. Future observations will confirm these detections and hopefully allow the characterization of these systems in more detail. The domain of Neptune-type and rocky planets will be drastically boosted in the near future with these detections. In particular, we expect to study a sufficient number of systems to discuss more deeply the emerging properties for these low mass planets Mayor *et al.,* 2009).

One of the most exciting possibilities offered by this large emerging population of low-mass planets with a short orbital period is the related high probability of transiting super-Earths among the candidates. If detected and then targeted in complementary observations, these transiting super-Earths would provide a tremendous contribution to the study of the expected diversity of the structure of low-mass planets.

10.10 Exomoonology: the habitability of satellites around other stars

It is no longer evident that the habitability zone, in which planets and their satellites may develop life, need be found in a limited region near the star. In the case of our solar system, the habitability zone has been taken to lie outside the orbit of Venus (0.7 AU) and beyond the orbit of Mars (1.5 AU). What we do know about Europa suggests that the characteristics needed for suites of extrasolar satellites around Jupiter-like planets to harbor life are (cf., Chapter 8):

- A non-circular orbit (caused by the gravitational interaction of the remaining satellites of the suite).
- The vicinity to a sufficiently large planet, so that tidal interaction may be effective in producing hydrothermal vents, favorable for bacterial ecosystems.

We should underline that all that is relevant is the presence of large Jupiter-like planets, not that the Jupiter-like planet be near the companion star. Traditionally the only concept that has led to estimating the boundaries of the zone of habitability has been the reasonable distance of the Earth-like planet from its star.

The easiest way to look for an exomoon (cf., Fig 10.3) is generally considered to be through closely watching the motion of the host planet, because the moon is too small to see directly. For example, as the Earth orbits the Sun, it exhibits a slight wobble due to the presence of our moon. In the animation below, you can see a Jupiter-like planet with an Earth-like moon. Notice how the Jupiter-like planet wobbles about a common centre. But even though we cannot see the exomoon directly, we can see the effect it has on the host planet. If this planet transits across its star, the planet passes in front of it once every orbital period and a dip in the amount of observed starlight takes place. For every transit, the planet's wobble makes the motion appear slightly differently.

Fig. 10.3. Artist impression of an exoplanet as seen from its exomoon (in the foreground)

These differences are signatures of a moon due to two changes:

- A change in planet's position
- A change in planet's velocity.

The change in position means that the transit light curve seems to shift about, in what is known as transit time variation, TTV (Sartoretti & Schneider, 1998). The duration of a transit is inversely proportional to the velocity of the planet. So since velocity is changing then transit duration is changing. This manifests itself with the transit duration variation, TDV (Kipping, 2009 a, b; Fossey *et al.,* 2009).

TTV and TDV always exhibit a 90-degree phase shift and this essentially gives astronomers a very unique signature to identify exomoons. If you try to use TTV by itself, you will run into the problem that a plethora of different physical phenomenon can cause TTV, not just moons. By using TTV and TDV together, you can finally say, that signal is definitely a moon. Combining the two approaches just described (TTV or TDV), allows the calculation of both the mass of the moon and the orbital distance of the moon. Consequently, though Earth-like planets have not yet been detected, and even though giant planets are not evident environments that may support life, the new extrasolar planets are already indicators of possible environments favorable to the origin of life. This conclusion follows from the generality of the arguments of the formation of Jovian planets from their corresponding subnebulae (cf., Chapter 2). From the mechanisms for the formation of satellites suggest that in the star cradles in the Orion nebula we should have a similar situation (cf., Fig. 10.4). New instruments will be needed that will improve on the current technology. A coronagraph is a telescope that produces an artificial total eclipse of the Sun inside the instrument, in order to be able to observe the Sun's corona and prominences (Encyclopædia Britannica, 2010).

Fig. 10.4 The Orion Nebula is a gas cloud, in which the younger stars are surrounded by disks of dust and gas, not unlike our own solar system in dimension

A round metal screen blocks the brightness of the Sun's photosphere (its central disk). A new class of similar instruments (called "stellar coronagraphs" to distinguish them from "solar" coronagraphs) has been developed in the area of the search for extrasolar planets. The spectral resolution or resolving power of a spectrograph, or, more generally, of a frequency spectrum, is a measure of its power to resolve features in the electromagnetic spectrum. It is usually defined by $R = \lambda / \Delta\lambda$, where $\Delta\lambda$ is the smallest difference in wavelengths that can be distinguished, at a wavelength of λ. A 1.5–2m class stellar coronagraph is suitable for blocking the star in order to study of giant planets and nearby super Earths (Schneider *et al.*, 2009). For detecting exomoons the technology must take giant steps (Table 10.2).

Table 10.2. Baseline distances needed for Exomoon direct imaging

Objectives	Required baseline at 600 nm
Exo-moon image	400m
Spectroscopy of a planetary transit image	645m
Astrometry of a planetary transit image	40km
Direct measurement of planetary size	20km
Image of continents or oceans	70km

(Williams and Gaidos, 2008, Schneider *et al.*, 2010)

An impressive success of these instruments was to directly observe a planet orbiting the nearby star Fomalhaut (cf., Fig 10.5).

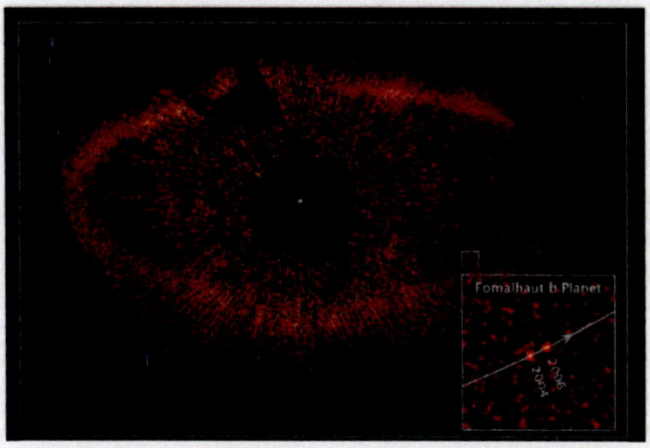

Fig. 10.5 This visible-light image from the Hubble shows the planet, Fomalhaut b, orbiting its parent star.

The planet could be seen clearly in these images taken by Hubble's Advanced Camera for Surveys' coronagraph (cf., Table 10.3)

Table 10.3 Some properties of Fomalhaut b

Properties	***Values***
Star	Fomalhaut
Mass	0.054 – 3 M_J
Distance	25 ± 0.1 ly
Detection method	Direct imaging
Orbital period	872 yrs
Semi major axis	~115 AU

(Kalas *et al.*, 2008).

The planet, called Fomalhaut b, orbits the 200-million-year-old star every 872 years. Fomalhaut is much hotter than our Sun and is 16 times as bright. Fomalhaut is vigorously burning hydrogen through nuclear fusion. This rate of hydrogen consumption limits its lifespan: the star will live for a mere 1 billion years, and the evolution of multicellular life is unlikely on any of its exoplanets, or exomoons.

10.11 Potentially habitable worlds: planets like Earth

The nearest to a planet like Earth so far is the set of super-Earths that we have already mentioned above. Indeed, none of them at the time of writing this book is similar in dimension to our own planet. CoROT discovered one small super-Earth: Corot-7b, which is less twice the Earth diameter d_E (it is ~ 1.7 d_E), but it too near its star to be habitable. The main issue is not to discover a planet with exactly equal to d_E. The real significant search is for a, Earth-like planet or even a super-Earth that lies in the habitable zone of its star, namely sufficiently separated to allow liquid water on its surface, just like the area bounded by the Venusian and Martian orbits in our own Solar System. In the vicinity of the Sun, stellar density can be determined from the various surveys of nearby stars and from estimates of their completeness. Wilhelm Gliese's catalog of stars closer than 65 light-years can be used for this purpose. It contains 1,049 stars in a volume with a radius of 65 light-years. The average calculated density is less than one-third the calculated density for stars known from other catalogs, yielding a stellar density of less than ~ 0.002 objects per cubic light-year. A dwarf star has average physical properties (luminosity, mass, and size) or even low values. These stars include most main-sequence stars (like the Sun). According to their temperature, their color can range from blue to red (with temperatures of a few thousand degrees. In the Gliese catalog the star labeled with the number 581 is a red dwarf. It is indeed surprising that in such a small sample—just over one thousand stars—one of them (Gliese 581) is gradually yielding the secrets of its extraordinary nature: it is the "Sun" of a solar system that will be the focus of astrobiology in the future; it is a stunning solar system.

At the time of writing our first neighboring cosmic cousin may consist of at least 6 planets. Labeled according to an alphabetical order as the distance from their Sun increases the new solar system is summarized in Table 10.4. Steven Vogt and co-workers at the Keck Observatory in Hawaii, using HARPS data have discovered that the year of Gliese 581 g lasts 37 days, all of which are spent in the habitable zone of that red dwarf (Vogt *et al.,* 2010). Like Europa in our Solar System, and the Moon, it always shows one face to that Sun. In the zone that separates day and night we should have about room temperature. We are witnessing this extraordinary discovery at the time of publishing the Second Edition of this book and, like all other epoch-making discoveries in the physical, earth and life sciences we, together with our readers, must expect a revision of the data to establish (or refute) one of the cornerstone discoveries in astrobiology that we have the fortune to report in the present publication.

Table 10.4 The planetary system of Gliese 581

Planets	***Mass of planet*** (M_E)
Planet b	16
Planet c	5
Planet d	6
Planet e	1.9
Planet f	7
Planet g	3

With these results from Gliese 581 and, in addition those coming from Kepler Space Telescope, especially from the planetary system around the yellow dwarf star Kepler 11 (cf., Sec. 10.8), the catalogue of known exoplanets is expected to explode way beyond the preliminary 500 that were known in the first decade of the 21st century.

Supplementary Reading

Boss, A. P. (1996) Extrasolar Planets, *Physics Today,* September, pp. 32-38.

Schneider, J. (1996) *Strategies for the search of life in the universe,* in Chela-Flores, J. and Raulin, F. (eds.), *Chemical Evolution: Physics of the Origin and Evolution of Life,* Kluwer Academic Publishers, Dordrecht, The Netherlands, pp. 73-84.

Tough, A., ed. (2000) *When SETI Succeds: The impact of high-information contact.* The Foundation for the Future. Washington.

References

Beichman, C. A., G. Bryden, T. N. Gautier, K. R. Stapelfeldt, M. W. Werner, K. Misselt, G. Rieke, J. Stansberry and D. Trilling (2005) An Excess Due to Small Grains around the Nearby K0 V Star HD 69830: Asteroid or Cometary Debris?, The *Astrophysical Journal* **626**, 1061.

Black, David C. (1999) *Extrasolar planets: Searching for other planetary systems.* In: "Encyclopaedia of the Solar System". P. R. Weissman, L.-A. McFadden and T. V. Johnson (eds.), Academic Press, San Diego, pp. 377-404.

Cosmovici, C. B., Pogrebenko, S., Montebugnoli, S. and Maccaferri, G. (2000) The 22 GHz Water Maser Line: a new diagnostic Tool for Extrasolar Planet Search, in "Bioastronomy 99" ASPC, vol. 213, pp.151-157.

Cosmovici, C.B., Pluchino, S., Salerno, E., Montebugnoli, S., Zoni, L. and Bartolini, M. (2008) Radio Search for Water in Exo-Planetary Systems, ASPC vol. 398, pp. 33-35

Davies, P. C. W. (1998) Did Earthlife come from Mars? in J. Chela-Flores and F. Raulin (eds.). *Chemical Evolution: Exobiology: Matter, Energy, and Information in the Origin and Evolution of Life in the Universe,* Kluwer Academic Publishers, Dordrecht, pp. 241-244.

De Duve, C. (1995) *Vital dust: Life as a cosmic imperative,* Basic Books, 1995, p. 294.

Encyclopædia Britannica (2010). Encyclopædia Britannica 2007 Ultimate Reference Suite. Chicago: Encyclopædia Britannica.

Fossey, S., Waldmann, I. & Kipping, D. (2009) Detection of a transit by the companion to HD 80606. *MNRAS,* 396, L16; [arXiv e-print

Fridlund, M., Eiroa, C., Henning, T., Herbst, T., Kaltenegger, L., Léger, A., Liseau, R., Lammer, H., Selsis, F., Beichman, C., Danchi, W., Lunine, J., Paresce, F., Penny, A., Quirrenbach, A., Röttgering, H., Schneider, J., Stam, D., Tinetti, G., and White, G.J. (2010b) A roadmap for the detection and characterization of other Earths. *Astrobiology* **10**:113–119.

Jakosky, B. (1998) *The Search for Life in Other Planets,* Cambridge University Press, London.

Kalas, P., Graham, J. R., Chiang, E., Fitzgerald, M. P., Clampin, M., Kite, E. S., Stapelfeldt, K., Marois, C., Krist, J.(2008) Optical Images of an Exosolar Planet 25 Light-Years from Earth. *Science* **322**, 1345-1348.

Kauffman, S. A. (1993) *The origins of order: Self-Organization and Selection in Evolution,* Oxford University Press, London.

Kipping, D., Fossey, S. & Campanella, G. (2009) On the detectability of habitable exomoons with Kepler-class photometry. MNRAS, 400, 398; [arXiv e-print]

Kipping, D. (2009a) Transit timing effects due to an exomoon II, 2009, MNRAS, 396, 1797; [arXiv e-print]

Kipping, D. (2009b) Transit timing effects due to an exomoon. MNRAS, 392, 181.

Lissauer, J. J. (1999) Three planets for Upsilon Andromedae, *Nature* **398**, 659-660.

Lissauer, J.J. *et al.* (2011) A closely packed system of low-mass, low-density planets transiting Kepler-11. *Nature* **470**, 53–58.

Marcy, G. (1998) Back in focus, *Nature* **391**, 127.

Mayor, M. and Queloz, D. (1995) A Jupiter-mass companion to a solar-type star. *Nature* **378**, 355-359.

Mayor, M., Queloz, D., Udry, S. and Halbwachs, J.-L. (1997) From brown dwarfs to planets, in C. B. Cosmovici, S. Bowyer and D. Werthimer (eds.), *Astronomical and Biochemical Origins and the Search for Life in the Universe,* Editrice Compositore, Bologna, pp. 313-330

Mayor, M., Udry, S., Lovis, C., Pepe, F., Queloz, D., Benz, W., Bertaux, J.-L., Bouchy, F., Mordasini, C., and Segransan, D. (2009) The HARPS search for southern extra-solar planets. XIII. A planetary system with 3 Super-Earths (4.2, 6.9, and 9.2 M□). *Astron. Astrophys.* **493**, 639–644.

Sagan, C. (1980) *Cosmos,* Random House, New York, p. 209.

Sartoretti, P. and Schneider, J. (1998) On the detection of satellites of extrasolar planets with the method of transits, *Astron. Astrophys. Suppl.* **134**, 553.

Schneider, J., Boccaletti, A., Mawet, D., Baudoz, P., Beuzit, J.-L., Doyon, R., Marley, M., Stam, D., Tinetti, G., Traub, W., Trauger, J., Aylward, A., Cho, J.Y-K., Keller, C.-U., Udry, S., and the SEE-COAST TEAM. (2009) The Super Earth Explorer: a coronagraphic off-axis space telescope. *Experimental Astronomy* **23**, 357–377.

Schneider, J., Tinetti, G., Karlsson, A., Gondoin, P., den Hartog, R., D'Arcio, L., Stankov, A.-M., Kilter, M., Erd, C., Beichman, C., Coulter, D., Danchi, W., Devirian, M., Johnston, K.J., Lawson, P., Lay, O.P., Lunine, J., and Kaltenegger, L. (2010) The search for worlds like our own. *Astrobiology* **10:**5–17.

Swain, M., Vasisht, G., Tinetti, G. (2008) The presence of methane in the atmosphere of an extrasolar planet. *Nature* **452**, 329–331

Tinetti, G., Vidal-Madjar, A., Liang, M.-C., et al. (2007) Water vapour in the atmosphere of a transiting extrasolar planet. *Nature* **448**, 169

Vogt, S. et al. (2010) *The Astrophysical Journal*, September.

Williams, D. and Gaidos, E. (2008) Detecting the glint of starlight on the oceans of distant planets. *Icarus* **195**, 927–937.

11

The search for the evolution of exointelligence in worlds around other stars

One of the greatest scientific achievements that have marked the Second Millennium is to begin the search of the place of Homo sapiens in the universe. Two great contributions in this search did occur in the 16th and 19th centuries.

The reader is advised to refer to the following entries in the Illustrated glossary:
Anthropocentrism, biogeocentrism, cell signaling, Drake (Frank), eukaryogenesis, the Fermi Paradox, geocentric, Heidmann (Jean), Kepler Mission, lipid bilayer. Ozma, Tarter (Jill).

11.1 What is our place in the universe?

The present chapter will begin to highlight three basic contributions that have shown us a vision of our own place in the universe. We shall postpone further comments of historical nature to Chapter 15. In Chapters 13 and 14 we have chosen to continue the scientific, philosophical and theological questions that serve as a frame to the main questions of astrobiology.

The pioneering work of Nicholas Copernicus revived the two thousand year old thesis of Aristarchus of Samos, the intuition of Giordano Bruno, and finally the Charles Darwin, respectively have begun, to a certain extent, to provide us with a balanced view of the position of the Earth in the cosmos, and the position of the genus *Homo* within the Earth biota.

On the other hand, what remains to be settled is to learn what place the Earth biota occupies in the cosmos. In other words, as we have emphasized elsewhere in this text, the distribution of life in the universe remains as the central question to elucidate by further research and astronomical observation. Having covered the extent to which our Solar System has been explored in Chapters 7, 8 and 9, we continued to review briefly what is known so far regarding how many sites in the known cosmos are likely to harbor life. In the search for inhabitable Earth like planets and corresponding moons, we have seen in Chapter 11 that this is only a question of time after the CoROT and Kepler searches have been given sufficient time. We now discuss the question of

J. Chela-Flores, *The Science of Astrobiology: A Personal View on Learning to Read the Book of Life*, Cellular Origin, Life in Extreme Habitats and Astrobiology 20, DOI 10.1007/978-94-007-1627-8_11,

whether the evolution of multicellularity, animals and intelligence is a natural outcome of the evolution of the cosmos.

11.2 Intelligent behavior is intimately related with the need to communicate

As mentioned in earlier, powerful searches have been implemented since the early 1960s for intelligent radio signals, a question whose discovery would fill the missing gap that all scientific research during the Second Millennium has been unable to answer Success in such a fundamental program would directly answer the question of whether we are alone in the universe.

The discipline is already almost 40 years old. Edward Mills Purcell was co-discoverer with Doc Ewen in 1951 of the 21-centimetre line, the first spectral line to be discovered by radio astronomy (microwave emission of hydrogen atoms in deep space). The serious systematic "search for extraterrestrial intelligence" (SETI) began in 1958 with the suggestion that a search be conducted at radio wavelengths emitted by instruments devised by intelligent beings that may have developed beyond our solar system (Cocconi and Morrison, 1959).

Intelligent behavior amongst evolved creatures seems to be intimately related with the need to communicate. Cocconi and Morrison suggested that a reasonable wavelength for communication might be that of the hydrogen line at twenty-one centimeters. Indeed, hydrogen is the most abundant element in the universe. It may be found in clouds and in interstellar space. When it absorbs some energy, it releases some of it by the emission of radio waves characterized by a wavelength of 21 cm. Due to its ubiquity the 21 cm line is now used as one of the principal means for exploring the structure of galaxies through radio observations. The eventual success of the SETI project seems inevitable. However, making the first contact with an extraterrestrial civilization has been delayed for some good reasons. They have been mentioned by one of the leading researchers, Paul Horowitz (cf., the BETA project) (Horowitz, 1996): "The hard part [of SETI] is the last step, which is intelligent life in the galaxy transmitting radio waves to us at the wavelength that we are expecting, and at a power level such that we can detect them. That is a lot of ifs."

11.3 The Drake Equation

The key to the central problem of Part 3 of the Book of Life is narrowed down to the problem of eukaryogenesis itself (Chapter 4). The Drake Equation may help us to drive the point home. The word "equation" need not deter any non-mathematical reader, since as Frank Drake pointed out, his equation was originally a simplification of an agenda for the points to be considered during the first meeting in 1961 in which SETI was to be discussed (cf., Fig. 11.1). *"I had the job of setting an agenda for the meeting [to discuss Project Ozma]. There was no one else to do it. So I sat down and thought, "What did we need to know about to discover life in space". Then I began listing the relevant points as they occurred to me... I looked at my list, thinking to arrange it somehow, perhaps in the order of the relative importance of the topics. But each one seemed to carry just as much weight as another in assessing the likelihood of success for any future Project Ozma. Then it hit me: The topics were not only of equal importance they were also utterly interdependent. Together they constituted a formula for determining the number of advanced, communicative civilizations that existed in space."* (Drake and Sobel, 1992).

Fig. 11.1 Frank Drake during his participation in the Fifth Trieste Conference in 1997, where he delivered the Cyril Ponnamperuma lecture

To summarize, Frank Drake reduced the agenda for the meeting as we have presented it in Table 11.1.

Table 11.1 The Drake Equation

$$N = R \times fp \times ne \times f_{\ell} \times fi \times fc \times L$$

N = the number of civilizations in our galaxy with which communication might be possible and
R^* = the average rate of star formation per year in our galaxy
fp = the fraction of those stars that have planets
ne = the average number of planets that can potentially support life per star that has planets
f_{ℓ} = the fraction of the above that actually go on to develop life at some point
fi = the fraction of the above that actually go on to develop intelligent life
fc = the fraction of civilizations that develop a technology that releases detectable signs of their existence into space
L = the length of time such civilizations release detectable signals into space.

The key parameter in the search for other civilizations is written as f_i, denoting the fraction of life-bearing planets or satellites, where biological evolution produces an intelligent species. We will consider in this chapter and the following one in some detail the significance o this parameter with our current knowledge of biology and with the present level of technological development.

Carl Sagan made an early reference regarding the transition from prokaryotes to eukaryotes in relation with the Drake Equation: His remarks were in the context of a discussion of the SETI projects at a 1971 conference (Sagan, 1973). In order to make Sagan's general comment more specific, we consider the Drake Equation, which is explained in Table 11.2.

Table 11.2 A simplified form of the Drake Equation

$$N = k f_i,$$

N is the number of civilizations capable of interstellar communication,
k is a constant of proportionality involving several factors that we need not discuss here f_i, denotes the fraction of life-bearing planets or satellites, where biological evolution produces an intelligent species.

It is useful to point out that the Drake parameter f_i is itself subject to the equation explained in Table 11.3. Our conjecture motivates the search *within our own solar system* for a key factor (f_e) in the distribution of life in the universe, including intelligent life.

Table 11.3 A simplified form for the equation for the f_i parameter.

$$f_i = k f_e ,$$

f_i, denotes the fraction of life-bearing planets or satellites, where biological evolution produces an intelligent species, k is a constant of proportionality involving several factors that we need not discuss here, f_e denotes the fraction of planets or satellites where eukaryogenesis occurs.

The extrapolation of the transition to multicellularity into an extraterrestrial environment is suggested by the selective advantage of organisms that go beyond the single-cell stage. Such organisms have the possibility of developing nervous systems and, eventually, brains and intelligence. If prokaryogenesis is possible in a short geological time frame, there are going to be evolutionary pressures on prokaryotes to evolve, due to symbiosis, horizontal gene transfer and natural selection (cf., Chapter 4).

In fact, these evolutionary mechanisms are going to provide strong selective advantage to those cells that can improve gene expression by compartmentalization of their genomes. (Larger genomes would be favored, since organisms with such genetic endowment would have better capacity for survival, and hence better ability to pass their genes to their progeny.) Whether the pathway to eukaryogenesis in an Europan-like environment, or elsewhere in the cosmos, has been followed, is clearly still an open question. We already know that f_e is non-vanishing on Earth. What is summarized in

Table 11.3 is that, instead of searching for intelligence, the more restricted search for eukaryogenesis is sufficient for understanding one of the main aspects of the postulated existence of extraterrestrial life, the basic assumption of any search for either extraterrestrial intelligent behavior, or the search for extraterrestrial eukaryotes.

As we have seen in Chapters 7 and 8, it is feasible to search for extraterrestrial microorganisms on Mars and Europa. But we must state clearly the theoretical bases that motivate such a search. To face the problems in the new science of the distribution of life in the universe, we could look for some hints at the problems of biology before Darwin. In Chapter 5 we recalled the significant step that Lyell took in his *Principles of Geology*. Lyell pictured a world constantly and slowly changing; in his view knowledge of the present conditions of climate, volcanic activity and Earth movements were sufficient for extrapolating back in time, in order to get insights into the same geologic questions in the distant past.

This 'doctrine' inspired Darwin when he successfully attempted to lay the scientific foundations of natural history. In volume 2 of *Principles of Geology*, Lyell accepted simple extrapolations into the past of the information we have in the present; but he did not take the crucial step to assume the same gradualism for living organisms as well. That crucial step was left for Darwin to take. In other words, we had to wait for the theory of evolution before we thought of all life on Earth in terms of a single tree of life; species change gradually (by the mechanism of natural selection) and are interconnected by the tree of life (the theory of common descent). Our generation has a similar undertaking, namely to provide a theoretical framework for the distribution of life in the universe. Today, thanks to molecular biology, we have isolated eukaryogenesis as the key phenomenon in the pathway that leads form bacteria to Man, intelligence and civilization. This insight, however, is restricted to the only tree of life that is known to us.

The SETI program attempts to make some progress in the missing theory of the distribution of life in the universe. Indeed, SETI directly searches for traces of intelligence of the same level of evolution that we have reached, or even beyond. The reasonable observation of SETI researchers is that, given the antiquity of the universe, there may be planets or satellites older than 5 billion years, still in orbit around a main sequence star that is not near its supernova stage.

The advantage of focusing on eukaryotes, as emphasized in this chapter, is that we can begin our search in our solar system. We assume as a working hypothesis that evolution of life in the universe can be explained only in terms of evolutionary forces that we experience today in our local environment. In other words, if other intelligence has developed on extrasolar planets, it must have gone through a similar "bottleneck" of eukaryogenesis. More precisely, life is not only a natural consequence of the laws of physics and chemistry, but once the living process has started, then the cellular plans, or blueprints, are also of universal validity: The simplest cellular blueprint (prokaryotic) will lead to a more complex cellular blueprint (eukaryotic). Eukaryogenesis will occur inexorably because of evolutionary pressures, driven by environmental changes in planets, or satellites, where conditions may be similar to Earth. Beyond eukaryogenesis and the first metazoans (cf., Chapter 12) the evolution of intelligence as searched in the SETI program is a rational scientific pursuit.

11.4 Progress in instrumentation

In the almost four decades that this search has been maintained, a great deal of progress in the necessary technology has been achieved (Drake and Sobel, 1992). After the first forty years no signal has been detected (Zuckerman, and Hart, 1995), but the equipment that is currently in use is about 100 million times more powerful than the equipment that was available in 1960. In fact, current searches with Project Phoenix, SERENDIP and BETA each are sampling 200 million frequency channels (cf., Table 11.1). The main enquiry may be formulated as the *Cocconi-Morrison question:* What waves would be most likely to be used by intelligent beings across the enormous interstellar distances that separate, for instance, an Earth-like planet or satellite in *Pegasi 51* (Mayor, 1996) from the Solar System? The answer to this question involves several factors, but the SETI 'window' is 1-10 GHz, in other words, this would correspond to wavelengths in the radio range. In this domain radio astronomers have already the equipment to face the problems raised by SETI research.

Phoenix arose form a previous NASA-funded project, the High Resolution Microwave Study, HRMS (Tarter, 1998; Lemarchand, 1998). However, private funding was the source that preserved this all-too-important research effort. From 1993 till the end of the decade the original NASA-funded equipment was preserved and much improved (cf., Table 11.4). The funding is managed by an independent corporation headed by Drake at the SETI Institute in Palo Alto, California. The Director of Research for the SETI Institute is Jill Tarter.

In the Phoenix Project, originally directed by Bernard M. Oliver (Oro, 1998), more than one radio telescope is used. The current director is Jill Tarter (cf., Fig. 11.2)

Fig. 11.2 Jill Tarter (in the middle row, seated, third from left to right) in the 1997 Trieste Conference

In the first phase of the project the main telescope was the 65-meter dish at Parkes, Australia; the complementary telescope measures 30 meters and is located at Mopra, Australia. The main objective was to observe target stars in the southern hemisphere over frequencies in the range 1-3 GHz. Then Phoenix operated at the NRAO Green Bank in West Virginia coupled with a facility at Woodbury, Georgia. At present the main objective is the northern hemisphere and the Arecibo radio telescope in Puerto Rico is being used. The search proposed initially to cover about 1000 target stars within 160 light years. The impressive sensitivity that has been achieved is capable of detecting the analogues of strong terrestrial radar signals.

Table 11.4 Some SETI Projects

	BETA	*META II*	*SERENDIP IV*	*Southern SERENDIP*	*Italian SERENDIP*	*Phoenix*
Site	Harvard	Buenos Aires	Arecibo, Puerto Rico	NSW, Australia	Bologna	Arecibo, Puerto Rico
Antenna diameter (m)	26	30	305	64	32	305
Channels (million)	250x8	8.4	168	4.2x2	4.2x2	28.7x2
Approximate sky coverage (%)	70	50	30	75	75	Targeted survey
Spectral resolution (hertz)	0.5	0.05	Down to 0.6	0.6	1.2	Down to 1

(Lemarchand, 1998)

META is an acronym for *Megachannel ExtraTerrestrial Assay*. In this project the basic idea is that there are no known processes in stellar dynamics that are capable of producing sharp lines as the one suggested by Cocconi and Morrison. If such a signal is received, then the natural conclusion is that an intelligent source is responsible for its emission. The META project used a 26-meter telescope at Harvard. Over a one-year period in cooperation with a radio telescope in Buenos Aires, Argentina, the complete celestial sphere is scanned for the presumably artificial signals that might be indicative of another intelligence. Over a period of 5 years several candidate signals were detected, but none confirmed. The BETA project has been developed in order to overcome the limitations of the previous META project (mainly the fact that signals are one-shot events that cannot be immediately be reconfirmed). Other initiatives are based in the United States (Berkeley), Italy (Medicina near Bologna cf., Fig. 11.3), and Australia (at the Parkes Observatory in New South Wales, NSW).

Fig 11.3 The VLBI 32m antenna in Medicina (Radioastronomical Station, Medicina, Bologne)

11.5 SETI on the Moon

As there is a background level of radio noise from radio, television, airplanes and telephones amongst others, it is natural to search for locations isolated form this impediment to SETI. One reasonable proposal is to move the research to the far side of the Moon in the 100 km diameter Saha crater (a 28 km diameter crater east of *Mare Smythii*), as advocated initially by the late Jean Heidmann from the Observatory of Paris (cf., Illustrated Glossary and Heidmann, 1996).

The position of the Saha crater (102° E, 2° S) permanently keeps it away from electromagnetic pollution, due to activities on Earth (cf., Fig. 11.4). None of the extremely ambitious engineering projects underlying this proposal seems beyond present technological capabilities, such as the construction of a 340 km road linking laboratories in *Mare Smythii* (at the equatorial level and at the Moon's limb) and crater.

The lava-like melt produced by impacts on the Moon can have a variety of morphologies. This texture could be the result of impact-melt coating boulders and other deposits on the floor of the crater. For evident reasons, it is likely that a base will be established on the near side of the Moon before the radio astronomy equipment may be taken to the far side. It is estimated that the Saha SETI base may be a reality by the second or third decade of this century. It seems possible to start establishing a small automatic radio telescope in the crater, which can be operated form the Earth by means of a radio station located in *Mare Smythii* (Genta, 2001).

Fig. 11.4 The floor of Saha E on the lunar farside. Image width is 1.08 km

11.6 From the first neuron to brains

We have divided the subject of astrobiology into three 'books', partly to emphasize two distinct aspects in which the scientific bases are on different footing. The fields of chemical evolution and Darwin's theory of evolution—time-honored sub-disciplines of astrobiology—support book 1. In Book 2, we discussed some relevant aspects of biological evolution. Finally, in Book 3, we still lack an underlying theory. We mentioned earlier, in Book 1, the transition form a simple prokaryotic blueprint of the Archean to the eukaryotic world of the Proterozoic, and Phanerozoic eons. We attempted to show that eukaryogenesis was probably the most transcendental step in the pathway that led from bacteria to Man. Once eukaryogenesis took place on Earth (Chapter 5), the steps leading up to multicellularity (Chapter 6), namely,

• Cell signaling and

• The organization into cooperative assemblies (tissues) were inevitable from the point of view of evolutionary advantage (cf., "The phenomenon of multicellularity", Sec. 6.3).

The onset of multicellularity is also due to the considerably larger genomes that were compatible with the eukaryotic blueprint. The densely packaged chromosomes in the cellular nucleus presented multiple options for opportunistic ways of passing genes to progeny, some of which allowed their carriers to be better adapted to the environment. Mitosis was a more advanced process of cellular division than simple prokaryotic fission. Such variety of options were the raw material for natural selection to improve upon the three billion year old single-cell strategy of life on Earth. The improvement was achieved within 30% of the single-cell 'era'. A full-organism strategy led subsequently to large-brained organisms and intelligence.

The first neuron could not arise at the prokaryotic level of development. A complex pattern of gene expression is required for a functional neuron; consequently, the first stage of a nervous system had to wait for the eukaryotic threshold to be crossed. The superior strategy of translation of the genetic message of nucleated cells permitted a variety of proteins to be inserted into the cellular membrane. This new stage in evolution provided all the channels necessary for the alteration of the ion concentration inside and outside the cell.

Indeed, early in the evolution of the eukaryotes we can conjecture that electrochemical imbalances were already being created by a disparity of electric charges on both sides of the cellular membrane. The details of the first steps towards multicellularity are not known in detail, but it is evident that in the struggle for survival, evolutionary pressures were going to encourage cells whose genomes codified proteins for cell signaling, as well as proteins that favored the formation of tissues of primitive multicellular organisms. We have not insisted in our discussions in Part 1 of the Book of Life on the complete understanding of all the intermediate steps that led from organic chemistry to life (still an open problem). In attempting to understand the distribution of life in the universe, we follow a similar strategy. In fact, we do not insist in filling in all the molecular biology details that are involved in the transition from the world of single-celled eukaryotes to multicellular organisms, neurons and brains.

11.7 Beyond geocentrism

Although there are still many questions to be answered, as we saw in Chapter 8, at present it seems possible (although not an easy matter) to penetrate the oceans of the iced Galilean satellites, provided it is confirmed that there are submerged oceans of liquid water. We can conceive experiments addressed specifically to the question of the search for extraterrestrial eukaryotes, although clearly many such experiments could be formulated. To sum up, verifying that f_e (cf., Table 11.4) is non-vanishing in one of the Galilean satellites goes a long way towards making contact with the initial steps towards other intelligence. It also would lay the foundation for the theory for the distribution of life in the universe. This chapter has summarized the efforts that science has made to free itself form the special way we see ourselves. The preliminary steps taken by science, philosophy and natural theology assumed that humans had a privileged position in the cosmos. With Copernicus geocentricism was abandoned.

Natural theology has incorporated this view, after an initial opposition to Galileo's inexorable conclusions (Chapter 8). After Darwin's theory of evolution was formulated, anthropocentrism was abandoned, although more than a century of research was needed, culminating with the solid genetic bases of neo-Darwinism. Natural theology is also beginning to incorporate this second step into its discussions (cf., Chapter 13). We have called, for lack of a better word in the English language, "biogeocentrism" the concept that life is exclusive to the Earth. The underlying difficulty for many scientists arises from a confrontation between the physical and biological sciences.

Firstly, everyone agrees that the Newton's theory of gravitation can be extrapolated without any difficulty throughout the universe, except for the small corrections required by Einstein's theory of relativity (cf., Chapter 1). One example was already discussed in Chapter 10 regarding the orbits of Jupiter-like planets. Secondly, the case of extrapolating the theory of biological evolution throughout the cosmos requires more care and is still an open problem.

Arguments against the hypothesis of biogeocentrism can now be formulated thanks to progress in our understanding of Darwinian evolution. The role of randomness has to be qualified since Darwin's time. Although chance is implicit in *The origin of Species* and Monod captured the essence of Darwinism with the suggestive contrast between chance and the necessary filtering of natural selection, we have also seen that molecular biology greatly constrains chance. Convergence will be a further factor to take into account, as we shall discuss in the next chapter. To sum up, Darwinian contingency is constrained and evolution tends to converge on similar solutions when natural selection acts on similar environments. Cosmochemistry and planetary science suggest that the environments where life can originate are limited. We already are gathering information on a significant number of Jupiter-like planets around stars arising from protonebulae that are likely to grant them an array of satellites. In our outer solar system this can be confirmed. Each of our giant planets has a large number of satellites. Such is the case in the Outer Solar System. Factors giving rise to atmospheres in Jovian satellites are known. Titan, for instance, has an atmosphere produced by outgassing, combined by seeding of volatiles by comets carrying a fraction of water ice. Evidence is leaning in favor of the existence of Jovian planets with masses larger than Jupiter; hence tidal heating responsible for Io's volcanic eruptions, could be even more efficient in other solar systems. On Europa it is not clear that tidal heating may produce hydrothermal vents capable of giving rise to life, unlike the case of satellites orbiting around large Jupiter-like planets.

To sum up, natural selection will be working on a finite number of similar environments. According to cosmochemistry, similar chemical elements will enter into pathways of chemical evolution. We have also learnt that there are no laws in chemical evolution that are specific to the Earth; it is reasonable to hypothesize that biological evolution will follow chemical evolution. Today Darwinism cannot be seen as a simple dichotomy between chance and necessity, but rather we must build into its fabric constrained chance and convergent evolution. Thus, in environments similar to Earth, we expect analogous pathways that lead from chemical evolution to the evolution of intelligent behavior.

11.8 The Fermi Paradox

Enrico Fermi (1901–1954) was an Italian physicist, who made significant contributions in quantum theory, nuclear and elementary particle physics (most significantly laying out the basis of the unified electroweak interactions that led to the work of Glashow, Salam and Weinberg that is reviewed in Chapter 15). Fermi questioned why, if a multitude of advanced extraterrestrial civilizations exist in our galaxy, the SETI project has not detected any trace of intelligent life after more that half century of search with

increasing technological sophistication. In an informal discussion in 1950, Fermi raised this question. Hence there are no publications due to him on this paradox (Zuckerman and Hart, 1995). The new science of astrobiology has brought a multidisciplinary approach to the Fermi paradox. There have been attempts to resolve the Fermi paradox, possibly the most trivial solution is to assume that we have not observed anything so far, due to the fact that there is nothing to observe. We are alone in the universe.
Alternatively some conjectures may be worth recalling here. It concerns the human brain. We need not assume that the human brain is the culmination of evolution. This viewpoint implies that there could be types of intelligence that we would not recognize if they would be vastly ahead of us (Rees, 2010).

A second approach is to assume that we are not alone but are the first multicellular entities that have sufficiently developed adequate technological. Possibly the most relevant solution to the Fermi Paradox that may be even pertinent for the current attempts to send humans to other bodies of the Solar System is that it might be impossible, or even difficult to implement appropriate space flights with a human crew. It could be much more risky that what is at present considered. For instance, cosmic rays or interstellar dust could militate against our presence outside essentially a low Earth orbit, and more generally, a low orbit of any inhabited exoplanet.

Some preliminary arguments have been put forward to try to rationalize the Fermi paradox. For instance, an in the Milky Way there could be some episodes within an evolution from an essentially lifeless planets or satellites, with exceptional pockets of microbial life to periods when complex life forms could evolve (Cirkovic and Vukotic, 2008). Such a scenario would rationalize the absence of evidence for other intelligences known to us by the usual means of bioastronomy. Other authors have argued in a different direction, namely, the potential risk of employing electromagnetic communications technology to announce Earth's presence to our cosmic companions, or to reply to a successful SETI detection. Such messaging to other exoplanets or satellites opens an alternative way of viewing the Fermi Paradox. It has been argued that a civilization attempting to detect radio evidence of other civilizations in the universe has to reveal its own presence. A scale of risk has been discussed (the "San Marino" Scale) attempting to argue in this direction (Almár and Shuch, 2007). Other authors are thinking along these lines, but on more certain mathematical bases (De Vladar, 2010). As we have seen in the previous section (and will discuss much more deeply in Chapters 13-14 once we place ourselves in a central position in the universe observational accuracy denies such "geocentrism". Beyond this aspect the science of astrobiology is centered on the eventual discovery of life in the Solar System, and thanks to the Kepler mission we have entered in the second decade of the 21st century, the discovery of life in other solar systems is gradually becoming feasible.

11.9 SETI: towards the future

The Very Large Array (VLA) is a radio astronomy observatory located some 80 km west of Socorro, New Mexico, USA. The VLA is a component of the National Radio Astronomy Observatory (NRAO). This is a multi-purpose instrument designed to allow investigations of many astronomical topics. Objects that are commonly studied include radio galaxies, quasars, pulsars, supernova remnants, gamma ray bursts, radio-emitting stars, the sun and planets, astrophysical masers, black holes, and the hydrogen gas that constitutes a large portion of the Milky Way galaxy as well as external galaxies. In 1989 the VLA was used to receive radio communications from the Voyager 2 spacecraft as it flew by Neptune.

The SETI project will benefit from a next stage in the growth of instrumentation by means of the wide field of view of the Allen Telescope Array (ATA), which is capable of enabling a survey comparable within a factor of a few of the current VLA. ATA is a pioneering centimeter-wavelength radio telescope that will produce science that cannot be done with any other instrument. The ATA is the first radio telescope that will simultaneously undertake the most comprehensive and sensitive SETI survey ever done, as well as the deepest and largest spectroscopic survey. A wide range of innovative technical developments enables the science of the ATA. The astronomy decadal panel, Astronomy and Astrophysics in the New Millennium, endorsed SETI and recognized the ATA (then, the 1 Hectare Telescope) as an important stepping-stone o the Square Kilometer Array (SKA), its highest ranked “moderate project” in radio astronomy. Located in Hat Creek, CA, formerly the site of the BIMA Array, the array currently consists of 42 antennas of 6.1-m diameter each with continuous frequency response from 0.5 to 11.2 GHz.

Science operations commenced in mid-2007. Highlights include large-scale continuum mosaics, a high-quality single-pointing image of the entire Andromeda galaxy in HI, and a 500-hour search for fast radio transients. The compact configuration of the ATA and the high density of antennas provides unprecedented snapshot imaging quality as well as a resolution that is intermediate between the most compact VLA configuration and that of the largest single dishes. The frequency range of the ATA permits sensitive observations at frequencies rarely studied.

Supplementary Reading

Tough, A. (ed.) (2000) *When SETI Succeeds: The impact of high-information contact.* A publication of The Foundation for the Future. Washington.

Billingham, J. *et al.* (eds.) (1994) *Social Implications of the Detection of an Extraterrestrial Civilization.* SETI Press.

Davies, P. (2010) *The Eerie Silence: Renewing Our Search for Alien Intelligence,* Houghton Mifflin Harcourt, 288 p.

Drake, F. and Sobel (1992) *Is there anyone out there? The scientific search for Extraterrestrial Intelligence,* D. Delacorte Press, New York.

Ekers, R. D.Cullers, D. K. Billingham, J. and Scheffer, L. K. (eds.) (2002) *SETI 2020.* SETI Press: Mountain View CA.

Heidmann, Jean (1996) *Intelligences extra-terrestres.* Editions Odile Jacob: Paris.

Hoover, R.B. (ed.), (1997) *Instruments, Methods and Missions for Investigation of Extraterrestrial Microorganisms,* Proc. SPIE, **3111.**

Lemarchand, G.A. (1992) *El Llamado de las Estrellas,* Lugar Cientifico, Buenos Aires.

Pinker, S. (1997) *How Mind Works*, W.W. Norton & Co. New York.

Fitch, W.T. (2010) *The Evolution of Language,* Cambridge University Press.

Raulin, F., Raulin-Cerceu, F. and Schneider, J. (1997) *La Bioastronomie,* Presse Universitaires de France, Paris.

Zuckerman, B. and Hart. M. H. (1995) *Extraterrestrials. Where are they?* 2nd Ed. Cambridge University Press.

References

Almár, I. and Shuch, H. P. (2007) The San Marino Scale: A new analytical tool for assessing transmission risk. *Acta Astronautica* **60,** 57–59.

Cirkovic, M. M. and Vukotic, B (2008) Astrobiological Phase Transition: Towards Resolution of Fermi's Paradox. *Orig Life Evol Biosph* (2008) **38**, 535–547.

Cocconi, G. and Morrison, P. (1959) Searching for interstellar communications, *Nature* **184**, 844-846.

DeVladar, H. P. (2010) Private communication.

Drake, F. and Sobel, D. (1992) *Is there anyone out there? The scientific search for Extraterrestrial Intelligence.* Delacorte Press, New York, pp. 51-52.

Genta, G. (2001) The Saha Crater Radioastronomic and SETI Obsservatory, in Chela-Flores, J., Owen, Tobias and Raulin, F. *The First Steps of Life in the Universe,* Kluwer Academic Publishers, Dordrecht.

Horowitz, P. (1996) cited in: *Is Anybody Out There?* Time Magazine February 5. pp. 42-50.

Heidmann, J. (1996) SETI from the Moon. A case for a XX Century SETI-Dedicated Lunar Farside Crater, in Chela-Flores, J. and Raulin, F. (eds.), (1996) loc. cit. Preface, ref. 18, pp. 343-353.

Lemarchand, G. A. (1998) Is there intelligent life out there? *Scientific American Presents Exploring Intelligence,* Quarterly, pp. 96-104.

Oro, J. (1998) The Abdus Salam Lecture, in Chela-Flores, J. and Raulin, F. (eds.), loc. Cit "Introduction", ref. 19, pp. 11-32.

Rees, M. (2010) Talking about Life. Conversations on Adstrobiology. Chapter 35. Cambridge University Press, Cambridge UK. p. 365.

Sagan, C. (1973) Discussion, in C. Sagan (ed.), *Communication with Extraterrestrial Intelligence (CETI),* The MIT Press, Cambridge, Mass., pp. 112-146.

Tarter, J. (1998) The search for intelligent life in the universe, *Commentarii* (Vatican City) **4**, N. 3., 305-310.

Zuckerman, B. and Hart, M. H. (1995) *Extraterrestrials. Where are they?,* (2nd ed.), Cambridge University Press, London.

THE BOOK OF LIFE

PART 4: THE DESTINY OF LIFE IN THE UNIVERSE

The cupola in the West Atrium of Saint Mark's Basilica in Venice

Since ancient times, eternal questions have been asked by humans, as we can clearly see when we enter the atrium of St. Mark's Basilica in Venice. As you go through St. Clement's Portal in its South Western corner, you may observe a small cupola, illustrating the initial events of the Book of Genesis. This is only part of a complete 13th century overview of the Torah (Old Testament), which continues with a cycle of five other cupolas along the western and northern sides of the Basilica. The Genesis cupola itself is divided into 24 scenes set on three concentric bands. The iconography follows an Alexandrian illuminated manuscript miniated during the 5th century, which is known as the "Cotton Bible", as parts of it were donated in the 18th century to the British Museum by Sir John Cotton. (This codex now belongs to the British Library in London, since its foundation in 1973.) The innermost band of the cupola, and part of the middle band, raise some of the questions that science has attempted to answer.

We have addressed some of these problems in the first and second parts of the Book of Life. The middle and outer bands of the Genesis cupola already touch on this event by raising the questions of the first appearance of fish, birds and, in separate scenes, mammals and humans. The topics we discuss in this fourth part of the Book of Life present to the readers what the science of astrobiology can provide to answer these ancient questions on the bases set by Newton, Galileo, Darwin together with the cutting-edge research of the 20th and 21st centuries. In Chapters 12 and 15 we confront the science of astrobiology with to frontiers with other sciences, astronomy (Chapter 12) and cosmology (Chapter 15). The Genesis cupola presents us with a further stimulating challenge: When we turn to Chapters 13 and 14 we prepare ourselves to discuss astrobiology within the context of culture in general—those branches of the humanities where similar questions have been raised. This is no idle exercise, since we may question our capacity for reading the complete Book of Life, Part 3 (distribution of life in the universe): it may depend on progress in cosmology, (Chapter 16).

12

Is the destiny of life inexorably linked with intelligence?

Throughout the book we have emphasized one possibility for evolution of life in the universe: this universality of life may be interpreted as the blueprint of the eukaryotic cell leading up to not a single tree of life (phylogenetic tree), but rather to a "universal forest of life". This is in sharp distinction with the view that ours is a "rare Earth". Clearly there are other alternatives, but only experiments carried out in appropriate solar system missions will give us the correct answer. We call the attention of the reader to the fact that experiments designed to select between the alternative pathways of evolution are possible. (It should be emphasized though that such experiments are both costly and slow in their implementation.)

The reader is advised to refer especially to the following entries in the Illustrated glossary: *Drake, exointelligence, flow cytometry. The reader should also refer to some of the recommendations suggested before reading the Introduction and Chapters 8 and 11.*

12.1 The origin of the neuron: a first step in the evolution of intelligence

What were the first steps towards the evolution of intelligence? What did the first neurons and nervous systems look like, and what advantages did they confer on the animals that possessed them? (Miller, 2009). These were questions the father of evolution, Charles Darwin, was ill equipped to address. Although comparative neuroanatomy dates back to ancient Greece, the tools of the trade had not been refined much by the mid-19th century. In Darwin's day, anatomists were limited to gross observations of brains; they knew relatively little about the workings of nerves themselves. Only around the time Darwin died in 1882 were scientists beginning to develop stains to label individual cells for more detailed postmortem neuroanatomical studies. Methods for investigating the electrical properties of individual neurons in living brain tissue were still decades away, to say nothing of techniques for investigating genes and genomes.

J. Chela-Flores, *The Science of Astrobiology: A Personal View on Learning to Read the Book of Life*, Cellular Origin, Life in Extreme Habitats and Astrobiology 20, DOI 10.1007/978-94-007-1627-8_12,

12.2 The origin of metazoans: a second step in the evolution of intelligence

The problem of the origin of metazoans, already discussed in Chapter 6, becomes a more urgent issue when seen form the point of view of the evolution of intelligence. By now it is clear that information on this crucial transition in the evolution of life can arise in a fourth pathway besides the three possibilities in the quest for simplicity outlined by Bonner in his classical book (Bonner, 2001).

In other words, Solar System exploration seems to be one way in the long-term to elucidate the simplicity of evolutionary development. This may be attempted on the Galilean moons of Jupiter where there is a possibility of detecting reliable biomarkers in the next decade with the Europa Jupiter System Mission in view of recent progress undertaken in order to succeed with penetrators on planetary or satellite surfaces (cf., Sec. 8.6). Mars is a second possibility in the inner Solar System, in spite of the difficulties faced by the fleet of past, present and future missions. The quest of information on evolution of biotas from planets around other stars does not seem to be feasible with present technology with direct visualization of living organisms on exoplanets. It has also seemed to be difficult after 50 years searching for extraterrestrial signals with the SETI project. We discuss a series of preliminary ideas for elucidating the origin of metazoans with instrumentation in potential payloads of feasible space missions to the Galilean moons.

12.3 Convergence and contingency in evolutionary biology

To begin with, some aspects of evolutionary behavior in animals are worth mentioning: Communication is a general social activity within species that have significant differences in brain organization (ants and humans); in ants communication has evolved by the emission of pheromones singly or in combination, and in various amounts, to say to other ants in effect, for instance: *danger,* or *follow me* (Wilson, 1998). In humans communication has ranged from drum beating in primitive populations to extensive use of the electromagnetic spectrum, as radio or TV signals (during the last century).

For the first time to the best of our knowledge, in a recent publication Frank Drake has yet added a fresh angle to our insights into the possible existence of life elsewhere in the Cosmos (Drake, 2001). He exposes in a conjecture that we shall refer to as the *Drake conjecture*, (the Drake conjecture is being formulated almost exactly 40 years after the formulation of the by now famous Drake Equation, cf., Chapter 11, Table 11.2): In simple terms the Drake conjecture is that there is a possibility that after an easy period of detection of an extraterrestrial civilization, progress of a given group of beings elsewhere, may be followed by a more silent period, because of the improvement of the means of transmitting information.

For instance, the current use of internet has created a definite trend towards diminishing the use of radio and TV signals, which after all are the indicators of the presence of other civilizations that may have acquired a technology similar to that of humans throughout most of the 20th century, prior to the massive onset of the internet and other means of transmitting digital information.

On the other hand, in this context it is worth remembering that cetaceans and primates, in spite of significant brain morphology, have a certain degree of *behavioral convergence*, such as a prolonged period of juvenile dependency, long life span, cooperative hunting, and reciprocal altruism. We shall use the pragmatic definition of intelligent behavior in those taxons that share such behavioral convergence with humans.

We shall now argue that terrestrial biota teaches us that multicellularity is inexorably linked with neurogenesis, which, in turn, is linked with the process of synaptogenesis, and finally with the formation of cerebral ganglions at the lowest levels of a given tree of life. Once protobrains are formed, brain evolution will be constrained by few mechanisms, prominent amongst these is brain organization across species in terms of, for instance, modules. It should be remarked, in addition, that critical clues have already been provided in developmental neurobiology in terms of molecular genetics, regarding the remarkable increase in brain size per body weight in humans (a concept that will be made accurate below). It concerns insights in terms of *homeotic genes,* whose expression pattern correlates with the repetition of similarly organized segments along the body.

Hence, we may consider brain evolution to be independent, to a certain extent, of historical contingency. Whether nature in an extraterrestrial context steers a predictable course, is clearly still an open question, but some hints from the basic laws of biology militate in its favor: These laws are natural selection and the existence of a common ancestor (a "cenancestor"). All the Earth biota can be interpreted as evidence in favor of the fact that, to a large extent, evolution is predictable and not contingent. The underlying question concerns the relative roles of adaptation, chance and history. This topic is subject to experimental tests. We shall assume that natural selection seems to be powerful enough to shape terrestrial organisms to similar ends, independent of historical contingency. In an extraterrestrial environment it could be argued that the evolutionary steps that led to human beings would probably never repeat themselves; but that is hardly the relevant point:

The role of contingency in evolution has little bearing on the emergence of a particular biological property.

Besides, it can be said in stronger terms that essentially evolutionary convergence can be viewed as a 're-run of the tape of evolution', with end results that are broadly predictable. (This question will be taken up again in Chapter 14, cf., *Can our intelligence be repeated elsewhere?*) The inevitability of the emergence of particular biological properties is a phenomenon that has been recognized by students of evolution for a long time. It is referred to as 'evolutionary convergence'. This may be illustrated with examples taken from malacology, and ornithology, as we have done in Chapter 13:

To discuss whether intelligent behavior is a logical consequence of the evolutionary process, we will recall the definition of astrobiology given in the Introduction: it is the subject that studies the origin, evolution, distribution and destiny of life in the universe. For clarifying our ideas, we may consider briefly, the overlap of these sub disciplines with neuroscience:

"The origin of life in the universe" is a well-studied discipline. The main lesson that we may derive from what we have learnt in the last 80 years, when the subject started with the experiments of Alexander Oparin, is that life's origin occurred on the early Earth, almost as soon as it could possibly do so. We shall return to this later, when we shall discuss the possibility that in just 200 million years after the formation of the Earth the conditions were suitable for the origin of life. This geologic discovery of early biofriendly environments should be seen together with the earliest evidence of microfossils, which suggests that just one billion years after the origin of the Earth, there were microorganisms of considerable evolution, such as cyanobacteria colonies, which are known as stromatolites.

"Evolution of life in the universe" is precisely within the scope of all biologists. For although life is still only known in one solar system of one galaxy—the Milky Way — we know a considerable amount, particularly in the topic of the evolution of intelligent behavior.

"Distribution of life in the universe" brings us closer to the subject of the possible universality of the evolution of intelligent behavior. For over 4 decades the SETI Institute under the direction, firstly of Drake, and presently of Jill Tarter, have made systematic progress in radio astronomy searching at various frequencies in the electromagnetic spectrum for signals from intelligent social groups of creatures that communicate amongst themselves. The technological progress achieved by the SETI researchers has been considerable, at present billions of channels can be processed simultaneously. There are no reproducible intelligent signals so far.

Finally, "destiny of life in the universe" does not overlap at all with neuroscience; it rather opens the door to interaction with other sectors of culture that normally does not interact directly with science itself. I have in mind philosophy and theology.

The main lesson we derive from the combined evidence of early geologic evolution of our planet, together with the earliest reliable microfossils, is that given a planet such as the Earth, the evolution of prokaryotic cells is intimately bound with the earliest stages of planetary evolution. We hope to have placed the topic 'evolution of intelligent behavior' within the wider frame of astrobiology. What I have to say draws a powerful lesson from the origin of life on Earth and lies somewhere between the evolution and distribution of life in the universe.

12.4 Biological evolution on other worlds

Let us discuss the rationale for the search for extraterrestrial signals of the evidence of intelligent behavior: We think that a central nervous system is bound to evolve, since in the phylogenetic tree, already at the level of the diploblastic animals (cnidarians, with the familiar example of jellyfish), there has been electrophysiological research demonstrating that such simple animals do have nervous nets. In Table 12.1, we should emphasize that the simpler multicellular arrangements, namely sponges have electrophysiological responses; but no primitive nervous systems have been demonstrated, for instance, similar to the nervous nets of the cnidarians. To sum up, almost as soon as some coordinated electrophysiological responses are possible in multicellular organisms, they have been demonstrated to exist.

We feel that brain is a sure thing to evolve. In the phylogenetic tree, the first appearance of a cerebral ganglion occurs very early, for instance in annelids (ancestors of common worms). Once again, as soon as the diploblastic/triploblastic barrier has been crossed, cerebral ganglions appear. The more difficult case to study is that of the first steps towards intelligent behavior and complex language. This is what the SETI project assumes to be occurring on other worlds. However, we already know that in humans the origin of language is probably a consequence of natural selection. There is a long debate on this issue, which started with Chomsky and was taken up by Pinker (cf., Chapter 6).

Without entering into the details of this argument, it seems reasonable to assume that natural selection may favor the appearance of language, once a sufficiently complex brain has appeared in a given phylogenetic tree.

Table 12.1 Some physiological responses at the lowest level of the phylogenetic tree

Organism	Physiological responses
Paramecium (protozoa)	Calcium channels in protozoan movements
Rhabdocalyptus dawsoni (sponge)	Ca- and Na-dependent channels
Aglantha digitale (jellyfish; cnidarian)	Action potentials (nervous nets)

(Villegas *et al.*, 2000)

The case of humans illustrates that the evolution of complex intelligent behavior, including the evolution of communication through language, rather than chemical signals (pheromones), has been favored by the emergence of mutually supportive social groups of creatures, at the later stages in the evolution of a complex brain.

12.5 Evolution of intelligent behavior in aquatic media

On Earth primates and dolphins present us with examples, which demonstrate that sharp increments in brain size can probably occur in different environments: terrestrial and aquatic. The dramatic increase in the last 2 million years (Myr) in relative brain size, compared to body weight (a concept that will be defined accurately below) has been considered as a hallmark of humans, and their special place in the universe.

There may be plausible mechanisms in terms of which the rapid increase of human brain size may be rationalized. Preliminary genetic arguments have been put forward. But independent of the molecular mechanisms that may be considered to be at the bases of hominid evolution, the anthropocentric heritage still within science gives undue attention to the last 2 Myr as a frame of reference for further discussions.

This may be inappropriate, since there is some evidence, which extends the favorable conditions for the origin of life on Earth back to 4,400 Myr BP. In other words, looking at the ascent of humans above other species, as an example of an unlikely, or a unique event, is at least misleading. Our brain has evolved remarkably only during the 0.5% of the history of life on Earth. We should look at other groups beyond the anthropoids, to which we belong, to verify whether within a brief period of geologic time, animals with different brain morphology, have gone through other episodes of a many-fold increase in brain size per body weight, above the average of similar sized animals. We shall return to the example of cetaceans that had a dramatic 9-fold increase in brain size per body weight over the relatively short period of the past 45 to 50 Myr. In fact, it is pertinent to dwell on this point a little longer. One of the correlates of intelligence is relative brain size. Across the barrier of species, we can assign a quantitative parameter, the encephalization quotient, EQ (Marino, 2000)). The EQ parameter reaches a value of 7 in *Homo sapiens* and 4.5 in some cetaceans; the parameter has lower values in non-human primates.

Encephalization is defined as the increase in brain size over and above that expected on the basis of body size. EQ is determined relative to a sample of species, so that its absolute value expresses the level of encephalization, relative to the rest of the species in the sample. This definition avoids the "chihuahua conundrum": These small dogs are much more encephalized than a bulldog, but are in no significant way smarter. This is why the definition of encephalization is with respect to the rest of the species.

The value for humans is to be interpreted as expressing that humans have, on average, a brain seven times larger than that of a mammal of similar body size. The point that is really relevant to the question of the evolution of intelligent behavior in a given species is that in the past 1-2 million years, the human EQ and that of cetaceans were equivalent (Marino, 1997).

12.6 Testing the evolution of microorganisms in the Solar System

This remark is pertinent, since it is only after the surveys of Voyager in the 1970s and of the Galileo Mission in the 1990s, that astrobiology began to take seriously the possibility of biological evolution in aquatic biotopes, other than terrestrial ones. Experimental tests of evolution of microorganisms in our solar system are of considerable interest in the oceans of the Solar System, or in isolated reservoirs on Mars. We proceed to list some key questions that remain to be cleared in the crucial experiments of evolution in the solar system. We mention here just a few of the outstanding difficulties: When we argue that evolution of a nervous system is an advantage for sensing the environment, we are only partly right. This is only true if there is need for motion or, more importantly, active predation. The truth of this statement is clear if one examines the other two multicellular kingdoms, the plants and fungi, in which motion (although it is present) is not evident.

David Attenborough has emphasized that swiftly moving creatures tend to regard plants as immobile organisms, rooted to the ground (Attenborough, 1995). But the production of new individuals leads to the extension of the domain of their species. Motion is normally invisible to the naked eye. However, the fact remains that we have illustrated some evidence that the first neurons, as well as primitive nervous systems in the phylogenetic tree of life, can be traced back to the simplest multicellular organisms. Such organisms evolved much earlier in time than the subsequent first appearance of plants and fungi. Granted that the argument of active predation excludes plants and fungi, Earth biology has shown that the first steps towards a nervous system predate the evolution of plants. Several phyla, such as Chordates, Mollusks and Arthropods can be traced back to the early Cambrian, earlier than the first appearance of the divisions of the Plant Kingdom. There are a few unifying threads in the process of cellular complexification: association, differentiation, patterning and reproduction, which have been discussed already. It is reasonable to assume that cells first got together as a result of chance mutations, which favored the multicellular association. In turn, they stayed together because they reproduced more successfully as a group, rather than as single cells. Today we can appreciate that slime molds can be seen as a model of how the first steps in multicellularity did occur. These organisms are eukaryotic and heterotrophic.

These unicellular microorganisms segregate a chemical—cyclic adenosine monophosphate (cAMP). This phenomenon leads to aggregation into a single macro-organism. Indeed, each unicellular eukaryote, on contact with cAMP, expresses new surface molecules with 'lock and key' possibilities that after they randomly come into contact, they remain locked to each other (De Duve, 1995). There are additional processes for holding the eukaryotes together: this occurs directly by the expression of surface cell-adhesion molecules (CAMs), which play a role of holding the cells together and also to extracellular "scaffoldings" built-up by means of substrate adhesion molecules (SAMs). Underlying this ability of the eukaryotes to form efficient and functional communities, there is a major role played by sexual reproduction, rather than the simpler reproduction processes of the prokaryotes (for instance by fission). But in the present context we shall not elaborate on this key aspect of evolution and diversification of the eukaryotes, as part of multicellular organisms. Amongst the earliest multicellular group that evolved simple nervous systems is Cnidaria. As summarized in Table 12.2, extensive work has been done with the jellyfish *Aglantha digitale,* in which action potentials have been characterized.

Table 12.2 The emergence of cerebral ganglions at the lowest level of the phylogenetic tree

Organism	Cerebral ganglion
Notoplana acticola (flat worms; Platyhelminthes)	Receives inputs form sensory organs and delivers outputs to muscles, via nerve filaments.
Ascaris lumbricoides (round worms; nematode)	Receives signals from sensory organs and sends output signals to muscles
Caenorhabditis elegans (round worms; nematode)	Receives signals from sensory organs and sends output signals to muscles

(Villegas *et al.*, 2000)

Hence, Cnidaria's primitive nervous system is well understood, but these are essentially diploblastic organisms (i.e., during development they just have endoderm and ectoderm, lacking a mesoderm). It is in the next step in the evolution of primitive animals, we should consider the triploblastic animals. They are animals that have a well-developed mesoderm. It is at this level in which primitive brains are first seen to evolve. In flat worms, there is one example, the *Notoplana acticola* that has a primitive brain (a cerebral ganglion); they receive inputs from sensory organs and delivering outputs to muscles, via nerve filaments. This has already been summarized in Table 12.2. The evolution of higher animals occurred explosively during the Cambrian, over 500 million years ago.

Soon enough after the emergence of the simplest triploblastic body plans, multiple phyla appeared in the Cambrian, which successfully persevered through subsequent geologic periods, right up to the present: we would like to underline particularly phyla in a major group of animals: Chordates, Mollusks and Arthropods. In all of them we find nervous systems with the capacity to support sensorial discrimination, learning, social behavior and communication. The property of communication through the spoken language had to await the advent of the vertebrates to reach the possibilities developed by humans that are being searched elsewhere in the universe.

In the above-mentioned phyla of the Animal Kingdom (Domain Eucarya), their nervous systems are considered well developed, according to their capacity for giving support to sensorial discrimination, learning, communication and social behavior. The lesson that we are continually learning from astrobiology is always one of unveiling environments that are not too dissimilar to the early Earth, when the evolutionary process of our phylogenetic tree started.

Some of these environments are nearby inside our own solar system, namely on Europa and Mars. But many other environments are also coming to our attention. Indeed, hundreds of planets were known to circle round nearby stars even before the era of CoROT and Kepler. They are large planets, comparable to our own Jupiter and some smaller ones, the super-Earths. In fact, it is not hopeless to raise the question: *Are there constraints on biological evolution in solar systems?* In the foreseeable future we are bound to discover earth-sized planets (cf., Sec. 10.8).

12.7 Precursors of the evolution of exointelligece—where should we look?

In a previous discussion we have attempted to answer the question raised in this section (DeVladar and Chela-Flores, 2011, where this topic is discussed more extensively). Under given conditions, it is indeed possible to safely speculate about the biomarkers that we could expect to find, even though our discussion must necessarily be restricted to the exploration of our own Solar System. This restricted attitude to research follows from what we have learnt in Chapter 10 about the search for Earth-like worlds around other stars that may bear the signs of life in atmospheric features (such as an ozone layer), but the detection of multicellularity, rather than life in general (microbial) would be limited to bioastronomical research, as the evolutionary pressures would tend to favor neuron multicellularity (brains), and hence the possibility of communication with high levels of technological development (cf., Chapter 11).

With a rather simple technology it may be possible to have a significant insight about the complexities that may be abundant in exogenous life. Life specimens might be trapped in the surface ice, which is in principle accessible to penetrators (cf., Chapter 8). A piece of ice containing some organismic tissue can be minced in such a way that the biological sample, if multicellular, is disrupted into smaller constituents, tentatively cell analogues. It would be hard at this point to perform any biochemical or calorimetric analyses, particularly in the Jovian System. Some experimental techniques may help us in the search for biomarkers either of microbial life, or higher levels of evolution, such as the onset of eukaryogenesis, multicellular prokaryotic communities, or even multicellular eukaryotic life (metazoans and higher levels of evolution).

Since we expect that the hypothetical organisms in Europa have metabolism based in sulfur, we require techniques that allow monitoring distinct chemical states of this element. Although sulfur compounds are very elusive, particularly *in vivo*, spectroscopy in the X-ray wavelengths (near 2.47 eV) shows a characteristic peak (termed *K-edge*) that allows discerning among distinct radical types involving this element (Rompel *et al.,* 1998). A few words are needed to introduce the reader to some of the concepts involved: X-ray absorption spectroscopy (XAS) is a widely used technique for determining the electronic structure of matter. The experiment is usually performed at synchrotron radiation sources, which provide intense and tunable X-ray beams. XAS data are obtained by tuning the photon energy using a crystalline monochromator to a range where core electrons can be excited (0.1-100 keV photon energy). In the standard tradition of quantum mechanics the name assigned to the edge depends upon the core electron which is excited: the principal quantum numbers n=1, 2, and 3, correspond to the K-, L-, and M-edges, respectively. The advantage of spectroscopic observation of sulfur is that it is non-invasive and therefore may allow coupling with a flow cytometer.

Cytometry is a technique for probing microscopic biological material, for instance cells. The sample is suspended in a stream of fluid and passing it by an electronic detection apparatus. This instrumentation allows simultaneous analysis of the chemical characteristics of up to thousands of particles per second. The specific case of interest is to identify biomarkers for the origin of metazoans in the Solar System. Cells go through a narrow channel that is to be monitored by an optical device that characterizes every particle. Conventionally, cytometers employ firstly, immunofluorescence—a technique used for light microscopy with a fluorescence microscope; it is used on biological samples. This technique uses the specificity of antibodies to their antigen to target fluorescent dyes to specific biomolecule targets within a cell, and therefore allows visualization of the distribution of the target molecule through the sample. Secondly, green fluorescent protein (GFP) is an alternative. GFP is a protein composed of 238 amino acid residues exhibiting bright green fluorescence when exposed to blue light. The flow cytometer allows categorizing, counting and even sorting the cells according to several chemical, or molecular properties. However, for astrobiology, both immunologically, or the use of GFPs are still challenges. X-ray absorption spectroscopy can, in principle, be developed as an alternative (Dalton *et al.,* 2003).

Laser-induced breakdown spectroscopy (LIBS) is a technique in which firstly we vaporize small quantities of material of the order of micrograms to nanograms (including biological matter) with the use of highly enregetic laser pulses. Secondly, by thermally exciting the resulting material into plasma, we can proceed to a spectroscopic analysis of the light emitted by the atoms. This has been a technique for the analysis of elements by retrieving a unique elemental fingerprint spectrum. Since chemical elements are known to emit light of a given frequency when excited to sufficiently high temperatures, LIBS suggests itself as a technique by means of which we could detect all elements in a given target. LIBS can be used to distinguish bacteria with fewer constraints that were previously thought to be possible. There are advantages when planning the exploration of the Solar System. For example, LIBS shows potential for development instrumentation with characteristics typical of LIBS, but in addition rapid *in situ* analysis is possible with little or no sample preparation and the feasibility of automated spectroscopic analysis (Multari *et al.*, 2010).

12.8 Can the origin of metazoans be detected elsewhere in the Solar System?

We have been arguing that the potential discovery of exogenous life, for example in Europa (Chapter 8), Mars (Chapter 7) or Enceladus (Chapter 9), is of immediate urgency and very likely to lead to positive results. We have two possible scenarios: that the exogenous life forms we find is unrelated to life on earth, or that it is related (e.g. by panspermia). In either case, one reason why the search for multicellularity in the solar system becomes urgent is because it might tell us about the "universality" of the traits that earth living creatures have developed.

The methods that we proposed above assume that all tissues are composed of cells that can be disrupted. Even in our own biota it is easy to think of counterexamples: (1) intertwined tissues (e.g. neural and gangliar), (2) pseudo continuous tissues where a compartmentalized, yet continuous cytoplasm hosts multiple nuclei (e.g. *Mycoplasma*), (3) progressively contained cells, like Russian dolls, where one cell encloses other cells, which in turn enclose other cells, etc. (e.g. *Volvox*) cannot be disrupted by those standard methods. Each of these would require a particular method of analysis. We can imagine many alternative ways of complex multicellular-like organizations, but most exciting would be to discover those arrangements that we did not even imagine, but for which detection is even harder to anticipate. In this chapter we have discussed the likelihood of finding multicellular life, in particular on Europa, and the challenges that lie on the pathway for the detection and analysis of simple metazoans. We have in mind technologies that could be contained in small probes that crush into the ice, called penetrators. These are significant innovations of the space technology. Life on Europa might be found in the cracks, or in the bottom of the ocean (cf., Fig. 12.1 from Greenberg 2002).

In either case this would be beyond the reach of the penetrators, and a direct observation, perhaps photographic, must depend on hydrobot/cryobot missions (cf., Fig. 8.6), which is at least decades away. Yet, the ice tectonics of Europa (Greenberg, 2005) stir and mix the ice that is to our favor, because it can make these samples accessible at any point in the surface, from which in principle ice samples can be extracted.

These samples can be analyzed using a combination of microscopy, tissue disruption and flow cytometry, and can provide information about the existence of multicellularity, their degree of differentiation ("level of complexity") and characteristics of individual cells. The cytometer could allow our first insight into the cell biology of exogenous metazoans. Once the first and second steps towards the evolution of intelligence are understood, we are left with a third step in the ascent of a neuron to multicellularity, the evolution of brains and eventually to intelligent recognizable behavior that will be exposed by communication through the different windows of the electromagnetic spectrum. This astronomical means was considered already in Chapter 11.

A relevant line of research has argued in favor of the emergence of multicellular life on Europa, not so much as we argue in this chapter and Chapter 8, based on the possible role of chemosynthesis triggering off the early anaerobic stages of microbial life (fed by hydrothermal vents in the bottom of the Europan ocean), but more on the possible role that exogenous oxygen may accumulate on the icy surface over geologic times and eventually feed the underlying ocean with sufficient oxygen to allow Earth-like aquatic metazoans and other possibly more advanced life (Hand *et al.*, 2007; Hand *et al.*, 2009; Greenberg, 2010). Radiolysis is the dissociation of molecules by radiation. It is the cleavage of one or several chemical bonds resulting from exposure to high-energy flux.

These processes have been shown to be relevant from the space weather conditions in the neighborhood of the Jovian satellite. This environment can provide oxidizing agents (oxidants), namely chemical compounds that readily transfer oxygen atoms, or even substances that gain electrons in certain types of chemical reactions (cf., redox in the Illustrated glossary).

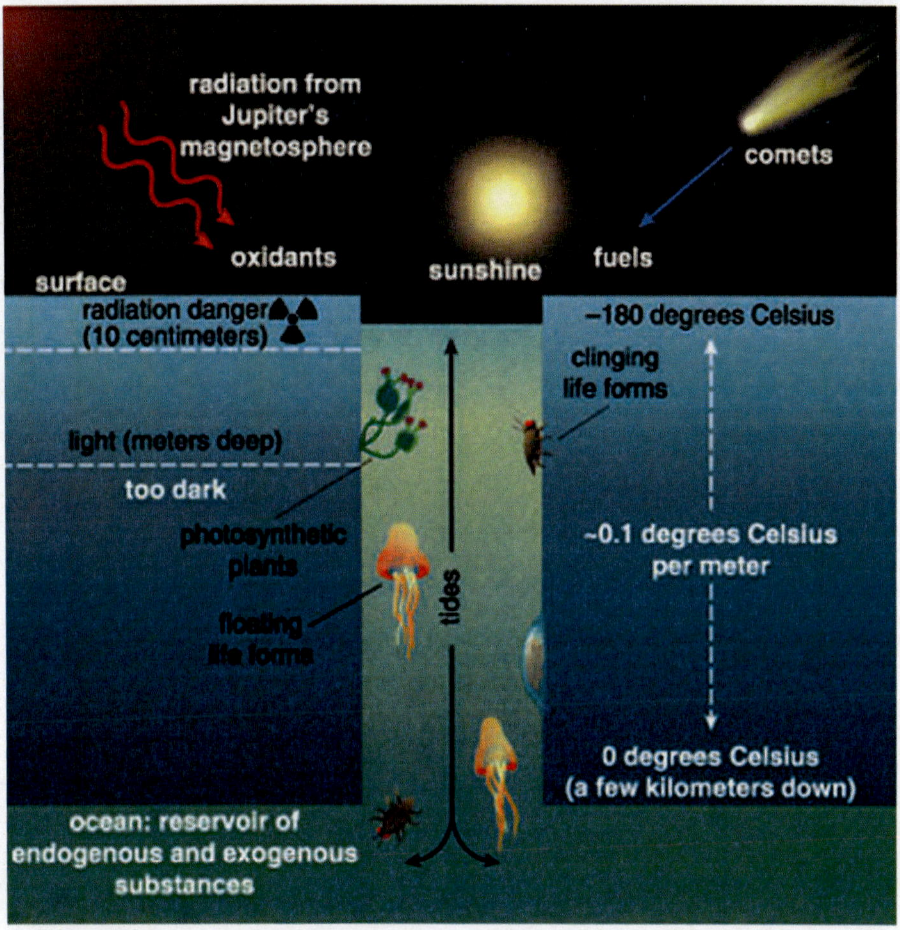

Fig. 12.1 One possibility is graphically explained for the presence of metazoans on the cracks of Europa's icy surface, or in the bottom of the ocean

Radiolytically produced oxidizing agents have been studied as candidates for accumulation and transport through the icy surface into the ocean of a biologically useful substance that could have led to a convergent evolution of aquatic metazoans as they did during the Earth Phanerozoic.

In fact the above-mentioned reasoning has led to the theoretical estimate that the rate of oxygen delivery by this means would support up to 3 billion kilograms of macrofauna (Greenberg, 2010). These possibilities raised by detailed calculations underline the relevance of the exploration of Europa in the next mission (EJSM, cf., Chapter 8) to provide the landers, or penetrators, with the correct instrumentation to probe for biomarkers that could expose such evolutionary developments. These tests are feasible, as we have argued in the above-mentioned earlier chapter. Our main point is that penetrators with *in-situ* micro-laboratories are realistic from the point of view of funding and, in addition, the relevant instrumentation has already been miniaturized.

12.9 Back to Europa for further constraining the Drake Equation

Just as we argued in Section 11.6 that some of the factors in the Drake Equation beyond those that Frank Drake introduced of astronomical nature (except for the duration of such civilizations due to war or natural catastrophes, meteoritic impacts or disease) can be studied to restrict the search for the number of civilizations N that may be present in the universe (cf., Tables 11.1, 11.2).

We now consider other means of restricting the Drake Equation in view of new technologies that may play a role in forthcoming exploration of the Solar System. Indeed, traces of seafloor microbial activity could, in principle, be found on the Europan icy surface for a variety of reasons. Amongst the many mechanisms that have been suggested, we underline the fluid-dynamics analysis of the Europan ocean (Thomson and Delaney, 2001). It invokes the possibility of plumes reaching the bottom of the icy surface from the seafloor, potentially carrying the microbial fractionated sulfur to within reach of surface probes. Indeed, the sulfurous material on the surface may be endogenous (Carlson *et al.*, 1999; Kargel *et al.*, 2000). The size of ice domes and movement of ice rafts on the surface of Europa are consistent with what one could expect if hydrothermal vent plumes in an ocean beneath the ice cause melting.

If the plumes arise from magma-heated regions of the seafloor, some of the sulfur non-ice elements observed on the surface of Europa could be remnants of the sulfur-particles that were processed on the seafloor and in the oceanic subsurface. They should be absent on Ganymede's surface, since the contact of a silicate core with the ocean would be missing and the longer distance from Jupiter would favor more cryovolcanism on Europa. To sum up, with Thomson and Delaney it is possible to interpret the non-water elements on the icy surface as the product of eruptions on the seafloor that were subsequently raised to the icy surface. The rotation of Europa and the weak stratification of its ocean are in principle capable of inducing hydrothermal vent plumes from dispersing, continuously rising like a cyclone through 100 kilometers of ocean to reach the base of the ice. This assumption is especially reasonable in the chaos-type features, such as melt-through structures that could be formed by rotationally confined oceanic plumes from heated regions on the seafloor. In other words, Thomson and Delaney apply plume dynamics to Europa from what we know on Earth, especially in our oceanic floors that have been fully explored, such as the Juan de Fuca Ridge off the west coast of Canada and the United States.

12.10 Convergence between SETI and the exploration of our Solar System

It is worth pursuing the exploration for life on Europa with the available technological developments, such as the penetrators for this aspect of the exploration of the Solar System would further reduce the uncertainties that are summarized by the Drake Equation (Gowen *et al.*, 2009, 2010). We will need some further concepts: In ice dynamics 'dilation' is a term associated with an opening process, where crust on either side is moved apart and new surface is created between them. Dilational areas, where cracks open wide—are called dilational "bands". These features are a common occurrence on the icy surface of Europa. These bands slope at about 5 degrees, or less. These broad tracks, tens of kilometers wide and hundreds of kilometers long, form when cracks in the ice shell open in response to the gravitational pull of Jupiter and the other Galilean moons. For landing, the penetrator should aim at a site that is not near levels that have been raised beyond an inclination of 40°. There is a large concentration of dilation features south of the equator from 180° for roughly 60°. For example, there is the "Sickle" dilation area at longitude 240° W, just below the equator, within a large area discovered by the Galileo NIMS team. It extends north of the equatorial region, between 0° and 30° N, and between the longitudes 240° and 270°. The best impact sites will be near polar (latitude > 75 degs), because this greatly increases frequency and volume of communications with the overhead orbiter.

In other words, one of the major discoveries of the Galileo Mission was this area, which for simplicity we call the *"large NIMS area"*. On this wide location there is an appreciable abundance of non-ice material. The interior of the Sickle band in this region is smooth and flat (Greenberg, 2005). Dilational areas could be smoother than the surrounding ice, because they formed gradually, unlike impact craters. Selecting these areas for future landings seems appropriate, keeping in mind that our paramount objective is to search for biomarkers with penetrators that should be deployed in the most favorable locations (cf., Fig. 12.2).

With respect to the case of Ganymede (and especially for testing the hypothesis for the origin of habitable ecosystems), some suggestions are possible at this stage for the best landing site for eventual JGO penetrators:

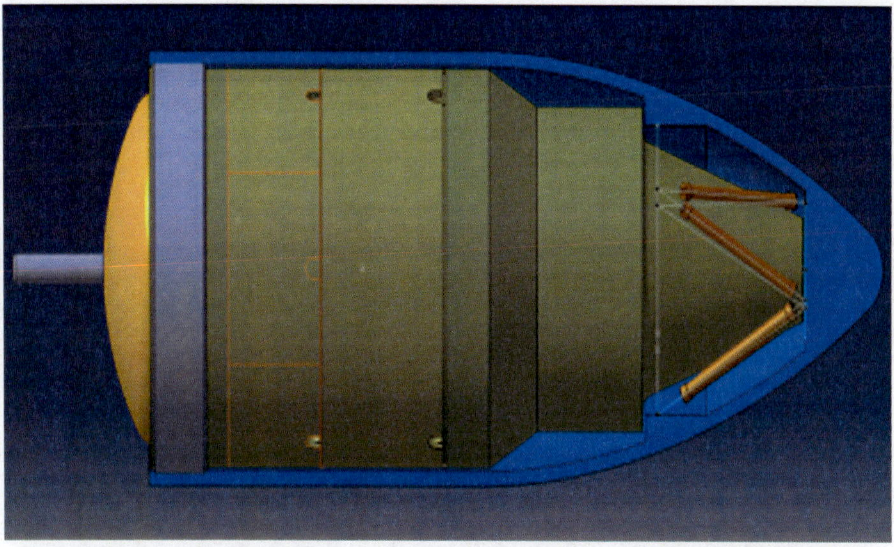

Fig12.2 Preliminary penetrator design (~11kg, length 37cm from nose to the release stud, diameter 15 cm)

During Galileo's G2 encounter a detailed examination of the small-scale characteristics of Ganymede's high latitude terrain was undertaken. This included a sequence of high-resolution SSI images. This effort was restricted to a region limited by (56° N, 174° W) to (65° N, 165° W) (Khurana *et al.,* 2007). These images not surprisingly showed few (if any) visible shadows, since small-scale slopes on Ganymede are generally less than 20° (Oberst *et al.,* 1999).

These small-scale slopes are compatible with the needs for a good landing site for the proposed JGO penetrators. Most attention on the non-ice components of the icy patches has been given to sulfate salts and sulfuric acid (Carlson *et al.*, 2002; McCord *et al.,* 1998, 1999; Dalton *et al.,* 2005). These interpretations of the sulfur-containing components of the icy patches have a most significant coincidence: the element sulfur continues to be present in both interpretations of the Galileo data. This remark helps to emphasize the main point: knowledge of the source of sulfur fractionation (either thermochemical sulfate reduction, bacterial sulfate reduction, or sulfur disproportionation) is of vital importance for defining a feasible and reliable biomarker. From the arguments discussed in this chapter a microbial ecosystem at the Europan seafloor and in its subsurface could indeed fractionate sulfur to a degree that is feasible to be detected by the instruments that are integrated in a penetrator, or landers (cf., Fig. 12.3). The penetrator's agility makes this instrument a leading candidate for the eventual payloads of both the JEO and JGO orbiters of EJSM. In addition, the UK supports current penetrator development for the exploration of Solar System bodies, including the Moon, Ganymede, Europa and Mars. Our main hypothesis that the origin of habitable ecosystems is linked to seafloor activity of hydrothermal vents requires a direct contact between the silicate core and a liquid water ocean, as it occurred in the early Earth, according to the geochemical evidence that we have retrieved from both the Pilbara craton and the Kaapvaal craton. There is also evidence for an ocean on Ganymede (McCord *et al.,* 2001), but it would not be in contact with its silicate core.

For the Galilean satellites that EJSM intends to explore, penetrators (supplied with mass spectrometry) should yield different results for fractionated sulfur according to our hypothesis for the origin of habitable ecosystems: The biogenically processed icy patches of Europa should give substantial depletions of ^{34}S, while Ganymede measurements should give significantly lower values for the depletion of ^{34}S.

Fig. 12.3 Artistic interpretation of a surface element penetrator in orbit around Europa.

In other words, diverging results—large minus $\delta^{34}S$ for Europa and small minus $\delta^{34}S$ for Ganymede (cf., Fig. 12.4)—would test the above hypothesis for the origin of habitable ecosystems.

Regarding the significant issue of selecting the most promising landing site, in the present work we have discussed some preliminary ideas implying that there are significant advantages from the point of view of the penetrator technology to prefer Sickle in the case of Europa (and possibly other similar flat dilational bands). Dilational bands would be steeped with sulfur if they were appropriately chosen (within the large 'NIMS' area).

If the search for biomarkers is focused on the NIMS area, non-ice chemical elements of the icy patches may bear fingerprints of life (biogeochemical ones). Non-ice elements, especially sulfur should be abundant on dilational bands chosen within this large area. Indeed, the disagreement of the two above-mentioned Galileo Mission groups does not question whether sulfur is present on the NIMS area (Carlson *et al.*, 2002; McCord *et al.,* 1998; Dalton *et al.,* 2005). Granted, there is some disagreement on the precise compounds that distort the absorption bands, since most likely they can be either sulfate salts, or alternatively sulfuric acid hydrates, but there is general consensus that sulfur is present in the molecular compounds that are present the icy surface.

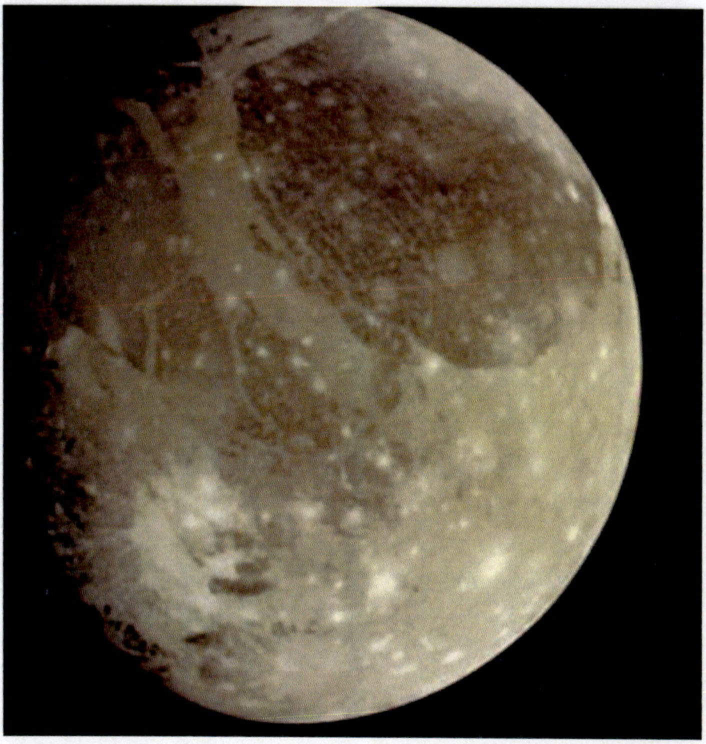

Fig. 12.4 Image of Ganymede, taken by the Galileo spacecraft in 1996. The satellite is illuminated from the right by the Sun

But, in any case, a landing site such as Sickle is appropriate for penetrators, due to the lack of significant elevations. The scientific return of taking advantage of the penetrator technology would be justified. Even if the sulfur patches turn out to have a small $\delta^{34}S$ parameter (with a modulus much smaller than -70 ‰, hence not necessarily biological), the nature of the chemically rich icy surface would be better understood at the geochemical level.

For the (essential) simultaneous exploration of EJSM with Ganymede's JGO, we have pointed out a suitable landing site. Further arguments discussed in favor of penetrator technology included the ability to provide measurements at globally spaced sites on planetary bodies, for example, a seismic network (Lunar-A) and, in an astrobiological context they are ideal instruments to carry out unambiguous biogeochemical search for biomarkers.

We conclude that the implementation of penetrators in future exploration of the Jupiter System is worthy of all the support that will be needed, both at the national, as well as at the international level. With the help of such appropriate instrumentation, we can face one of the most transcendental questions in astrobiology, namely the discovery of a habitable ecosystem elsewhere in our Solar System, which would constrain further research programs that seem at first sight to be distant from each other, namely the bioastronomy SETI program and the direct exploration of our Solar System, especially the Jovian system with its intriguing icy moons that were discussed in Chapter 8.

Supplementary Reading

Deacon, T. W. (1997) *The Symbolic Species The Co-evolution of Language and the Brain,* W. W. Norton & Co., New York.

Drake, F. and Sobel (1992) *Is there anyone out there? The scientific search for Extraterrestrial Intelligence,* D. Delacorte Press, New York.

Ekers, R. D.Cullers, D. K. Billingham, J. and Scheffer, L. K. (eds.) (2002) *SETI 2020.* SETI Press: Mountain View CA.

Greenberg, R. (2005). *Europa, the Ocean Moon*. Springer, Berlin

Wilson, E. O. (1998) *Consilience The Unity of knowledge,* Alfred A. Knopf, New York.

References

Attenborough, D. (1995) *The Life of Plants.* BBC Books: London.

Bonner, J. T. (2001) *First Signals: The Evolution of Multicellular Development.* Princeton University Press, USA.

Carlson, R. W., Johnson, R. E. and Anderson, M. S. (1999). Sulfuric Acid on Europa and the Radiolytic Sulfur Cycle. *Science* **286**, 97-99.

Carlson, R. W., Anderson, M. S., Johnson, R. E., Schulman, M. B. and Yavrouian, A. H. (2002). Sulfuric Acid Production on Europa: The Radiolysis of Sulfur in Water Ice. *Icarus* **157,** 456-463.

Dalton, J.B., Rakesh, M., Kagawa, H.K., Chan, S.L. And Jamieson, C.S. (2003) Near-infrared detection of potential evidence for microscopic organisms on Europa. *Astrobiology* **3**(3):505-529.

Dalton, J.B., Prieto-Ballesteros, O., Kargel, J.S., Jamieson, C.S., Jolivet, J. and Quinn, R. (2005) Spectral comparison of heavily hydrated salts with disrupted terrains on Europa. *Icarus* **177**, 472-490.

De Duve, C. (1995) *Vital Dust Life as a cosmic imperative,* Basic Books, New York.

De Vladar, H. P. and Chela-Flores, J. (2011) Can the evolution of multicellularity be anticipated in the exploration of the Solar System? in: *Genesis: Origin of Life on Earth and Planets.* Arnold Hanslmeier, Stephan Kempe, Joseph Seckbach (eds.) Cellular Origin and Life in Extreme Habitats and Astrobiology, Springer: Dordrecht, The Netherlands (in preparation).

Drake, F. (2001) *New Paradigms for SETI,* in Chela-Flores, J., Owen, T. and Raulin, F. (2001). *The First Steps of Life in the Universe.* Proceedings of the Sixth Trieste Conference on Chemical Evolution. Trieste, Italy, 18-22 September, 2000. Kluwer Academic Publishers: Dordrecht, The Netherlands.

Gowen, R., Smith, A., Ambrosi, R., Prieto Ballesteros, O., Barber, S., Barnes, D., Braithwaite, C., Bridges, J., Brown, P., Church, P., Collinson, G., Coates, A., Collins, G., Crawford, I., Dehant, V., Dougherty, M., Chela-Flores, J., Fortes, D., Fraser, G., Gao, Y., Grande, M., Griffiths, A., Grindrod, P., Gurvits, L., Hagermann, A., van Hoolst, T., Hussmann, H., Jaumann, R., Jones, A., Jones, G., Joy, K., Karatekin, O., Kargl, G., Macagnano, A., Mukherjee, A., Muller, P., Palomba, E., Pike, T., Proud, B., Pullen, D., Raulin, F., Richter, L., Ryden, K., Sheridan, S., Sims, M., Sohl, F., Snape, J., Stevens, P., Sykes, J., Tong, V., Stevenson, T., Karl, W., Wilson, L., Wright, I., Zarnecki, J. (2009). Looking for Astrobiological Signatures with Penetrators on Europa. In: *Physical and Engineering Sciences Exploratory Workshops*, W08-115: Biosignatures on Exoplanets; The Identity Of Life□, 22-26 June 2009, Mulhouse, France□.
http://www.ictp.it/~chelaf/ESFsummary.pdf

Gowen, R. A., Smith, A., Fortes, A.D., Barber, S., Brown, P., Church, P., Collinson, G., Coates, A. J., Collins, G., Crawford, I. A., Dehant, V., Chela-Flores, J., Griffiths, A. D., Grindrod, P.M., Gurvits, L.I., Hagermann, A, Hussmann, H., Jaumann, R., Jones, A.P., Joy. A. Sephton, K.H., Karatekin, O., Miljkovic, K., Palomba, E., Pike, W.T., Prieto-Ballesteros, O, Raulin, F., Sephton, M. A., Sheridan, M S., Sims, M., Storrie-Lombardi, M. C., Ambrosi, R., Fielding, J, Fraser, G., Gao, Y., Jones, G. H., Kargl, Karl, W. J., Macagnano, A., Mukherjee, A., Muller, J.P., Phipps, A., Pullan, D., Richter, L., Sohl, F., Snape, J., Sykes, J., Wells, N. (2010). Penetrators for in situ sub-surface investigations of Europa, *Advances in Space Research,* doi: 10.1016/j.asr.2010.06.026.

Greenberg R. (2002) Tides and the Biosphere of Europa. *American Scientist* **90**, 48-55

Greenberg, R. (2005) *Europa, the Ocean Moon.* Springer, Berlin, p. 136.

Greenberg, R. (2010) Transport rates of radiolytic substances into Europa's ocean: implications for the Potential Origin and Maintenance of Life. *Astrobiology* **10**, 275-83.

Hand, K.P., Carlson, R.W., and Chyba, C.F. (2007) Energy, chemical disequilibrium, and geological constraints on Europa. *Astrobiology* **7**,1006–1022.

Hand, K. P., Chyba, C.F., J.C. Priscu, Carlson, R.W. & K.H. Nealson (2009) Astrobiology and the Potential for Life on Europa. In: *Europa*. R. Pappalardo, W. McKinnon and K. Khurana (eds.). Univ. of AZ Press, pp. 589-629.

Kargel, J. S., Kaye, J. Z., Head, III, J. W., Marion, G. M., Sassen, R., Crowley, J. K., Ballesteros, O. P., Grant, S. A., and Hogenboom, David L. (2000) Europa's Crust and Ocean: Origin, Composition, and the Prospects for Life. *Icarus* **148**, 226-265.

Marino, L. (1997) *Brain-behavior relations in primates and cetaceans: Implications for the ubiquity of factors leading to the evolution of complex intelligence,* in: *Astronomical and Biochemical Origins and the Search for Life in the Universe.* Eds. C.B. Cosmovici, S. Bowyer and D. Werthimer. Editrice Compositore: Bologna. pp. 553-560.

Marino, L. (2000) *Turning the empirical corner on F : The probability of complex intelligence,* in *A New Era in Astronomy,* Lemarchand G.A. and Meech K. (eds.), ASP Conference Series, San Francisco, USA, **213,** 431-435.

McCord, T. B., Hansen, G. B., Clark, R. N., Martin, P. D., Hibbitts, C. A., Fanale, F. P., Granahan, J. C., Segura, N. M., Matson, D. L., Johnson, T. V., Carlson, R. W., Smythe, W. D., Danielson, G. E. and the NIMS team (1998) Non-water-ice constituents in the surface material of the icy Galilean satellites from the Galileo near-infrared mapping spectrometer investigation, *Jour. Geophys. Res.* **103**(E4), pp. 8603-8626.

McCord, T. B., Hansen, G. B., Matson, D. L., Johnson, T. V., Crowley, J. K., Fanale, F. P., Carlson, R. W., Smythe, W. D., Martin, P. D., Hibbitts, C. A., Granahan, J. C., Ocampo, A. and the NIMS team (1999). Hydrated salt minerals on Europa's Surface from the Galileo near-infrared mapping spectrometer (NIMS) investigation. *J. Geophys. Res.* **104**, 11827-11851.

McCord, T. B., Hansen, G. B., Hibbitts, C. A. (2001). Hydrated Salt Minerals on Ganymede's Surface: Evidence of an Ocean Below. *Science* **292**, 1523 – 1525.

Miller, G (2009) On the Origin of The Nervous System. *Science* 325, 24 - 26.

Multari R.A., Cremers D. A., Dupre J.M., Gustafson J.E. (2010) The use of laser-induced breakdown spectroscopy for distinguishing between bacterial pathogen species and strains. *Appl Spectrosc.* **64**(7), 750-759.

Oberst, J., Schreiner, B., Giese, B., Neukum, G., Head, J.W., Pappalardo, R.T. and Helfenstein, P. (1999). The distribution of bright and dark material on Ganymede in relation to surface elevation and slopes. *Icarus* **140,** 283–293.

Rompel A., Cinco, R.M., Latimer, M.J., McDermott, A.E., Guiles, R.D., Quintanilha, A., Krauss, R.M., Sauer, K., Yachandra, V.K. and Klein, M.P. (1998) Sulfur K-edge x-ray absorption spectroscopy: A spectroscopic tool to examine the redox state of S-containing metabolites *in vivo*. *Proc. Natl. Acad. Sci. USA* **95**, 6122–6127.

Thomson, R. E. and Delaney, J. R. (2001). Evidence for a Weakly Stratified Europan Ocean Sustained by Seafloor Heat Flux. *Jour. Geophys. Res.* **106,** 12,355-12,365.

Villegas, R, Castillo, C. and Villegas, G.M. (2000) *The origin of the neuron: The first neuron in the phylogenetic tree of life,* in: Chela-Flores, J., Lemarchand, G.A. and Oro (eds.) *Astrobiology from the big bang to civilization.* Kluwer Academic Publishers: Dordrecht, The Netherlands. pp. 195-211.

Wilson, E. O. (1998) *Consilience The unity of knowledge***.** Alfred A. Knopf: New York, p. 70.

13

Cultural frontiers of astrobiology

An answer to the fundamental question of the relation man/universe requires a broad cultural discussion. It was characteristic of the Enlightenment, the movement of ideas current during the 18th century that distrusted tradition in cultural matters. Truth was to be approached through reason. At the end of that period Auguste Comte (1798-1857) founded a movement that advocated that intellectual activities should be confined to observable facts. The reason why this movement was called "positivism" is that Comte called observable facts "positive". This point of view was developed much later by a group of philosophers working in Vienna at the beginning of the 20th century. They were known as the "Vienna Circle". They maintained that scientific knowledge is the only kind of factual knowledge.

The reader is advised to refer especially to the following entries in the Illustrated glossary: *Biogeocentrism, contingency, convergent evolution, Coyne (George), Enlightenment, Logical positivism, natural theology,*

13.1 The frontiers of science, philosophy and theology

Some of the deepest questions that have persistently remained with us since biblical times are:

What is the origin of the universe?
What is it made of?
What is its ultimate destiny?
How did life, in general, and how did humans originate?
Are we alone in the cosmos?

Research in astrobiology has provided arguments that Pasteur longed for during his lifetime. Indeed, we have attempted to be persuasive, in the sense that during the 20th century there was considerable progress in understanding the evolution of intelligent behavior on Earth. At this stage it may be argued that our task as a scientist should be independent of that of the other approaches to the question of our origins, for Russell, 1991): "Philosophy is something intermediate between theology and science. Like theology, it consists of speculations on matters as to which definite knowledge has, so far, been unascertainable; but like science it appeals to human reason rather than authority, whether that of tradition or that of revelation."

J. Chela-Flores, *The Science of Astrobiology: A Personal View on Learning to Read the Book of Life*, Cellular Origin, Life in Extreme Habitats and Astrobiology 20, DOI 10.1007/978-94-007-1627-8_13,

On the other hand, it may be useful to be aware that this view of the role of science has also been discussed from another point of view; for instance (Townes, 1995): "If science and religion are so broadly similar, and not arbitrarily limited in their domains, they should at some time clearly converge. "

The idea that is reflected in the last citation is that both science and religion are concerned with the common understanding of life in the universe. Since they largely address the same questions, both of these aspects of human culture should at some point converge. With subsequent progress in science, philosophy and theology, convergence is therefore unavoidable. There does not seem to be any evident convergence in culture at present, but the status of the present relationship between the three disciplines: science, philosophy and theology (relevant to an eventual integrated view of understanding the evolution of intelligent behavior on Earth), has recently been comprehensively discussed (John Paul II, 1992): "Contemporary culture demands a constant effort of synthesis of knowledge and of an integration of our understanding... but if the specialization is not balanced by an effort aiming to pay attention to relationships in our understanding, there is a great risk of arriving at a "splintered culture", which would in fact be the negation of the true culture. "

We may avoid a splintered culture by bringing closer to each other various approaches regarding the origin, evolution and distribution of life in the universe. This is an area of research in which we should expect substantial progress to occur in the future. This is particularly likely, due to the large funds potentially available for space missions (cf., Chapter 1). Moreover, we are bound to converge towards better answers, due to the deep insights that all of us, scientists, philosophers and natural theologians, can collectively provide in the future. However, in spite of the impressive progress of science, it is important to underline that many of the fundamental questions in astrobiology still remain unanswered.

Therefore, before beginning our task it may be prudent to recall that cosmology suggests that there is sufficient time available for science to progress to a stage in which further, and possibly better attempts than are now possible, will be made, in order to search for answers to some of the deepest questions, which humans have been asking themselves from time immemorial (cf., the introductory pages of the Book of Life, Part 4. Up to the present partial answers have often led to controversy between scientists, philosophers and theologians. We should be made aware of the limited scope of the scientific method, a point that has been stressed. Recognizing a limit to the applicability of present day science Bertrand Russell, the British Literature Nobel Laureate, philosopher and mathematician expressed that (Russell, 1991): *Almost all the questions of most interest to speculative minds are such as science cannot answer.*

13.2 Positivism and the Vienna Circle

The Vienna Circle maintained that all traditional doctrines are to be rejected as meaningless. They went beyond positivism in maintaining that the ultimate basis of all knowledge rests on experiment. Since they were also considering the unification of science and were using mathematical logic in their formulation, the Viennese version of extreme positivism came to be known as "Logical positivism". Although some scientists have adopted this philosophy, either consciously or unconsciously, the fact remains that modern science begins with Galileo, who initiated the tradition of formulating theories based on observation and experiments. No underlying philosophy was adopted then, or need to be adopted now, beyond the dialogue theory/experiment.

Positivism avoided all considerations of ultimate issues, including those of metaphysics and religion. However, as anticipated by Russell, the reduction of all knowledge to science is a matter that debate has not yet settled. Issues of first causes and ultimate ends are precisely topics relevant to the subject matter of this book. To 19th-century scientists, the problems of the origin and distribution of life in the universe were issues that were to be excluded from the scientific discourse.

In the three Books into which the subject of astrobiology has been divided, we have attempted to show that these problems are approachable by scientific methods. These subjects have a consistent history of valuable efforts by some of the best scientific minds of the 20th century. The long list of such scientists began with Alexander Oparin, John Haldane and included many others, some of which are listed in the name-index at the end of the present work. In view of the progress that has been achieved, extreme aspects of first causes and ultimate ends are naturally inserted in the science of astrobiology; yet, neither of the two problems (origins and distribution of life) is solidly set on scientific bases: it has been impossible to synthesize a living organism so far, and no signal from an extraterrestrial civilization has yet made contact with the highly sensitive radio telescopes that have been discussed in Chapter 11. Due to this unsettled state of affairs, it seems unavoidable that a reasonable collective approach to the deepest questions in astrobiology should be encouraged by all sectors of culture. Regarding the limits of science, inspired by Russell's philosophical outlook, probably a pertinent question is whether man is really what he seems to the astronomer, a tiny lump of impure carbon and water impotently crawling on a small planet of no special significance. To address such questions, we must take a step back and ask ourselves what is the place of humankind amongst the Earth biota.

13.3 Position of humans in the totality of all earthly species

From the perspective of biology, human beings represent only a single species among four thousand mammals. Yet, this is a small number when compared with the 30 million species that are expected to constitute the whole of the Earth biota. One aspect of this bewildering abundance of species of which humans are only one, has led to the metaphor (Gould, 1991): "If the history of evolution were to be repeated, such an alternative world would teem with myriad forms of life, but certainly not with humans."

Our main concern is not the origin and evolution of our own species. Our main concern is rather the likelihood that the main attributes of man would raise again, if the history of evolution starts all over again elsewhere, not in a hypothetical Earth that would be miraculously reconstructed. We are mainly concerned with the repetition of biological evolution in an extrasolar planet, or satellite, that may have had all the environmental conditions appropriate for life.

The main question is whether the attributes of man are repeatable. Such attributes are, for example, a large brain and consciousness. These features of man evolved from lower primates over the last 5 to 6 million years. This is a short period of time compared to the evolution of other organisms, such as mollusks, which have survived since the Lower Cambrian. Indeed, their first appearance occurred 500 million years ago, about 100 times earlier than the first appearance of man.

13.4 Is evolution more than a hypothesis?

We have already reviewed some arguments that suggest that the problem of the position of man in the cosmos critically depends on the evolution of microorganisms up to the level in which eukaryogenesis occurred. This forces upon us the question of the position of the eukaryotic cell in the cosmos, as the main focus of our attention. I feel that such a radical break with the past has some implications in our understanding of the origin and destiny of man. Nevertheless, none of these arguments lie outside the scope of the question raised in the Papal Message to the Pontifical Academy of Sciences (John Paul II (1996a,b): "*New knowledge has led to the recognition that the theory of evolution is no longer a mere hypothesis*."

In spite of this important step, the acceptance of evolution has not led to a consensus amongst scientists, either on its mechanism, or on its implications. Nevertheless, in spite of this shortcoming, we shall base our subsequent arguments on Darwin's theory of evolution. This leads us to a discussion of the implications that such a search might imply for the dialogue between science and natural theology.

Constraints imposed by philosophy and theology on our view of life mostly favors a special place of man in the universe. In the case of philosophy there is a continued quest for the impact of technological progress on the future of mankind. We have already encountered the perspective of cultural evolution on the breakdown of a straight coupling between chance and necessity, namely, the continued accumulation of mutations that may favor the adaptation to changing environmental conditions.

A separate question concerns the changes in our theological outlook that may follow the incorporation of knowledge of the place of earthly biota in the cosmos. Would there be problems in the traditional monotheistic view of Deity as being confined with the affairs of man? It could be argued that the Deity is omnipotent and can be concerned with the affairs of many intelligent species that may exist in the universe. Yet the question remains whether the original image of God as portrayed in the Scriptures would be acceptable if SETI, or a more restricted search in our own solar system, were to confront us with parallel evolution in other worlds. The question of the impact of extraterrestrial life on our culture, a discipline referred to as "astrotheology", has been reviewed previously (Dick, 1998).

13.5 What is specific to a human being?

A first contact with extraterrestrial life would confront us with new problems to be solved in biology. For example, a more extensive view of taxonomy would be needed. We would have to learn to classify new organisms. This would be within the domain of scientific enquiry. In parallel to the Christian tradition, a position that has been discussed since the Enlightenment regards the confrontation between science and religion. It maintains that God acts only in the beginning, creating the universe and the laws of nature. This thesis is called deism, usually taken to imply that God leaves universal evolution to its own laws, without intervening, once the process of creation has taken place. On the other hand, within the Christian tradition an approach towards integration has been advocated in the relationship between science and theology. It concerns the problem of biological evolution. Like Darwin, John Paul II, while referring to the living world, for good reasons has not put the main emphasis on first causes and ultimate ends. At the beginning of this chapter we emphasized that for the first time within science there is a branch, namely astrobiology, which makes first causes and ultimate ends its own subject matter.

We identify as a first cause the origin of life in the universe; the distribution of life in the universe may be identified as an ultimate end. With respect to human beings there is much ground to cover yet in the road of convergence between science and religion. Once again, the Papal Message refers to remaining points still to be discussed: "With man we find ourselves in the presence of an ontological difference, an ontological leap, one could say. However, does not the posing of such ontological discontinuity run counter to that physical continuity which seems to be the main thread of research into evolution in the field of physics and chemistry? Consideration of the method used in the various branches of knowledge makes it possible to reconcile two points of view that would seem irreconcilable. The sciences of observation describe and measure the multiple manifestations of life with increasing precision and correlate them with the time line. The moment of transition to the spiritual cannot be the object of this kind of observation, which nevertheless can discover at the experimental level a series of very valuable signs indicating what is specific to the human being."

One example is provided by one aspect of the human being that is exclusively discussed in the context of natural theology, namely, "the moment of transition to the spiritual which cannot be the object of [scientific] observation".

We have already touched upon the concept of evolutionary convergence at the beginning of this chapter: we should not consider the "rewinding of the evolutionary clock" as a thought-experiment, but as a real possibility that may have occurred on extrasolar planets. Evolution may not produce man again, but within the scope of science we can discuss the possible convergence of some of the attributes that are characteristic of human beings. For instance, we discussed language and intelligence (cf., Chapter 6), two attributes that are of extreme importance for the search of extraterrestrial life.

What questions would a first contact with extraterrestrial life imply for both science and religion? We have endeavored to demonstrate that contact need not come only at the level of a fully intelligent message (Chapter 11); contact could come first in the form of detecting the first cellular steps towards intelligence (Chapter 12). There remains a difficulty of addressing those attributes of human beings that are raised in theology, but not in science, namely the spiritual dimension, a point that has been clearly stated by the American astronomer George Coyne SJ (Coyne, 1998 and Fig. 13.1).

While we are still not in a position to answer this question, we have endeavored to gather a number of efforts within science that suggest that contact with life elsewhere cannot be excluded in the future. Such an experience would give us a unique opportunity; it would provide us with a solid point of reference on which to base original discussions of the implications on all the attributes of human beings. In such discussions the participants should be scientists and natural theologians.

Facing the discussion of this possibility now is neither premature nor idle: Exploration on Earth in the 15th century led to the difficulty of widening the horizons of the accepted attributes of man.

Fig. 13.1 George Coyne SJ lecturing at the Sixth Trieste Conference, 2000

The confrontation of Europeans with the American natives proved to be traumatic. In retrospect, the dialogue that took place in Valladolid, Spain, between Bartolomé de las Casas, the former Bishop of Chiapas (Mexico), and the learned Juan Ginés de Sepúlveda, is still of considerable interest. The question of the attributes that characterize man was raised on that occasion. We are still not ready to decide which attributes make us human till we reach consensus on what is our position, first on the tree of life and, subsequently, when we understand what the position of our tree of life is in the universe.

13.6 Are there trends in evolution?

In order to decide what is the position of our tree of life in the universe, we must appeal to science. First of all, we should put aside some philosophical objections that have been deeply rooted in the literature. We have touched on this question earlier, when we considered constraints on chance and evolutionary convergence; putting together these two aspects of evolution with its intrinsic randomness suggests that there are trends within evolutionary history that might reflect the existence of general principles in the evolution of increasingly larger and more complex forms in the Earth biota (Carroll, 2001) including the brain (Krubitzer, 1995).

In previous chapters we have already mentioned Monod's book *Chance and Necessity*. The author overemphasized the role of 'pure chance' in evolution. He excluded the role that evolutionary convergence may have had in the evolution of life on Earth. Basing himself on these arguments, Monod concluded that trends in biological evolution must be rejected. This question is not merely philosophical, although its philosophical implications are important. The question of evolutionary trends is relevant to the subject of astrobiology and in particular to bioastronomy. For we have learnt in Chapter 11 that there has been an enormous technological revolution in the capability of scanning the celestial sphere for traces of ongoing communication amongst creatures that are the product of evolution of intelligent behavior elsewhere. In concluding this chapter we underline the fact that chance at the molecular level in terms of mutations in the genome, does not exclude organisms from exhibiting trends at a higher level of organization.

We have already given an example of common trends at a higher level of organization in our illustrations of evolutionary convergence, which were taken form malacology and ornithology. Another aspect of the question of the existence of trends in evolution is also relevant to natural theology, as it has been discussed extensively (Peacocke, 1988). This seems to be an appropriate place to leave the subject of deeper implications of the search for extraterrestrial life at this particular crossroad of astrobiology, philosophy and theology.

13.7 Cultural implications of discovering extraterrestrial life

Often astrobiologists have to face the criticism of doing science without a subject matter. Currently, we do not have a definition of life that is unanimously accepted, but there are many potential cultural inputs in the subject of astrobiology. This is in fact an open question in both philosophy and science (Aretxaga and Chela-Flores, 2011 where these questions are discussed more extensively).

Our present knowledge of life on Earth allows us to elaborate a reasonably well-accepted explanatory model. Putting these remarks together with our recent insights into life in extreme environments induce us to think that a similar phenomenon—life—could have taken place elsewhere in the universe in the favorable conditions that the increasing repertoire of exoplanets begin to suggest. Extrapolating elsewhere in the universe our present understanding for the emergence of life does provide us with an object for astrobiology, since it allows us:

• To set some theoretical bases for the possible existence of the main object of our research (a second Genesis).

• To suggest the elaboration of a coherent and realistic strategy, namely with possibilities of success, of the search and obtaining reliable data of extant, extinct life or simply the identification of reliable biomarkers. Consequently, astrobiology can be understood as the *scientific exploration* of the universe searching for a second Genesis (Aretxaga, 2008).

As we have seen in Book 3 (Chapters 7-9), we are gradually discovering other places in the universe where such conditions exist suggesting that it may be possible to detect the presence of bacterial life. As a consequence of the exciting discoveries of the Cassini Mission (cf., Sec. 9.2) that come after several epoch making discoveries related to the Jovian moons by the Galileo mission (Chapter 8), our general view of the conditions favorable for a second Genesis in our own Solar System, our cosmic backyard, have been greatly enhanced. The conditions for habitability are more widespread than were thought to be possible at the end of the 20th century.

To sum up the earlier part of this book, on our Solar System there are several cases of some interest, where life can reasonably expected to be discovered:

• Evolutionary tests are also suitable in potential water sources under the Martian surface, such as in its north pole. These environments may contain microorganisms, an exciting possibility that was raised by the Allan Hill meteorite (cf., Chapter 7). Although the possibility of having detected life in this meteorite was not subsequently confirmed, it was nevertheless an important contribution that stimulated discussion about the possibility of life elsewhere in our Solar System. Cases of possible habitability to test in the future have been discussed in the literature (Chela-Flores, 2010).
• Europa may be a typical biofriendly body amongst extra-solar planets. The main problem is how to select appropriate experiments in situ, after surface landers are able to filter melt water from Europa's frozen surface.

From the above examples we have learnt that searching for a second Genesis is feasible in the foreseeable future. The ultimate aim of a biology experiment in Solar System exploration is to develop robotic tests that are compatible with the necessarily reduced dimension of landers. In the case of a Martian mission to subterranean pockets of liquid water we find a possibility that has attracted wide attention both by scientists and by the popular press. For this purpose miniaturization of instrumentation is essential. Many difficulties though are inherent in the eventual design of a test that would intend to identify microbes, robotically, in any extraterrestrial environment. This question begins to be important, in view of the decisions that have to be made in the selection of biological experiments that should be performed in situ on Europa and on Mars.

The discovery on Earth of unicellular microorganisms (eukaryotes) in an analogous habitat to Mars—the Tinto River, Spain—suggests the possibility of the presence of eukaryotic microorganisms elsewhere in our Solar System. Since there is no evidence for inhibiting the origin and evolution of such microorganisms beyond the Earth in analogous habitats, it seems reasonable to discuss the hypothesis of the universality of eukaryogenesis (Chapter 5). Finally, eukaryoticity seems to be a necessary condition for the emergence of a neuron (with the capability of setting up a series of action potentials that characterize it) and eventually multicellular complex nervous systems seems inevitable. The lowest multicellular organisms where action potentials have been documented are the eukaryotic sponges (Chapter 12). In multicellular prokaryotes, such as stromatolitic colonies of cyanobacteria action potentials seem unnecessary, and in fact have not been detected to the best of our knowledge. These arguments reinforce the hypothesis that has been assumed by radio astronomers anyway that there will be other intelligences that would give away their presence by radio waves emitted from some exoplanet or exomoon (the SETI project).

13.8 Relevance and implications of discovering life elsewhere

Both the implications and the degree of importance of the eventual discovery of a second Genesis will depend on the type of life that is discovered. Considering our human condition, the discovery with the largest impact and relevance at all levels of our society would be the discovery of intelligent life, especially if it were technologically advanced. But the majority of scientists would be of the opinion that the discovery of the simplest extraterrestrial microbial life would be an event of historic proportions. In the present work we shall distinguish implications at different levels.

In this context we have two different aspects:

• The setting up of different protocols for our actions to be taken for communicating the news to society in general, as well as preventing undesirable ethical, political, or scientific consequences, both for life on Earth as well as for the life that would be discovered (Race, 2008).

• Confirming the universality of biology, sharpening the definition of life, and the fall of biogeocentism: The sharp distinction between chance (contingency) and necessity (natural selection as the main driving force in evolution) is relevant for astrobiology. For this reason, it is important to document the phenomenon of evolutionary convergence at all levels, in the ascent from stardust to brain evolution.

The universality of biochemistry suggests that in Solar System missions, biomarkers should be selected from standard biochemistry. Given the importance of deciding whether the evolution of intelligent behavior has followed a convergent evolutionary pathway, and given the intrinsic difficulty of testing these ideas directly, we can alternatively begin testing the lowest stages of the evolutionary pathway within the Solar System. Within a few years we will be in a position to search directly for evolutionary biomarkers on Europa, the Jovian satellite.

The preliminary steps taken by science, philosophy and natural theology assumed that humans had a privileged position in the cosmos. With Copernicus geocentricism was abandoned. Natural theology has incorporated this view, after an initial opposition to Galileo's inexorable conclusions. After Darwin's theory of evolution was formulated, anthropocentrism was abandoned, although more than a century of research was needed, culminating with the solid genetic bases of neo-Darwinism. Natural theology is also beginning to incorporate this second step into its discussions. We have called, for lack of a better word in the English language, 'biogeocentrism' the concept that life is exclusive to the Earth (Aretxaga, 2004; Chela-Flores, 2001, 2009). The underlying difficulty for many scientists arises from a confrontation between the physical and biological sciences. Firstly, everyone agrees that the Newton's theory of gravitation can be extrapolated without any difficulty throughout the universe, except for the small corrections required by Einstein's theory of relativity. One example regards the orbits of Jupiter-like planets.

Secondly, the case of extrapolating the theory of biological evolution throughout the cosmos requires more care and is still an open problem. Arguments against the hypothesis of biogeocentrism can now be formulated thanks to progress in our understanding of Darwinian evolution

13.9 Consequences of other intelligent behavior

The consequences and implications of the specific nature of life to be discovered elsewhere in the Solar System, or eventually on exoplanets or exomoons are the main topics of this section. They shall depend on the type of life and the stage of the evolution of the life to be identified (Race, 2008). Taking as a point of reference the terrestrial model, we can conceive the following possibilities: unicellular life (both prokaryotic and eukaryotic), multicellular life capable of coordination via the emission of action potentials, multicellular life that has reached the stage of intelligent behaviour, and possibly a stage further where the intelligent behavior manifests itself at the social and cultural levels which we commonly associate with civilization. We should not rule out other possibilities, and Dick has discussed some further possibilities, such as post-biological entities (Dick, 2008).

While the discovery of eukaryotic microorganisms would confirm the hypothesis of the universality of eukaryogenesis, reinforcing the possibility of the existence of intelligent life elsewhere in the universe, the detection of technologically advanced civilizations would have more profound consequences, for instance the incorporation of the category of *person* assigned to deserving beings different from humans (Aretxaga, 2006). We should also consider the possibility that the messages received by Earth-bound radio telescopes could provide information on other forms of thinking and comprehending reality in its multiple aspects, such as science, philosophy, art or religion that would induce us to revise our theories and knowledge related to the origin, structure and function of mind and intelligence, not forgetting the origin, evolution and contents of its most significant creations: society and culture (Billingham, 1994; NASA, 2000; Tough, 2000).

Faced by these possibilities, the Spanish theologian E. Miret-Magdalena (Miret-Magdalena, 2008) recommends an open mind in searching all the positive common elements that might be learnt after the discovery, especially including the idea of the new beings of God, and other ideas of religious significance. These reflections led him to suggest a cosmic ecumenism. In this context it is also worth mentioning the concept of *cosmotheology* (Dick, 2000, 2005).

Supplementary Reading

Colombo, R., Giorello, G. and Sindoni, E. (eds.), (1999) *Origine della Vita Intelligente nell'Universo (Origin of intelligent life in the universe),* Edizioni New Press, Como.

Chela-Flores, J. (2008) Astrobiological reflections on faith and reason. The Issues of Agnosticism, Relativism and Natural Selection. In: *Divine Action and Natural Selection: Science, Faith and Evolution.* J. Seckbach and R. Gordon (eds.) World Scientific Publishers, Singapore, pp. 48-63.

Chela-Flores, J. and Seckbach, J. (2008) Divine Action and Evolution by Natural Selection A Possible and Necessary Dialogue. In: *Divine Action and Natural Selection: Science, Faith and Evolution.* J. Seckbach and R. Gordon (eds.) World Scientific Publishers, Singapore, pp. 1034-1048.

Martini, C. M. (1999) *X Cattedra dei non credenti: Orizzonti e limiti della scienza,* Raffaello Cortina Editore, Milano.

Russell, R. J., Murphy, N. and Isham, C.J. (eds.), (1996) *Quantum Cosmology and the Laws of Nature. Scientific Perspectives on Divine Action,* (2nd ed.), Vatican Observatory Foundation, Vatican City State.

Russell, R.J., Murphy, N. and Peacocke, A.R. (eds.), (1995) *Chaos and Complexity. Scientific Perspectives on Divine Action,* Vatican Observatory Foundation, Vatican City State.

Russell, R. J., Stoeger, W. R. and Ayala, F. J. (eds.), (1998) *Evolutionary and Molecular Biology: Scientific Perspectives on Divine Action,* Vatican Observatory and the Center for Theology and the Natural Sciences (a joint publication),Vatican City State/Berkeley, California.

Seckbach, J. and Chela-Flores, J. (2008) Preface 1: Where did we come from? In: Divine Action and Natural Selection: Science, Faith and Evolution. J. Seckbach and R. Gordon (eds.) World Scientific Publishers, Singapore, p. 30.

Tough, Tough, Allen, ed. (2000) *When SETI Succeeds: The impact of high information content,* Foundation for the Future: Washington.

References

Aretxaga, R. (2004) Astrobiology and Biocentrism, in *Life in the Universe*, J. Seckbach, J. Chela-Flores, T. Owen and F. Raulin (eds.), Series: Cellular Origin and Life in Extreme Habitats and Astrobiology **7**, Kluwer Academic Publishers, Dordrecht, The Netherlands, 345-348.

Atetxaga, R. (2006) La ciencia astrobiológica: Un nuevo reto para el humanismo del siglo XXI, *Letras de Deusto* **36**, No 110, January-March 2008, 12-20.

Aretxaga, R. (2008) Astrobiología: entre la ciencia y la exploración, *Letras de Deusto* **38**, No 118, January-March 2008, 13-27.

Aretxaga-Burgos, R. and Chela-Flores, J. (2011) Cultural Implications of the Search and Eventual Discovery of a Second Genesis. In: *Origins: Genesis, Evolution and Diversity of Life* (2nd Edition), J. Seckbach and R. Gordon (eds.). Cellular Origin and Life in Extreme Habitats and Astrobiology, Springer, Dordrecht, The Netherlands. Submitted by invitation.

Billingham, J., Heins R., Milne, D., Doyle, S., Klein, M., Heilbron, J., Ashkenazi, M., Michaud, M., Lutz, J. y Shostak, S. (eds.) (1994) *Social Implications of the Detection of an Extraterrestrial Civilization.* California: SETI Press, SETI Institute.

Carroll, S.B. (2001) Chance and necessity: the evolution of morphological complexity and diversity, *Nature* **409**, 1102-1109.

Chela-Flores, J. (2001) *The New Science of Astrobiology from Genesis of the Living Cell to Evolution of Intelligent Behavior in the Universe*, Kluwer Academic Publishers, Dordrecht, The Netherlands.

Chela-Flores, J. (2009). *A Second Genesis: Stepping stones towards the intelligibility of nature.* World Scientific Publishers, Singapore, 248 pp.

Chela-Flores, J. (2010) Instrumentation for the search of habitable ecosystems in the future exploration of Europa and Ganymede. *International Journal of Astrobiology* **9** (2), 101-108.

Coyne, SJ, G. V. (1998) *The concept of matter and materialism in the origin and evolution of life* in Chela-Flores, J. and Raulin, F. (Eds.), *Chemical Evolution: Exobiology. Matter, Energy, and Information in the Origin and Evolution of Life in the Universe,* Kluwer Academic Publishers, Dordrecht, pp. 71-80

Dick, S. J. (1998) *Life on Other Worlds,* CUP, London, pp. 245-256.

Dick, S. J. (2000) Cosmotheology, in Steven J. Dick (ed.), *Many Worlds: The New Universe, Extraterrestrial Life, and its Theological Implications*, Templeton Press, Philadelphia.

Dick, S. J. (2005) Cosmotheology revisited: theological implications of extraterrestrial life, *Cons-Ciências: actas do Fórum Internacional Ciência, Religião e Consciência*. Porto, Ediçōes Universidade Fernando Pessoa, CTEC, 2005, pp. 287-301.

Dick, S. J. (2008) The Postbiological Universe, *Acta Astronautica* **62**, No 8-9, April-May 2008, 499-504.

Gould, S. J. (1991) *Wonderful life. The Burgess Shale and the Nature of History.* Penguin Books, London, pp. 48-52.

John Paul II (1992) Discorso di Giovanni Paolo II alla Pontificia Accademia delle Scienze. L'Osservatore Romano, 1 November, p. 8.

John Paul II (1996a) Papal Message to the Pontifical Academy, *Commentarii* **4**, N. 3. Vatican City, 1997, pp. 15-20.

John Paul II (1996b) Papal Message to the Pontificial Academy of Sciences of 22 October 1996, *L'Osservatore Romano Weekly Edition.* N. 44, 30 October, p.3 and p. 7. (A translation from the official version in the French Language.)

Krubitzer, L. (1995) The organization of neocortex in mammals: are species differences really so different? *Trends in Neuroscience* **18**, 408-417.

Miret-Magdalena, E. (2008) Ciencia astrobiológica y pensamiento cristiano, *Letras de Deusto* **38**, No 118, January-March 2008, 58-67.

NASA Technical Memorandum: *Workshop on the Societal Implications of Astrobiology. Final Report*, 15 December 2000.

Peacocke, A. (1988) Biological evolution - a positive theological appraisal, in *Evolutionary and Molecular Biology: Scientific Perspectives on Divine Action.* R. J. Russell, W. R. Stoeger and F. J. Ayala, Editors. Vatican City State/Berkeley, California: Vatican Observatory and the Center for Theology and the Natural Sciences, pp. 357-376.

Race, Margaret S. (2008) Communicating about the discovery of extraterrestrial life: Different searches, different issues, *Acta Astronautica* **62,** 71–78.

Russell, B. (1991) *History of Western Philosophy and its Connection with Political and Social Circumstances from the Earliest Times to the Present Day,* Routledge, London, p.13.

Tough, A. (ed.) (2000) *When SETI Succeeds: The Impact of High-Information Contact.* Foundation for the Future, Washington, USA.

Townes, C. H. (1995) *Making Waves*, AIP, Woodbury, NY. pp. 157-167.

14

When astrobiology meets philosophy

In the 19th century not even the advent of theory of evolution could set speculative minds at rest regarding life's origins, either from the point of view of philosophy, or with the support of the scientific method. However, both philosophy and astrobiology are concerned with fundamental questions regarding the position of humans in the universe. But since the methods and training of both philosophers and astrobiologists are so different it is tactful to address in the present chapter the philosophical foundations and the main issues that are involved in the common frontier between these two sectors of modern culture. Our aim is to discuss the relevant issues when astrobiology meets philosophy.

The reader is advised to refer especially to the following entries in the Illustrated glossary: *Biogeocentrism, convergent evolution, De Duve (Christian) existentialism, Lyell (Charles), natural theology, reductionism, rationalism, Teilhard de Chardin (Pierre), Weinberg (Steven).*

14.1 Pasteur, Darwin and Wallace

The research of Louis Pasteur had brought upon the scientific community the urgency of rationalizing the question of the origin of life. It became evident, due to Pasteur's own work, that the concept of spontaneous generation was untenable. In 1860 Pasteur wrote to a friend on the "impenetrable mystery of life and death" (Holmes, 1961): "There is so much passion and so much obscurity on both sides that it will require nothing less than the cogency of an arithmetical demonstration to convince my adversaries of my conclusions."

When Pasteur was writing these lines, Charles Darwin had just published "*The origin of Species*". It was only with the publication of this fundamental work that the basic questions on the nature of life would be seen in their proper perspective. In fact, the fundamental question of "how life might have been breathed into matter", had to wait for two events that were to be well separated in time: First of all, Darwin had to have his work widely read and discussed by the scientific community. The publication of his theory of evolution had taken place the previous year, although it had been maturing ever since his trip on *The Beagle*. Darwin's publication was motivated by Alfred Russell Wallace (1823-1913), whose work was the fruit of travels as a naturalist

J. Chela-Flores, *The Science of Astrobiology: A Personal View on Learning to Read the Book of Life*, Cellular Origin, Life in Extreme Habitats and Astrobiology 20, DOI 10.1007/978-94-007-1627-8_14,

on the Amazon (1848-1852) and in the Indo-Australian archipelago (1854-1862), where he independently had developed a version of evolution by natural selection.

Secondly, the first steps in the understanding of the spontaneous appearance of life form basic chemical compounds had to wait for another hundred years. It was not until the advent of chemical evolution, based on the work of Oparin, Miller and others in the middle of last century, when scientists suggested the possibility of life spontaneously appearing on Earth somewhere in the remote past.

On the basis of the work of Darwin, Wallace, Oparin and the pioneers of chemical evolution that we have mentioned in Chapter 1, philosophers of all schools were to be in a position to discuss one of the deepest questions that science has raised up to the present time, namely, *What is life and how did it first appear on Earth?*

14.2 A philosophical issue: design in biology

We cannot repeat often enough in the present book that astrobiology's second objective is the *evolution* of life in the universe. Life on Earth is one certain example of life that can be understood in terms of evolution by means of natural selection. Hence, any account of the philosophical issues of the theory of evolution will be pertinent to astrobiology. For this reason we should take a closer look several issues that have been discussed for a considerable time. Amongst them we would like to highlight just one, for the purpose of illustration:

The question of design in biology has a long history going back to ancient Greece. However, in modern times we may begin with the work of William Paley (1743-1805) who was an Archdeacon, and Doctor of Divinity at Cambridge University. His writings were highly respected in the Anglican order. His *Horae Paulinae* was written in 1790 specifically to prove the historicity of the New Testament. Another famous book was *View of the Evidences of Christianity* (1794), a text that was standard reading amongst undergraduates during Charles Darwin early university education. However, his best-remembered book is *Natural Theology,* which played an important role in the early stages of the establishment of Darwin's arguments. Paley presented some observations from nature that was meant to prove not only the existence of a grand design, but more importantly, in his 'Natural Theology' Paley attempted to prove the existence of an intelligent designer. The famous quotation that follows is at the beginning of his book: "Suppose I found a watch upon the ground, and it should be enquired how the watch happened to be in that place...When we come to inspect the watch, we perceive that its several parts are framed and put together for a purpose...the inference, we think, is inevitable, that the watch must have had a maker..."

This argument can be traced back to classical times, but Paley's defense of it in modern times was influential in the 19th century dialogue between science, philosophy and theology. One of the fundamental steps in the ascent on man towards an understanding of his position in the universe (a key to understanding the present state of astrobiology) has been the realization that natural selection is indeed a creative process that can account for the appearance of genuine novelty, independent of a single act of creation, but more as a gradual accumulation of small successes in the evolution of living organisms. This is a point that has been defended by many of the founders of Darwinism, most recently others, who refer to an analogy that is artistic creation (Ayala, 1998). The creative power of natural selection arises, according to Jacques

Monod to an interaction between chance and necessity, a phrase that became familiar thanks to his very popular book *"Chance and Necessity"*. Consider, for instance, a painter who mixes and distributes pigments over a canvas. The artist does not create the canvas and pigments, but the painting is the creation of the artist. A random mixture of pigments could not have created Leonardo's *Mona Lisa*, or at least the probability is infinitesimally small. This underlines the fact that natural selection is like the painter-it is not a random process. The complicated anatomy of the human eye for instance, is the result of a non-random process: natural selection.

14.3 Towards a common interest in astrobiology and the humanities

We have seen in the Introduction and Chapter 1 that the present time is one of expansion of the number of people that are interested in the problem of the origin of life. Theologians have been deterred to get involved in this fascinating field mainly due to some reservations that can be traced back a long time to the fundamental question of how to read the Holy Books of the three monotheistic religions of the world, namely the Old Testament of the Holy Bible that is shared by Judaism, Christianity and Islam.

In Chapter 1 we also attempted to illustrate that in the Book of Genesis there are questions raised that are of interest to theologians, philosophers, scientists and artists. To extend the fortunate phrase of Lord Snow, we may refer to these four groups as the 'four cultures'. In this terminology, it is hardly surprising that the fourth culture (art) should have shown interest in Genesis.

The very rich iconography of Christianity and Judaism was a permanent source of funding for artists throughout the rise of Western civilization. The reason why the third culture (science) has been interested in the evolution of life is self-evident. Science took one of its most transcendental steps towards understanding the complexity of the biosphere when Darwin and Wallace and formulated natural selection as a mechanism for evolution. Progress since the publication of *The Origin of Species* has been considerable. The second culture, philosophy, has been intricately connected with the development of Darwinism. Philosophers such as Karl Popper have meditated deeply on the philosophical implications of the Theory of Evolution.

We wish to dwell at a certain length on the interest that the first culture, theology, has had on the question of the origin and evolution of life on Earth and the possibility of there being life elsewhere in the universe. Traditionally, there has been a certain caution of theologians with respect to the questions that we have discussed at some length in this work. Saint Augustine touched on this question in his book *"The City of God"* with respect to possible conflicts that may arise from a literal reading of the Bible (St. Augustine, 1984). In our own time a clear position was expressed at the Pontifical Academy of Sciences, which had met to discuss the origin and evolution of life. In this message, the subject of the present book was defined as John Paul II, 1996): "A basic theme which greatly interests the Church, as revelation contains teachings concerning the nature and origins of man".

Raising the Augustinian conundrum of whether scientifically reached conclusions, and those contained in revelation on the origin of life, seem in contradiction with each other: "In what direction should we seek their solution? We know in effect that truth cannot contradict truth". This position has opened the way to a fruitful dialogue between scientists and theologians, two cultures, which are not too distant from each other (Polkinghorne, 1996). New understanding is arising between cultures that were once distant from each other. Such progress can only open the way to enhancing the subject of the origin, evolution and distribution of life in the universe, a field of research that is bound to continue its robust growth.

14.4 Is there life elsewhere in the universe?

It is useful to reflect on the philosophical consequences of the *biogeocentric point of view* that has been advocated in the past by both evolutionists (Mayr, 1995) and molecular biologists (Monod, 1972). Mayr went as far as assigning extraterrestrial life an "improbability of astronomical dimensions". On the other hand, Monod believes that: "The present structure of the biosphere certainly does not exclude the possibility that the decisive event occurred only once". The biogeocentric position has been maintained from the point of view of paleobiology (Conway-Morris, 1998): "If indeed we are alone and unique, and this possibility, however implausible, cannot yet be refuted, then we have special responsibilities".

Almost a century and a half separates us from the beginning of philosophical thinking in modern times. In fact, Sir Charles Lyell was concerned with the position of man amongst all living organisms. As we saw in Chapter 5, it seems that the underlying difficulty in the dialogue of faith and reason is still related to inserting Lyell's "ugly facts" of Darwinian evolution into our culture. Darwin was prudent enough to avoid ideological issues. He even avoided philosophical issues. Bertrand Russell in Religion and Science inevitably raised these issues subsequently: "From evolution, so far as our present knowledge shows, no ultimately optimistic philosophy can be validly inferred" (Russell, 1997).

Darwin himself concentrated on the narrow, but transcendental problem of establishing the theory of evolution of life on Earth and prudently postponed the wider issue of the position of man in the universe. As we mentioned in Chapter 13, at the beginning of a new era in space research (Burrows, 1998), a much wider problem remains to be solved, namely we still do not know what is the position of our tree of life in the forest of life. However, we may still have to wait for preliminary progress in the still undeveloped science of the distribution of life in the universe. One of the main difficulties is deciding the position of man in our parochial tree of life. General consensus on this issue has been missing since Lyell's time. Clearly, if we are not alone in the universe, there are some unavoidable theological and philosophical consequences for which we still have no answer (cf., Chapter 13). However, we are convinced that discovering extraterrestrial life would induce a fruitful dialogue between sectors of culture that are not always willing to approach one another. For this reason we feel that the problem of extraterrestrial life is one of the most important questions ever raised.

We hope to have conveyed to the reader the multiple hints from science that such discoveries are likely to take place sometime in the future. The implications are likely to have an impact in our culture requiring adjustments possibly more radical than those arising form the evidence that humans descend from microorganisms.

Recalling the two main subtopics of astrobiology that we have been discussing in previous pages, namely, distribution and destiny of life in the universe, what we have argued is that if evolution experiments were to be successful, the science of the distribution of life in the universe would lie on solid scientific bases. This concludes the first of the two topics I wished to comment upon. Given such bases for the distribution of life in the universe, it does not seem premature to include in our discussions other sectors of our society. Keeping this in mind, we move on to the second topic of the present work, the destiny of life in the universe.

Indeed, if biology experiments such as the ones we have discussed in this book succeed in the future, they would undoubtedly have a significant impact in our culture, not just in our scientific outlook. The influence of the new knowledge will also be felt while discussing the deep questions raised in Chapter 13. These questions may be discussed not just against a background of our particular evolutionary line, which has been followed up by life on Earth. Such questions ought to be discussed already in terms of the many evolutionary lines that are hinted at by astrobiology. In fact, the intercultural dialogue is urgent for various reasons, which are strongly suggested by astrobiology:

• A human-level type of intelligent behavior may be widespread in the cosmos,
• Within the context of astrobiology it is clear that our human descent does go back all the way to microorganisms and,
• Ultimately, our origins go back to star dust.

Yet, these three items lie outside our cultural patrimony. But it is not an easy task to integrate our knowledge. In this work, my main overall thesis has been that such an integration is not only possible, but it is also timely, and necessary. As science progresses, the dialogue with religion is enriched, and previous apparent controversies disappear. The moment we accept that the frontiers of science are clearly delineated, all controversy disappears. For science is a means of formulating reasonable hypotheses that are only retained until our experimental ability (or our observations) improve, so that we can go on to make better hypotheses that will make nature more comprehensible.

Science is justified to remain outside the scope of theology, since science aims to develop technologies that make our lives better, and it gives us a better understanding of nature. Theology, on the other hand, helps us to understand and rationalize our traditions that ultimately we believe are based on revelation, which is clearly beyond the experimental method that constrains all scientific activities.

14.5 Can our intelligence be repeated elsewhere?

In an extraterrestrial environment the evolutionary steps that led to human beings would probably never repeat themselves. However, the possibility remains that a human level of intelligence may be favored when the combined effect of natural selection and cultural evolution are taken together. This is independent of the particular details of the phylogenetic tree that may lead to the intelligent (non human) organism. Conway-Morris in *The Crucible of Creation* states it briefly: the role of contingency in evolution has little bearing on the emergence of a particular biological property (Conway-Morris, 1998). To illustrate the inevitability of the emergence of particular biological properties we may use examples of convergent evolution, a phenomenon that has been recognized by students of evolution for a long time (Tucker Abbott, 1989).

In the phylum of mollusks, the shells of both the camaenid snail from the Philippines, and the helminthoglyptid snail from Central America resemble the members of European helcid snails. These distant species (they are grouped in different Families), in spite of having quite different internal anatomies, have grown to resemble each other outwardly over generations in response to their environment. In spite of considerable anatomical diversity, mollusks from these distant families have tended to resemble in a particular biological property, namely, their external calcareous shell.

Swallows provide a second example (Passeriformes). This group is often confused with swifts (Apodiformes), but is not related to them. Members of these two orders differ widely in anatomy and their similarities are the result of convergent evolution on different stocks that have become adapted to the same life styles in ecosystems that are similar to both species.

In the light of these examples, the question of whether our intelligence is unrepeatable goes beyond biology and the geological factors mentioned in the metaphor on the repetition of the history of evolution. Indeed, the question is rather one in the domain of the space sciences; in which the radio astronomers have led the way with the SETI project (Chapter 11). The question of whether we are alone in the cosmos concerns astrometry measurements for the search for extrasolar planets. This activity has led to the current revolutionary view that planets of the Solar System are not unique environments that may be conducive to the origin and evolution of life in the universe (Chapter 10)· The presence of a dozen planets in the cosmic neighborhood of the Sun, does not answer positively the question whether we are alone in the universe, but extrasolar planets increase the possible sites where life, given the right ingredients, may evolve. Progress in observational astronomy in the foreseeable future with new projects will begin to provide some answers, such as the possible presence of water and oxygen.

A separate question, much closer to our capability to perform practical experiments, concerns the search for microbial life in the Solar System. So, we believe it is appropriate to shift our attention away from "attempting a full and coherent account of the phenomenon of man" (Teilhard de Chardin, 1965 and Illustrated Glossary). Instead we should focus our attention at the level of a single cell. Indeed, we feel that the progress of molecular biology forces upon us a search for a full and coherent account of eukaryogenesis, the first transcendental transition in terrestrial evolution at the cellular level which led to intelligence. The cosmic search for extraterrestrial intelligence ought, in our opinion, begin with a single step, namely, the search for the first cellular transition on the pathway to multicellularity, and inevitably to brains (due to their selective advantage). This emphasis on the eukaryotic cell as a "cosmic imperative" has been called the phenomenon of the eukaryotic cell (Chela-Flores, 1997; 1998)· The task of understanding the origins of the eukaryotic cell is not easy (De Duve, 1995). But let us at least dwell on clarifying the terms being used.

14.6 Some of the larger issues when philosophy meets astrobiology

Although astrobiology is far from reaching its maturity, it lies squarely within the frontiers of traditional research (cf., Aretxaga and Chela-Flores, 2011 for a more comprehensive review of the topics we discuss below). Some of the deepest questions raised within astrobiology lie close to those raised within the humanities. From the point of view of philosophy and theology, it is conceivable to view conventional science as one aspect of a wider empiricism that would take into account such facts as the intelligibility of the universe. Some humanists feel that a search for a rich empiricism could possibly rationalise what lies beyond the scientific approach.

The implications of a second Genesis are likely to be relevant, not only to astrobiology, but also ethics and especially for philosophy and theology. Such a discovery would provide significant additions to our insights on the intelligibility of nature. Intelligible means the capability of being understood, or comprehended. Alternatively, intelligible can signify to be apprehensible by the intellect alone. A third aspect of intelligible is related to something that is beyond perception. An "intelligible universe" can be the starting point of a prolonged and systematic discussion amongst astrobiologists, as well as philosophers and theologians (Chela-Flores, 2009).

The Belgian Nobel Laureate Christian de Duve, at the end of a review on the origin and evolution of life, asks himself the question: "*What does it all mean?*" not only in science (De Duve, 2002), but also in philosophy (Dear, 2006) and in theology (Russell, 2001). The intelligibility of the universe raises questions that lie on the frontier between science and the humanities; and we need an open all-embracing approach.

Considerations of intelligibility return to the often quoted, but less frequently debated statement of Steven Weinberg, the American Physics Nobel Laureate: *The more the universe seems comprehensible, the more it also seems pointless* (Weinberg, 1977 and Fig. 14.1).

Fig 14.1 Steven Weinberg during a talk at the ICTP

Such pessimism seems to leave no room for a first Genesis on Earth as a special case of a wide and meaningful phenomenon of second genesis elsewhere. We should clarify that the quoted statement reflected a specific philosophical trend, or attitude, characteristic of the first half of last century. Hence, if that were the case, the statement should not discourage the general dialogue at the frontier of science and religion. The common approach from both ends of the academic discourse to discovering a second Genesis anywhere else in the universe should contribute, as we shall see below, to the progress of the philosophy and theology. Reciprocally, the questions raised in any field of the humanities arising from the discovery of a second Genesis could, in turn, enrich the search for the place of humans in the universe. The above quotation can be best understood in the context of a philosophical trend called existentialism.

But we should go further back in time to come to grips with the larger philosophical issues of a second Genesis. Indeed, Western philosophy began raising the question of the intelligibility of nature, in such a manner that the first Greek philosophers were called physicists. These first thinkers conceived the natural universe as an ordered set of all the beings (*cosmos*) that arose from a first principle (*arche*). An inexorable dynamic guiding principle made it possible to transform the *arche* in *cosmos*. In this scenario the human being considered himself as a privileged manifestation of the natural universe: in which the structure and ordering principle of everything in existence becomes intelligible. To know consisted, therefore, in discovering the structure of nature, as well as comprehending the reason for such structure.

During the Middle Ages Western thought was constrained by the Christian faith. In this context, higher knowledge was *wisdom*, which consisted in discovering the structure of reality, and at the same time and intimately related, in confirming that such structure had its basis in God's Intelligent Action. The natural universe was considered God's *creation*, which implied that the universe was intelligible and had a meaning. During the Renaissance and through the scientific revolution of the 16th to 18th centuries, Western reason went through a process of replacing the contribution of faith by a contribution of observation and experimentation, thus giving rise to the scientific method. During this long and laborious process, the assumption that the universe was intelligible and meaningful was firmly maintained. This remark rationalizes the Galileo statement that nature was a book written in mathematical language.

Supplementary Reading

Chela-Flores, J. (2009). *A Second Genesis: Stepping stones towards the intelligibility of nature.* World Scientific Publishers, Singapore, 248 pp.

Colombo, R., Giorello, G. and Sindoni, E. (1999) *Origine della Vita Intelligente nell'Universo (Origin of intelligent life in the universe).* Edizioni New Press, Como.

Conway-Morris, S. (1998) *The Curcible of Creation. The Burgess Shale and the Rise of Animals,* Oxford University Press: New York, pp. 9-14.

Eligio, Padre, Rigamonti, G. and Sindoni, E. (1997) *Scienza, Filosofia e Teologia di Fronte alla Nascita dell'Universo (Reflections on the Birth of the Universe: Science, Philosophy and Theology).* New Press, Como.

Davies, Paul (1994) *Siamo Soli? Implicazioni filosofiche della scoperta della vita extraterrestre,* Editori Laterza, Roma.

Drake, F. and Sobel, D. (1992) *Is there anyone out there? The scientific search for Extraterrestrial Intelligence,* Delacorte Press, New York, pp. 45-64.

Impey, C. and Petrry, C. (eds.). (2002) *Science and Theology: Ruminations on the Cosmos.* Vatican Observatory and Templeton Foundation.

John Paul II (1997) *The Holy Father Message to the Pontifical Academy, Commentari* **4,** No. 3, Part I, The origin and early evolution of life, Pontificia Academia Scientiarvm, Vatican City, pp. 15-20.

Mayor, M., Queloz, D., Udry, S. and Halbwachs, J.-L. (1997) From Brown Dwarfs to planets, in C.B. Cosmovici, S. Bowyer and D. Werthimer (eds.), *Astronomical and Biochemical Origins and the Search for Life in the Universe,* Editrice Compositore, Bologna, pp. 313-330.

Russell, R. J., Stoeger, W. R., (SJ). and Ayala, F. J. (eds.) (1998) *Evolutionary and Molecular Biology: Scientific Perspectives on Divine Action.*Vatican City State/Berkeley, California: Vatican Observatory and the Center for Theology and the Natural Sciences

Russell, R.J., Stoeger (SJ), W.R., and Coyne (SJ), G.V. (eds.), (1995) *Physics, Philosophy and Theology. A common quest for understanding,* (2nd. ed.) Vatican Observatory Foundation: Vatican City State.

References

Aretxaga-Burgos, R. and Chela-Flores, J. (2011) Cultural Implications of the Search and Eventual Discovery of a Second Genesis. In: *Origins: Genesis, Evolution and Diversity of Life* (2nd Edition), J. Seckbach and R. Gordon (eds.). Cellular Origin and Life in Extreme Habitats and Astrobiology, Springer, Dordrecht, The Netherlands. Submitted by invitation.

Ayala, F.J. (1998) Darwin's Devolution: Design without Designer, in *Evolutionary and Molecular Biology: Scientific Perspectives on Divine Action***,** Robert John Russell, William R. Stoeger, S.J. and Francisco J. Ayala (eds.), Vatican City State/Berkeley, California: Vatican Observatory and the Center for Theology and the Natural Sciences. pp. 101-116.

Burrows, W. E. (1998) *This New Ocean: The Story of the First Space Age,* Random House, London.

Chela-Flores, J. (1997) Cosmological models and appearance of intelligent life on Earth: The phenomenon of the eukaryotic cell, in Padre Eligio, G. Giorello, G. Rigamonti and E. Sindoni (eds.), *Reflections on the birth of the Universe: Science, Philosophy and Theology,* Edizioni New Press, Como, pp. 337-373.

Chela-Flores, J. (1998) The Phenomenon of the Eukaryotic Cell, in R. J. Russell, W. R. Stoeger, and F. J. Ayala (eds), *Evolutionary and Molecular Biology: Scientific Perspectives on Divine Action,* Vatican City State/Berkeley, California and Vatican Observatory and the Center for Theology and the Natural Sciences, pp. 79-99.

Chela-Flores, J. (2009). *A Second Genesis: Stepping stones towards the intelligibility of nature.* World Scientific Publishers, Singapore, 248 pp.

Conway-Morris, S. (1998) *The Curcible of Creation. The Burgess Shale and the Rise of Animals,* Oxford University Press: New York, pp. 222-223.

Dear, P. (2006) *The Intelligibility of Nature How Science Makes Sense of the World,* The University of Chicago Press, Chicago, USA, 254 p.

De Duve, C. (1995) *Vital dust: Life as a cosmic imperative,* Basic Books, New York, pp. 160-168.

De Duve, C. (2002) *Life Evolving Molecules Mind and Meaning.* New York, OUP.

Henderson, L. J. (1913) *The Fitness of the Environment An Enquiry into the Biological Significance of the Properties of Matter*, Peter Smith, Gloucester, Mass., 1970, p. 312.

John Paul II (1992) Discorso di Giovanni Paolo II alla Pontificia Accademia delle Scienze. *L'Osservatore Romano,* 1st November, p. 8.

Mayr, E. (1995) The search for extraterrestrial intelligence in Extraterrestrials. Where are they?, in B. Zuckerman and M.H. Hart (eds.), (2nd. ed.), Cambridge University Press, London, pp. 152-156.

Monod, J. (1972) *Chance and Necessity An Essay on the Natural Philosophy of Modern Biology,* Collins, London, p. 136.

Polkinghorne, J. (1996) *Scientists as theologians,* SPCK, London.

Russell, B. (1997) *Religion and Science,* Oxford University Press, New York, pp. 49-81.

Snow, C. P. (1993) *The Two Cultures*, Cambridge University Press.
Teilhard de Chardin, P. (1965) *The phenomenon of man,* Fontana Books, London, p. 33.
Tucker Abbott, R. (1989) *Compendium of landshells,* American Malacologists, Melbourne, Florida, USA, pp. 7-8;
Weinberg, S. (1977) *The First Three Minutes*, Fontana/Collins, London.

15

Why we may be unable to read the complete Book of Life

The current incomplete understanding of the evolution of the cosmos sets a limit to astrobiology's major aim of understanding the last two chapters of the Books of Life, namely the distribution and destiny of life in the universe. To illustrate this fundamental limitation we review briefly the present search for a complete theory of the basic forces of nature that could give us a deeper understanding of both cosmic evolution and the ultimate aims of astrobiology. This chapter should produce most rewarding reading. However, the reader should be prepared to do a reasonable amount of preparation by consulting the suggested items in the Illustrated glossary:

Dispersion relations, cross section, Dirac (Paul), electroweak interactions, Feynman diagrams, Glashow (Sheldon), Heisenberg (Werner), Kant (Immanuel), M- theory, Mandelstam (Stanley), Nambu (Yoichiro), Planck (Max), Plato, potential scattering, quantum chromodynamics, quantum electrodynamics, quantum mechanics, Regge poles and trajectories, Regge (Tulio), Salam, scattering amplitude, Schwinger (Julian), standard theory, string theory, Veneziano (Gabriel), Weinberg (Steven).

15.1 Evolution of the universe and its contents including the living process

After our preliminary contact with cosmology in Chapter 1 we need to look a little closer at current developments of the study of the fundamental forces, since such studies are themselves work in progress. Due to some of the limitations of the search for unification of the fundamental forces, corresponding limitations begin to emerge in our own general program of astrobiology.

In order to understand why these two scientific approaches to the evolution of the universe and its contents (including the living process), it is inevitable to invite our readers to pursue the ideas that are discussed in this chapter. They should not be discouraged by the intrinsic difficulties of a science that unlike astrobiology is based in more mathematical language that is typical of other scientific areas. For this reason, as mentioned above, we encourage the readers to follow our invitation to precede the reading of this chapter by browsing through the *Illustrated glossary*.

Inflation is the assumed rapid (exponential) expansion of the early universe by a factor of at least 10^{78} in volume. This lasted from 10^{-36} seconds after the big bang to sometime between 10^{-33} and 10^{-32} seconds. Following the inflationary period, the

J. Chela-Flores, *The Science of Astrobiology: A Personal View on Learning to Read the Book of Life*, Cellular Origin, Life in Extreme Habitats and Astrobiology 20, DOI 10.1007/978-94-007-1627-8_15,

universe continues to expand. Andre Linde envisions a vast cosmos in a model in which the current universal expansion is a bubble in an infinitely old "superuniverse" (Linde, 1982). The scale parameter R (cf., Chapter 1) evolves as a function of time t in such a cosmology, but as in the model of Alan Guth (Guth, 1981), the Linde model differs from the Friedmann solution, already discussed in Chapter 1, in the first instants of cosmic evolution:

While the possibility of an exponential expansion of the early universe had been noted before, it was Guth who realized that inflation would solve some of the major problems confronting the big bang cosmology. Difficulties with the original inflationary model were recognized by Guth and others, and were overcome with the introduction of "new" inflation by Linde and Steinhardt (with Andreas Albrecht and Steinhardt, 1982). Linde went on to propose other promising versions of inflationary theory, such as chaotic inflation (cf., Fig. 15.1).

The greatest success of inflationary theory has been in accounting for the existence of inhomogeneities in the universe and predicting their spectrum. In other words, in this model the universe we know is assumed to be a bubble amongst bubbles, which are eternally appearing and breeding new universes. Inflation makes a number of predictions that have been confirmed by observation and is now considered part of the standard big bang cosmology.

Fig. 15.1 From left to right K.R. Sreenivasan (the Abdus Salam ICTP Director), Alan Guth, Paul Steinhardt and Andrei Linde during the award of the 2002 Dirac Medal for the development of the concept of inflation in cosmology

Depending on the way the word "standard" is used, both the Linde and Guth models can be considered standard big bang models.

15.2 Constraints on the universality of biology

We have failed so far to have a unified physical theory that would give reliable constraints on the nature of cosmic evolution everywhere. We will consider this question that is vital for the eventual growth of our understanding of the universality of biology. We first return to cosmic evolution based on Einstein's theory of general relativity, which was reviewed in Chapter 1 and in the earlier book (Chela-Flores, 2009), on which this section is based. While in classical Greek philosophy Thales of Miletus and Plato made attempts to understand the cosmos, the earliest scientific accounts were developed within the gravitational theory of Sir Isaac Newton, notably by Newton himself and later by and Emmanuel Kant. The contemporary cosmogonical account has been tested in space by probes of the main space agencies of Europe and the United States. They have successfully tested the bases of Albert Einstein's theory of gravitation.

From Chapter 2 we have seen how measurements of the CMB yield some information on the earliest stages of the universe. Our present understanding of the composition of the universe in terms of dark matter and dark energy, on the other hand remains a challenge for the future. Our insights into dark matter have arisen as follows. All visible and microscopic form of matter is made up of protons, neutrons and electrons. Protons and neutrons are bound together into nuclei. Atoms are nuclei surrounded by electrons. For instance, hydrogen is composed of one proton and one electron. Helium is composed of two protons, two neutrons and two electrons. Carbon is composed of six protons, six neutrons and six electrons. Heavier elements, such as sulfur, iron, lead and uranium, contain even larger numbers of protons, neutrons and electrons. The universe is not composed entirely of this "baryonic matter". There is evidence that suggests there is an additional form of matter that has been called "dark matter" (Quinn and Nir, 2008).

This concept arose thanks to the insights of Fritz Zwicky (1898-1974). He was a Bulgarian astrophysicist, who held a professorial position at the California Institute of Technology from 1925 till 1968. He was responsible for introducing the concept of missing mass. In collaboration with Walter Baade he based his observations on the Coma Cluster that is some 300 million light years away. This is one of two clusters that lie in the constellation of Coma Berenices (the other is the Virgo Cluster). It is a remarkable cluster as it contains over a thousand galaxies with an average distance between them that is about three times smaller than the distance between the Milky Way and Andromeda.

Zwicky was the first to realize that there should be a great deal of invisible matter to make the cluster stable. Such invisible matter is conjectured to be one form of what we now call dark matter. Indeed, the mass inferred for galaxies, including our own, is roughly ten times larger than the mass that can be associated with stars, gas and dust in a Galaxy. This mass discrepancy has been confirmed by observations of the effect of gravitational lens, the bending of light predicted by the theory of general relativity. By measuring how the foreground cluster distorts the background galaxies, we can measure the mass in the cluster. The mass in the cluster is more than five times larger than the inferred mass in visible stars, gas and dust (baryonic matter). The remaining mass contribution of the cluster arises from matter that, as we mentioned above, we call dark matter that is matter exerting a gravitational attraction, but does not emit nor absorb light, as can be inferred form the Hubble Space Telescope images from which these studies have been conducted. For theoretical reasons some candidates for dark matter have been suggested: brown dwarfs, supermassive black holes and even new forms of matter.

Secondly, dark energy is still another challenge. In 1998 two groups studying supernovae showed that a fraction of energy of the cosmos is accelerating the expansion of the Universe (Reiss *et al.*, 1998; Perlmutter *et al.*, 1999). Subsequent work revealed that dark energy may make up about 70 percent of the Universe, but our current understanding of cosmology based on the general theory of relativity is unable to explain this unknown form of energy.

Our hope for progress is based on the many missions that the major space agencies are planning. For example, a more accurate gravitational theory would emerge from the Laser Interferometer Space Antenna (LISA), if this mission were approved. This was initially jointly sponsored by ESA, as a Cornerstone mission in their "Cosmic Vision Program", and NASA's Structure and Evolution of the Universe 2003 Roadmap, "Beyond Einstein: From the Big Bang to Black Holes." LISA would test the Theory of General Relativity, probe the early Universe, and last but not least, LISA will search for gravitational waves. LISA is a space-based gravitational-wave observatory, capable of detecting waves generated by binaries within the Milky Way, and by waves generated by massive black holes in other galaxies. LISA would use a system of laser interferometry by directly detecting and measuring gravitational waves.

15.3 Extending Newton's mechanics into the microscopic domain

The realization by Max Planck that Newton's mechanics fails at the atomic size eventually led to a new mechanics called quantum mechanics for the following reasons: Planck originally suggested in 1900 that to understand the phenomenon of color change when objects get hotter (technically known as the black-body radiation), it is forced upon us to assume that the surface of the hot body emits light in little bundles called "quanta". This led to the present science describing the behavior of matter and light on the atomic, nuclear and subnuclear levels.

Niels Bohr (1885-1962) was a Danish physicist who went beyond Planck by applying the concept of discrete energy states to the atomic nucleus. He received the Nobel Prize for Physics in 1922. His contribution is referred to in the literature as the Bohr theory of the atom, which was superseded by subsequent work of Heisenberg and Schrödinger, but was nevertheless a major factor in the evolution of our understanding of the microscopic world. Indeed, in 1913 Niels Bohr had introduced a model of the atom by going beyond Planck in an attempt in the realm of microscopic physics to describe an atom as a small, positively charged nucleus at the centre of electrons bound in circular orbits, in perfect analogy with the Solar System.

The crucial difference being that electrostatic forces replace the law of gravitation. It began to be taken seriously as the Rydberg formula for the spectral emission lines of atomic hydrogen was understood theoretically. Not only did the Bohr model explain the reason for the structure of the Rydberg formula, it also provided a justification for its empirical results in terms of fundamental physical constants using the quantum concept of Max Planck. Later on, in 1925 though there appeared two original approaches that were compatible with the increasing knowledge of the microscopic world since the early efforts of Planck and subsequently Niels Bohr:

• Matrix mechanics, proposed by Werner Heisenberg and other German scientists (cf., Fig. 15.2).

• Wave mechanics was developed by the Austrian physicist Erwin Schrödinger (1887-1961). He showed that, in spite of apparent dissimilarities, the matrix mechanics and his own wave mechanics are equivalent mathematically. He shared the 1933 Nobel Prize for Physics with Paul Dirac. In 1927, he succeeded Max Planck at the Friedrich Wilhelm University in Berlin. In 1940 he received a personal invitation from Ireland's Taoiseach Éamon de Valera to help establish an Institute for Advanced Studies in Dublin (DIAS). He became the Director of the School for Theoretical Physics, DIAS, where he remained for 17 years. In 1944, he wrote *What is Life?*, which suggested the concept of a complex molecule with the genetic code for living organisms.

According to both James D. Watson (in *DNA, the Secret of Life)* and Francis Crick, (in *What Mad Pursuit)* Schrödinger's book gave them inspiration that eventually led to the discovery of the DNA. The Schrodinger Equation is the basic contribution. This is an equation that describes how the microscopic (quantum) state of a physical system changes in time. The quantum state is also called a wave function, which is a complete description of a microscopic system.

Fig. 15.2 Werner Heisenberg lecturing during the Symposium on Contemporary Physics, Trieste, June 1968

15.4 From quantum electrodynamics to quantum chromodynamics

Our insights into the evolution of the universe are based on the interaction of the elementary particles of the subnuclear world. The theories that allow us to probe the initial instants of the evolution of the universe go back to the great progress of the early 20th century with the quantum theories. Indeed, quantum mechanics coupled with the special theory of relativity of Albert Einstein was the achievement of Paul Dirac in 1926 (cf., Fig. 15.3).

Fig. 15.3 Julian Schwinger first on the left in the front row listening to Paul Dirac during the Symposium on Contemporary Physics, Trieste, June 1968

His theory received the name of quantum field theory. The basic difficulties that atomic structure presented earlier at the onset of the revolution of Niels Bohr, Werner Heisenberg and others, received satisfactory treatment. This progress in understanding the atomic world of electrons and the electromagnetic field and the electrons amongst themselves led to the Physics Nobel Prize to Richard Feynman, Sinitiro Tomonaga and Julian Schwinger in 1965. The Dirac Equation describes electrons reconciling quantum mechanics and the theory of special relativity.

The theory was developed in the 1940s based on the idea that charged particles (electrons and positrons) interact by emitting and absorbing photons, the particles of light that transmit electromagnetic forces. These photons cannot be detected in any way because their existence violates the conservation of energy and momentum- they are called virtual particles.

The particle exchange is interpreted as the particle interaction, since they change momenta as they release or absorb the photonic energy. Photons also can be emitted in an observable state. Quantum electrodynamics (QED) has become a model for other quantum field theories, including the electroweak interaction and the theories that incorporate the strong nuclear force (cf., Sec. 15.5).

15.5 The unification of the elementary forces

The interaction that is responsible for the decay of the neutron into a proton, an electron and a neutrino ("beta" decay) was called the weak force: It is five orders of magnitude smaller than the electromagnetic interaction. Sheldon Glashow, Abdus Salam and Steven Weinberg proposed a theory that unified both of these interactions, which we refer to as the electroweak interaction (cf., Figs. 4.3, 14.1 and 15.4).

Fig 15.4 Sheldon Glashow during a talk at the ICTP in 1986

The electroweak theory predicted the existence of the Higgs boson, a massive elementary particle in terms of which we can understand the difference between the massless photon, which mediates electromagnetism, and the massive W and Z particles that mediate the weak force. With a field theory of the strong nuclear force that keeps the nucleus together (quantum chromodynamics QCD), these theories constitute the Standard Model.

Instead of the emission and subsequent absorption of photons, the strong force acts upon quarks that are bound together in the protons and neutrons of the atomic nucleus. By analogy with QED, in QCD quarks are the constituent elements of particles that are subject to the nuclear force, he so-called "hadrons". The equivalent of the electric charge is called color (hence the name of the theory as QCD). Since the electron has instead an electric charge, it is not subject to the strong nuclear interaction.

There were intrinsic difficulties due to the presence of infinite values that were obtained when physical quantities were evaluated, including the electric charge. Fundamental progress in this direction was due to work in the 1950s of Abdus Salam and independently Freeman Dyson and others under the name of "Renormalization Theory". These difficulties reappeared beyond QED especially when the other nuclear forces were taken into account, as in QCD. In these circumstances some scientists began to move away from the pathway that Dirac had started, but perhaps beyond the difficulty with infinities there was a more stimulating influence for progress, namely observational and experimental difficulties that had to wait for a new generation of instrumentation.

15. 6 Hopes for clarification from new instrumentation

In addition the instrumentation from the physics of high energies may also offer further help in improving our understanding of the microscopic world at distances much smaller than those envisaged by quantum mechanics. The progress, as it has always happened in the physical sciences is now due to much more sophisticated instrumentation.

At a depth of about 100 m experiments are able to collide two beams of particles head-on recreating the conditions in the Universe moments after the big bang. Showers of new particles are revealing new physics beyond the standard model. The main aspects of the Large Hadron Collider (LHC) experiment are the LHC detectors and the LHC particle beams. They are named LHCb, Alice, Atlas and the Compact Muon Solenoid (CMS). While LHCb and Alice are designed to investigate specific physical phenomena, Atlas and CMS are designated "general purpose" detectors. They will both aim to identify the elusive Higgs boson important for the validity of the standard model. Its microscopic description incorporates the interactions of three sub-atomic forces: they include the force that holds the atomic nucleus together, the force responsible for nuclear beta decay and electromagnetism.

With the progress of observational cosmology, we now realize that the standard model cannot incorporate gravity. It is also restricted to ordinary matter, which makes up only a small part of the Universe, excluding implicitly dark matter. The LHC aims to add a missing ingredient of the standard model: A hypothetical subnuclear particle that explains why all other particles have mass. According to the standard model, particles acquire their mass by means of a mechanisms theorized by Peter Higgs in 1964 with a mechanism that explains how two types of particles, massless like everything else immediately after the big bang, came to acquire different masses as the universe cooled down. The Higgs particle could shed light on the still-to-be understood nature of dark energy. With the help of Atlas and CMS will search for new kinds of particles. New forces and new types of particles may be the constituents of dark matter. One of the primary motivations for LHC is to try to produce this matter in Geneva. Since the universe was very dense and hot in the early moments following the big bang, the universe itself was a "particle accelerator". Dark matter may be made of weakly interacting massive particles that were produced shortly after the big bang. These areas of subnuclear physics that still represent challenges to our understanding of the microscopic world have encouraged new physical concepts that go back a few years to the work of Stanley Mandelstam, Tullio Regge, Gabriel Veneziano and Miguel Angel Virasoro, amongst others.

15.7 Influential steps other than quantum field theory

In 1937 John Archibald Wheeler‘s 1937 introduced the S-matrix. Four years later Werner Heisenberg went on to suggest this approach to the strong nuclear force as a principle of the interactions of the fundamental particles. It was considered to be an alternative to quantum field theory as the basic principle of elementary particle physics. Its renaissance in the 1960s was due to several scientists, including Mandelstam. S-matrix theory avoided the notion of space and time by replacing it with abstract mathematical properties of the S-matrix by relating the infinite past to the infinite future in one step, without intermediate steps, as in the Feynman diagrams.

Stanley Mandelstam, a South African physicist introduced the relativistically invariant variables into particle physics in 1958 (cf., Fig. 15.5 and Mandelstam, 1958).

Fig. 15.5 the South African physicist Stanley Mandelstam during the award of the 1992 Dirac Medal in Trieste (1992)

This contribution turned out to be a convenient coordinate system for formulating his double dispersion relations for scattering amplitudes (called the "Mandelstam Representation"). This result was a central tool in the program that attempted to formulate a consistent theory of infinitely many particles with certain quantum properties. Starting with Mandelstam in 1958, theoreticians began to prefer looking directly to the measurable parameters in the collisions (cross-sections) in the scattering at high energies. This led to revive an approach called the S-Matrix Theory in which the mathematical theory of complex variables played a leading role, instead of proceeding along the lines started by Paul Dirac, with quantum field theory. A series of original discoveries were at the frontier of mathematics and physics. In the 1960s these developments led to concepts, which were to influence the first decades of the 21st century.

This approach differed from the Dirac field theory highlighting analytic properties of scattering amplitudes as attractive features that could be understood fully in a similar theory called "potential scattering", essentially in the non-relativistic approach of quantum mechanics that Erwin Schrodinger had initiated in 1926 with his wave equation (Chela-Flores, 1967; 1970).

There were some criticisms at the time (Mandelstam, 1963; Feynman, 1965). Feynman pointed out that with Mandelstam Representation we cannot discuss any production reaction, and cannot accommodate virtual states with more than two particle; one must throw away terms without being able to assess their importance.

15.8 String theory as a basis for reading the Book of Life

In this section we shall discuss further the implications of the most accepted theory that has emerged from the efforts of unifying all the fundamental forces of nature following the significant contribution of Mandelstam and other physicists.

It was pointed out that for scattering of unequal mass particles there might be failures (Goldberger and Jones, 1966) leading to corrections for the scattering amplitude in terms of the singular mathematical behavior extrapolated from its physical values to complex values. Such singularities are well studied in mathematical analysis. They are called poles. The relevance of these singularities is distinguished with the name of the physicist responsible for the idea. They are called "Regge poles". The Italian physicist Tullio Regge had studied the dynamics of a particle that is governed by the Schrodinger equation (cf., Fig. 15.6).

This original way of viewing physics allowed him to study the scattering of microscopic particles, which is referred to as "potential scattering" (cf., the Illustrated glossary). This original way of viewing the work of Schrodinger led to the discovery that the microscopic collisions can be thought of in terms of smooth mathematical properties of one of its variables. In technical terms the scattering amplitude was described precisely as an analytic function of the angular momentum (cf., Illustrated glossary).

Regge showed that the position of the poles determines rates of growth of the amplitude in the region of large values of one of the variable he had singled out (the angular momentum). The work of Regge was not only an inspiration, but it was also timely: it was focused on the fashionable principles of the S-matrix theory that were dominant in the 1960s, especially analyticity in the framework that Mandelstam had proposed earlier. In spite of these criticisms the contribution of Mandelstam and Regge was basic to the modern understanding of relativistic particle scattering and theories that emerged subsequently: the string theories discussed below. The combined influence of Mandelstam and Regge became evident with Gabriel Veneziano, an Italian physicist working at the European Organization for Nuclear Research (CERN).

Fig. 15.6 Tullio Regge, an Italian physicist who was responsible for the development of a theory of the strong nuclear force during his participation in the ICTP 40th Anniversary Conference, 4 October 2004

Veneziano contributed a key breakthrough in 1968 with his realization that a single mathematical function (the Euler beta function), was capable of explaining much of the data on the strong force that was being collected at various particle accelerators around the world (Veneziano, 1968, cf., Fig. 15.7). A few years later, three physicists—Leonard Susskind of Stanford University, Holger Nielsen of the Niels Bohr Institute, and Yoichiro Nambu of the University of Chicago—significantly amplified Veneziano's insight by showing that the mathematics underlying his proposal described the vibrational motion of minuscule filaments of energy that resemble tiny strands of string, inspiring the name string theory.

In fact, they provided a physical interpretation in terms of an infinite number of simple harmonic oscillators describing the motion of an extended one-dimensional string. This gave rise eventually to the "string theories", which attempt to encompass the gravitational force in theoretical frameworks that go beyond the efforts to unify QED and QCD (the "grand unified theories abbreviated as GUTs).

String theories intend to unify quantum mechanics and Einstein's general theory of relativity. They assumed that the electrons and quarks are one-dimensional oscillating lines ("strings"), having only the dimension of length.

Fig. 15.7 Gabriel Veneziano lecturing at the ICTP, Trieste

While vibrating the observed particles acquire their quantum properties. String theory has been influenced by the discovery of various duality relations (mathematical transformations connecting, in this case, what appeared to be mathematically distinct string theories). These theories have been given specific names (type I, type IIA, type IIB, HE and HO). In other words duality links these versions to one another and to eleven-dimensional particle theory. This led to the conjecture by Edward Witten that the five theories are really aspects of a single underlying theory, which was given the name 'M theory'. (M theory is also used to describe the unknown theory of which the eleven-dimensional particle theory is just the low energy limit).

As mentioned in the last section string theories had extra dimensions beyond space and time and also forced the introduction of massless particles that had not been foreseen in previous attempts to unify the strong nuclear force and the electroweak interaction ("Grand Unified Theory", GUT). But string theory made a significant step forward in the search for a reliable theory of the fundamental forces, by incorporating the force of gravitation through these new features that were needed for allowing consistency. This search for a complete theory of elementary forces made available a new set of vibrations for the interpretation of the elementary particles, but more significantly it allowed for the introduction of the graviton (the particle mediating the gravitational force just as the photon mediates the electromagnetic interaction). As mentioned above, string theory has been extended trying to reconcile general relativity and quantum mechanics into a whole family of different theories.

15.9 Questions that lie in the frontier where astrobiology meets philosophy

Given this stage of development of cosmology, one specific question that concerns the main theme of this book is: How did the universe evolved in the distant past? Is it possible to read the Third Part of the Book of Life? Our observable universe, where galaxies have emerged with stars such as the Sun with life on one (or more) of its planets, may not be the only universe where we can build a complete astrobiological description by means of reproducible observations. More importantly, we should address the question: With our current understanding of cosmology can astrobiology complete its program of discovering the distribution and nature of life in the cosmos, especially in the remote past? In the final chapter we shall return to these questions that lie in the frontier where astrobiology meets philosophy.

Supplementary Reading

De Grasse Tyson, N. and Goldsmith, D. (2004) *Origins Fourteen Billion Years of Cosmic Evolution*. W.W. Norton & Company, New York.

De Grasse Tyson, N. (2007) *Death by a Black Hole and Other Cosmic Quandries,* W.W. Norton & Company, New York.

Gribbin, J. (2009) *In search of the multiverse*. London, Allen Lane, 228 pp.

Quinn, H. R. and Nir, Y. (2008) *The Mystery of the Missing Antimatter*, Princeton University Press, Princeton, N. J.

Yau, S.-T. and Nadis, S. (2010) The shape of inner space: string theory and the geometry of the universe's hidden dimensions New York, NY: Basic Books.

References

Albrecht, A. and Steinhardt, P. J. (1982) Cosmology For Grand Unified Theories With Radiatively Induced Symmetry Breaking. *Phys. Rev. Lett.* **48**, 1220.

Caldwell, R.R. and Marc Kamionkowski, M. (2001) Echoes from the Big Bang, *Scientific American* January 2001, pp. 28-33.

Chela-Flores (1967) *A study of analytic properties of scattering amplitudes*. M. Phil Thesis. University of London. 125pp.

Chela-Flores, J. (1970) Mandelstam Representation in Potential Scattering. *J. Math. Phys.* **11,** 2013-2015.

Chela-Flores, J. (2009) *A Second Genesis: Stepping stones towards the intelligibility of nature*. World Scientific Publishers, Singapore, 248 pp.

Feynman, R. P. (1965) *Symmetries in elementary particle physics*. A. Zichichi (ed.) New York, Academic Press.

Goldberger, M. L. and Jones, C. E. (1966) Consistency Questions Raised by Simultaneous Mandelstam and Angular-Momentum analyticity. *Phys. Rev. Lett.* **17**, 105–107.

Guth, A. H. (1981) The Inflationary Universe: A Possible Solution to the Horizon and Flatness Problems, *Phys. Rev. D* **23**, 347.

Linde, A. (1982) A New Inflationary Universe Scenario: A Possible Solution Of The Horizon, Flatness, Homogeneity, Isotropy And Primordial Monopole Problems, *Phys. Lett. B* **108**, 389.

Mandelstam, S (1958) Determination of the Pion-Nucleon Scattering Amplitude from Dispersion Relations and Unitarity. General Theory. *Phys. Rev.* **112**, 1344–1360.

Mandelstam, S (1963) Cuts in the Angular-Momentum Plane-I. *Nuovo Cimento* **30**, 1127-1147.

Perlmutter, S., G. Aldering, G. Goldhaber, R.A. Knop, P. Nugent, P.G. Castro, S. Deustua, S. Fabbro, A. Goobar, D.E. Groom, I. M. Hook, A.G. Kim, M.Y. Kim, J.C. Lee, N.J. Nunes, R. Pain, C.R. Pennypacker, R. Quimby, C. Lidman, R.S. Ellis, M. Irwin, R.G. McMahon, P. Ruiz-Lapuente, N. Walton, B. Schaefer, B.J. Boyle, A.V. Filippenko, T. Matheson, A.S. Fruchter, N. Panagia, H.J.M. Newberg, W.J. Couch (1999) Measurements of Omega and Lambda from 42 high redshift supernovae. *Astrophysical J.* **517**, 565–86.

Veneziano, G. (1968) Construction of Crossing Symmetric, Regge-Behaved Amplitude for Linearly Rising Trajectories. *Nuovo Cimento* **57A**, 190-197.

16

An intelligible universe with the science of astrobiology

Intelligibility of the universe is a question where humanists, especially philosophers can interact with astrobiologists. We have seen in Part 3 of the Book of Life that one aspect of intelligibility can be achieved by searching for the distribution of life in the Solar System, even before Kepler, CoROT and other instruments can provide definite answers. In this final chapter we discuss intelligibility in the domain of philosophers to attempt to pose proper questions in the science of astrobiology. Once we accept that a definite answer to the distribution of life in the universe intrinsically a question that even with the best instruments cannot be answered completely at present. We must wait until cosmology progresses further.

The reader is advised to refer especially to the following entries in the Illustrated glossary: *Biogeocentrism, Dyson (Freeman), Einstein (Albert), enlightenment, existentialism, Feferman (Solomon), Green (Brian), Hegel (Georg Wilhelm Friedrich), Kant (Immanuel), logical positivism, M-theory, Maxwell (James), realism, reductionism, string theory, Wilson (Edward), Witten (Edward).*

16.1 The meaning of reality

Even if considerable progress occurs in the near future in understanding the origin and evolution of life in the universe (parts 1 and 2 of the Book of Life), in order to succeed in the last two objectives of astrobiology the distribution and destiny of life in the universe, we have to work together with cosmologists for eventually giving definite answers. It is possible that reality depends on the solution of the equations (cf., Sec. 16.8), rather than on the fundamental equations themselves. In a problem of modern physics the philosophical question of realism has been relevant for the theory of quantum gravity, where we could accept the statement:

"No data, no theory, no philosophy."

This is not at all trivial statement. It requires a careful dialogue with philosophers (Butterfield and Isham, 2001). The distribution of life in the universe is in a similar position requiring some philosophical meditation. Indeed, we have no data (exolife), no theory (as the nature of extraterrestrial life is purely a question of space exploration), but as with the above-mentioned case of quantum gravity, we have demonstrated in the

J. Chela-Flores, *The Science of Astrobiology: A Personal View on Learning to Read the Book of Life*, Cellular Origin, Life in Extreme Habitats and Astrobiology 20, DOI 10.1007/978-94-007-1627-8_16,

previous chapter (and will do so in part of the present one) that the specific knowledge of philosophy is an asset. Thus the question of realism in philosophy is suddenly highlighted in the frontier of astrobiology. But in modern philosophy realism, in contrast with idealism, is the view that physical objects exist independently of being perceived. We can say that it is the viewpoint that grants objects of knowledge an existence that is independent of whether they are being perceived or even thought about. These insights have been forced upon philosophy due to the earlier thinking of idealists.

We feel that for simplicity we must return deeper into philosophy (cf., Aretxaga-Burgos and Chela-Flores, 2011, where some of the topics of this section have been discussed). In fact, we begin with the great effort of rationalism, which is the doctrine of a group of philosophers including Descartes, Spinoza and Leibniz. They attempted to construct an entirely deductive philosophy in the manner of geometry, which demonstrates how firmly rooted was the idea that reality was intelligible and meaningful, as well as the power and self-sufficiency of reason to discover both aspects of reality. Empiricism transformed radically how to face this question by considering that the only source of knowledge was the sensorial experience. Hence, everything that was beyond the experience of the senses will not be the subject of knowledge, but instead it will be in the domain of belief. Since the meaning of reality cannot be experienced empirically, it shall not be the subject of knowledge. It follows that metaphysics, the discipline that concerns itself with the meaning of reality, is not knowledge and should therefore be discarded.

16.2 The philosophy of Kant and Hegel

Kant maintained that human knowledge is only possible through science. The German philosopher coincides with the empiricists in so far as human knowledge cannot go beyond the frontiers of experience but differed from them since granted that all of human knowledge arises from experience, not for this reason its source is necessarily experience.

For Kant, science should only be concerned with *phenomena*, since only these have an intelligible structure that is the property that provides cognoscible (i.e., knowable) patterns of the human being—*the a priori forms of sensibility* (space and time), as well as the *pure concepts* (the categories). In other words, the intelligibility that the human being encounters in reality is the one that he himself has inserted in it. The totality of *phenomena*, instead, is not the subject of the experience of the senses that is to say is not a *phenomenon*; hence there cannot be the concern of science. Since metaphysics—the subject that reflects on the significance of reality as a whole—is not science, it shall not be considered as knowledge. Nevertheless, Kant does not consider that metaphysics can be discarded, but instead it is an inevitable tendency of reason to exceed the frontiers of experience (*phenomena*) in search of the *unconditioned*, namely the discovery of laws and principles gradually with more generality for a unified explanation of *phenomena*.

Hegel's philosophy is the last all-embracing Western philosophical system that searches the significance and comprehension of the whole of reality. According to Hegel, nature is the manifestation of God turned as matter, and everything that takes place in the universe is nothing but the transformation of matter into spirit and, finally, a transformation into spirit that is self-conscious (*Absolute Spirit*). In this scenario, human reason is the place where God thinks about Himself. Knowledge consists in the discovery and comprehension of the phases of this process and of its final result. It is only from the conclusion of this process, from the point of view of the totality of what has taken place that each phase or element of the process acquires its meaning. With Hegel, the intelligibility and meaning of the universe are within reach of human reason.

16.3 Is the universe a well-determined ordered system?

The above-mentioned thinkers had maintained that the cosmos is a well determined ordered system, and hence intelligible to all observers. In that framework there was no motivation for viewing the origin and evolution of the universe and life as being absurd, or pointless. But strong objections were raised to this line of thinking. Outstanding among these points of view were those due to Darwin and to the probability calculus. Darwin developed an explanation of the origin of humans based on natural selection and the underlying random mutations. This approach made it unnecessary, within the scope of the scientific method, to appeal to metaphysical concepts—that cannot be rejected by experiments—such as Divine Action. The development of a mathematical formulation of randomness and its increasing application to a wide range of scientific disciplines, such as physics and biology, allowed the substitution of the concept of necessity—determinism—for the concept of chance—probability—in order to explain both the existence of phenomena as well as their type of structure.

On the other hand, the new physics had multiple roots: discoveries of Boltzmann (thermodynamics), Maxwell (electromagnetism), Planck (quantum of energy), Einstein (the theory of relativity), Schrödinger (wave mechanics), or Heisenberg (the principle of uncertainty), just to quote a few, has required since the 20th century to go beyond the Newtonian approach. The new physics suggests a fresh approach to intelligibility of the universe that differs from traditional understanding and rationalization.

In the present context existentialism considered the human life as a contingent phenomenon, namely both its existence as the events that take place during its course happen at random. Life in the universe is not intelligible (pointless and absurd). Sartre maintained that existentialism is an attempt to live logically in a universe that is ultimately absurd. Another eloquent supporter of this doctrine was the Literature Nobel Laureate Albert Camus. In the mid-20th century Camus, through writings addressed the isolation of man in what he considered to be an alien universe. At the end of the line of intellectuals that were under the influence of existentialism, in the above quotation Weinberg reflects a view of the universe to which he was constrained by the adopted philosophic trend that influenced his generation.

In the existentialist view of the universe there still remains some hope for the concept of a meaningful universe, a living universe where intelligibility could be approached with the hope of the eventual emergence of a future "theory of everything" (cf., Chapter 15).

In this proposed all-embracing future theory we would hopefully discover the fundamental laws of nature in terms of a set of equations. Then, all phenomena should follow from these equations (the hope being that chemistry and biology could also be deduced). This is an extreme form of reductionism, *not an inevitable choice*, given the many insights of current progress in science as a whole. Since the Enlightenment (cf., Chapter 13) the ever-increasing growth of science has encouraged reductionism. The ideas that physical bodies are collections of atoms, or that thoughts are combinations of sense impressions are forms of reductionism.

16.4 Logical positivism and consilience

Philosophers have held two very general forms of reductionism in the 20th century. Firstly, logical positivists have maintained that expressions referring to existing things or to states of affairs are definable in terms of directly observable objects, or sense-data, and, hence, that any statement of fact is equivalent to some set of empirically verifiable statements. In particular, it has been held that the theoretical entities of science are definable in terms of observable physical things, so that scientific laws are equivalent to combinations of observation reports.

Secondly, proponents of the unity of science: Two Cultures (Snow, 1993), Edward Wilson's "consilience" (Wilson, 1998) have held the position that the theoretical entities of particular sciences, such as biology or psychology, are definable in terms of more basic science, such as physics. In other words, laws of these sciences can be explained in terms of more basic science.

The logical positivist version of reductionism also implies the unity of science insofar as the definability of the theoretical entities of the various sciences in terms of the observable would constitute the common basis of all scientific laws. Although this version of reductionism is no longer widely accepted, primarily because of the difficulty of giving a satisfactory characterization of the distinction between theoretical and observational statements in science, the question of the reducibility of one science to another remains controversial.

The reductionist dream has been supported by preliminary sets of successful equations that have embodied general phenomena at the most disparate scales (both microscopic and macroscopic). Today we recognize such efforts by assigning the equations the surnames of their authors: Newton, Einstein, Dirac, Schrödinger, Glashow, Salam and Weinberg. We are still at a very early stage in the comprehension of life in the universe. When the open question of the intelligibility of the universe is posed in a wider cultural context, including the earth sciences and astrobiology, Reductionism's restricted view becomes more evident. These arguments force upon us the following question:

Are we in need of facing a new synthesis of science and metaphysics that would allow us to reconcile them and lead to their joint progress?

One possible answer could be that a new future synthesis would have its basis in a physical science that would conceive life as a natural tendency of the cosmic evolution. If this was the pathway to follow, then we should also ask: Which role would astrobiology play in the systematic search for meaning beyond Hegel's Absolute Idealism?

In other words, what would be the implications of a new philosophical approach for the meaning of human life? The eventual success of astrobiology in its search for life elsewhere in the universe would force upon us a reconsideration of the question of intelligibility and comprehension of nature, a subject that is the very nucleus of Western thought, as we have attempted to demonstrate in this Chapter.

16.5 Are we approaching the end of biogeocentrism?

From this new perspective we could profit by returning to concepts that can be retraced to Lawrence J. Henderson, especially those of fitness and biocentrism: The fitness of the cosmos for the origin and evolution of life is discussed in Henderson's *"The Fitness of the Environment"* (Henderson, 1913), where the shift form two separate discussions, one for the cosmic evolution, and the other for biological evolution (Darwinism) begin to merge. This influential book was published almost a century ago, not as a response to the crisis in the systematic search for meaning that we have attempted to review in this chapter. Henderson was a graduate of Harvard and a professor of the University (Chela-Flores, 2007). Henderson's main interests ranged widely: he was a physiologist, chemist, biologist, philosopher and sociologist. He discussed the question of teleology in biochemistry to give some rationale to the question of fitness of the environment for the evolution of life. For many chemical compounds he discussed the difficulties that the evolution of life would have encountered had these compounds not been freely available in the environment. Water was one example. Its search, even today, is a main objective of the exploration of the Solar System. Henderson concludes that:

> *The properties of matter and the course of cosmic evolution are now seen to be intimately related to the structure of the living being and to its activities; they become, therefore, far more important in biology than has been previously suspected. For the whole evolutionary process, both cosmic and organic, is one, and the biologist may now rightly regard the universe in its very essence as biocentric.*

We have argued in favor of fitness of the cosmos for the origin and evolution of life without touching on the question of teleology. In this sense, we approach the subject without restricting ourselves exclusively to biological evolution in the universe, but rather we also include the evolution of the structure of the cosmos itself. The contrast between different views on the intelligibility of the universe, as discussed in this chapter, illustrates the need for philosophical understanding of meaning with fresh approaches in philosophy at the frontier of astrobiology.

We need especially to understand the influence of philosophical doctrines that will tend to encourage any future constructive dialogue between astrobiology and the humanities— especially philosophy—pointing towards unified two-culture knowledge, aiming at consilience. As mentioned in Sec. 11.7 this philosophically well-motivated new worldview has a close frontier with the exciting science that is to come in our search for life in our Solar System and other solar systems.

16.6 Is reductionism inevitable in the physical and in the life sciences?

This question is relevant for the main conclusion that inherent in cosmology there is an impediment for reading the third chapter of the Book of Life: the distribution of life in the universe. Reductionism refers to any doctrine that claims to reduce the apparently more complex phenomena to the less so (Flew, 1979). This will be referred to as the weak form of reductionism; its long string of successes in science, particularly in physics is one of the success stories best known to scientists, as well as to a large sector of the educated layman. These successes can probably be best illustrated with Isaac

Newton's theory of gravitation; with the introduction of quantum mechanics by Max Planck, and finally with the unified electroweak theory (cf., Chapter 15).

On the other hand, there is a second meaning for the word reductionism (Russell, 2001): It is the belief that human behavior can be reduced to or interpreted in terms of the lower animals, and that ultimately it can be reduced to the physical laws controlling the behavior of inanimate matter. For convenience this form of reductionism will be referred in this text as its strong version.

For example, in sociobiology and behavior genetics we have followed the tradition of the giants of physics, who restricted themselves to the weak form of reductionism). These groups of life scientists have extrapolated reductionism into its strong form. Indeed, the basis of human social behavior has been studied, in order to determine the relation between genetic constraints and their cultural expression. Some opponents have even referred to this approach as "genocentrism". Further applications of strong reductionism have led to controversy, as clearly illustrated by Robert Russell in his review on *"Life in the universe: Philosophical and Theological Issues"* (Russell, 2001). In particular, Russell discusses the arguments that have been put in favor of interpreting the capacity and content of human morality as products of evolution.

In this context we may recall that the physicists Freeman Dyson and Brian Green were invited to an open discussion in which the main thrust of the debate was whether science is ending its course, or whether we are far from achieving this aim. Dyson has made a significant impact on quantum field theory that was discussed in Chapter 15, but more importantly his wide range of cultural interests have led to important recognitions in the humanities, such as the Templeton Award. Dyson, in the debate with Green has raised a fundamental philosophical question: it is not evident that reductionism is inevitable in all scientific pursuits (Dyson, 2004, 2010). The reductionist dream of searching for an ultimate theory that explains the whole of natural phenomena is not inevitable. Within this Dysonian point of view we would not insist on the unified theories mentioned in Chapter 15, with the implications that we discussed for philosophical issues such as realism and, especially, the possibility that in different regions in earlier phases of the universe the quantum state has adopted a different configuration, leading to different details of the elementary particles. In such conditions life, as we understand it, becomes a phenomenon that we cannot understand if the constituents of atoms and subnuclear phenomena have not universally maintained their identity since the beginning of the universe.

The main point we wish to bring to the reader's attention is not the detailed argument that Dyson presents in his 2004 article or in his book, but rather the argument that the physical sciences are inexhaustible. The American philosopher Solomon Feferman made this remark (Feferman, 1998, 2004) when he discussed the type of mathematics that underlies the natural language of the physical sciences. The implications of Gödel's Incompleteness Theorem, on which Dyson based his arguments is not needed. Rather the Feferman argument concerns the remark that no matter which axiomatic system S is taken to underlay the physical sciences, there is a potential infinity of propositions that can be demonstrated in S, but scientists can only discuss a finite number. Hence the Feferman conclusion is that science indeed is inexhaustible, suggesting a cautious note on the unfinished program of the physical sciences in the search for an ultimate theory of nature.

16.7 Difficulties reading Part 3 of the Book of Life

The theoretical understanding of the fundamental forces of Nature may have to be revised, in order to take into account careful measurements of the velocities of very distant galaxies as defined by their stars in their late stage of evolution, namely, exploding stars that have exhausted their nuclear fuel. These important dying stars are normally called supernovae (cf., Chapter 1 and Reiss *et al.,* 1998). The value of these velocities may be interpreted as some evidence for an accelerating expansion of the universe. We have to learn whether the constant that Einstein introduced into his equations of gravitation (the 'cosmological constant'), purely on theoretical grounds, may represent some form of gravitational repulsion, rather than attraction. We should dwell on this question a little longer. Only a small fraction of the matter in the universe is in the form of the familiar chemical elements found in the Periodic Table. It is assumed that a large proportion of the cosmic matter consists of 'dark matter', whose composition consists of particles that play a role in the sub-nuclear interactions, mostly foreign to our everyday experience. The term 'dark matter' is not a misnomer, for the subnuclear particles that contribute to it, do not interact with light.

However, a remarkable aspect of cosmic matter is emerging: the sum total of the standard chemical elements and the dark matter make up only half of the matter content of the universe. The remaining fraction of cosmic matter has been referred to as dark energy with the astonishing property that its gravity is repulsive, rather than attractive. A possibility that has to be considered seriously in the future is that the repulsive gravity may dominate the overall evolution of the universe. This could lead to ever increasing rates of expansion. If this was to be the future of our cosmos, then future of life in the universe may hold some surprises. Eschatological considerations may have to be revisited. Will our universe be biofriendly? But this has taken us far enough into the general picture of the origin and evolution of the universe and within this framework we may begin to consider how life was inserted in the universe in the first place.

Ever since Hubble's discovery of an expanding universe (cf., Chapter 1 and Illustrated Glossary: "Hubble"), some galaxies that we now see may be out of sight in billions of years in view of the above-mentioned accelerating nature of the expanding universe. Consequently, looking back in time it is possible that some regions of space may be already similarly out of sight. String theory allows the possibility that in different regions, where the quantum state has adopted a different configuration, leading to different details of the elementary particles (Witten, 2010 and Fig. 16.1).

This lack of consensus, as to what theoretical formalism drives the evolution of the cosmos presents us a few basic difficulties not only to cosmology, but especially to the science of astrobiology: some of these challenges are not too far from the general philosophical issues that lie at the frontier of science and the humanities (extensively discussed in Chapters 13 and 14):

- It is possible that aspects of what we observe are not uniquely determined by the way the laws of nature are, but rather by which solutions of the equations of nature are relevant in the region we are able to see today (which clearly is not the whole of the cosmos as understood by the current physical theories that we have briefly sketched in this chapter).
- It is possible that reality depends on the solution of the equations, rather than on the fundamental equations themselves.

Both of these issues are central to the eventual outcome of what we have described in Part 3 of the Book Life, namely the objective of astrobiology to understand the distribution of life in a universe that is still hiding some of its secrets. We see the future development hand in hand of cosmology and astrobiology with great excitement and enthusiasm, as we can now appreciate that much wonderful research work lies ahead of us before we can with confidence assert that astrobiology is the science that can eventually succeed in the study of:

- The origin of life in the universe,
- The evolution of life in the universe,
- The distribution of life in the universe, and
- The destiny of life in the universe.

Fig. 16.1 The American physicist Edward Witten lecturing at the ICTP with Abdus Salam in the audience (front row, first from the right). Witten made a major contribution to the understanding of the interactions in the standard model and gravitation with his formulation of M-Theory

16.8 Perhaps the Third Part of the Book of Life can be intelligible

From our point of view astrobiology has included the distribution of life in the universe as the third chapter of The Book of Life. To complete the program that will eventually lead to a proper understanding of the distribution of life in the universe we have to base ourselves in a complete understanding of cosmology. Dropping the reductionist dream of physicists, as advocated by Dyson, signifies that the difficulties with reality that were described at the end of Chapter 15 are not compulsory at the present stage of development of science in general and cosmology in particular. In other words, the intuitive ideas that astrobiologists are currently pursuing are not confronted with the deeper aspects of philosophy, namely the reality of the phenomena that will be searched by astronomers when the search for life in our galaxy or in extragalactic environments become accessible to our astronomical instrumentation throughout the complete electromagnetic spectrum, not only in the visible range of observation but even further in infrared-, ultraviolet-, X ray- and gamma ray astronomy.

The eventual unification of gravitation and quantum mechanics (for instance as conceived by the M theory discussed in Chapter 15) could in principle be reconsidered in the final choice of cosmology that should be taken into account for discussing the evolution and distribution of life in the universe. Gravitational waves are, in principle, detectable by a new generation of detectors in solar orbits, such as LISA (a proposal of the European Space Agency). Hence, we could eventually interpret gravitational waves as classical waves and not as collections of gravitons. We would have two successful theories supporting cosmological discussions: quantum mechanics for the atomic and subatomic phenomena and General Relativity for the description of gravitation. There are two different worlds in Dyson's reasoning: the classical world of gravitation and the independent quantum world of the atomic and subatomic domains.

In our approach to these fundamental questions from the physical sciences, there is an implication that is relevant for astrobiology in the reading of the third chapter of the Book of Life. All our efforts to detect life in the Solar System (Chapters 7-9) and around other stars (Chapters 10-12) would be based on classical cosmology, namely on a cosmology that is based on Albert Einstein's theory of gravitation. The deep philosophical issues that were raised at the end of Chapter 15 concern reality of the physical world. While we persevere with the reductionist dream of unification of the fundamental forces, we have to wait for the profound difficulties that the physical sciences have raised. The Dyson viewpoint discarding reductionism allows us for the moment to proceed with our reading the complete third chapter of the Book of Life.

Supplementary Reading

Callender, C. and Huggett, N. (2001) *Physics Meets Philosophy at the Planck Scale*, Cambridge University Press, Cambridge.

Flew, A. ed. (1979) *A dictionary of philosophy*. Pan Books, London.

Hinneells, J.R. (ed.) (1984) *The Penguin dictionary of Religions*. Penguin Books, London.

Livingstone, E.A. (ed.) (1977) *The Concise Oxford dictionary of the Christian Church*. Oxford University Press, London.

Rovatti, P.A. (ed.), (1990) *Dizionario Bompiani dei Filosofi Contemporanei*. Bompiani, Milan.

References

Aretxaga-Burgos, R. and Chela-Flores, J. (2011) Cultural Implications of the Search and Eventual Discovery of a Second Genesis. In: *Origins: Genesis, Evolution and Diversity of Life* (2nd Edition), J. Seckbach and R. Gordon (eds.). Cellular Origin and Life in Extreme Habitats and Astrobiology, Springer, Dordrecht, The Netherlands. Submitted by invitation.

Butterfield, J. and Isham, C. (2001) Spacetime and the philosophical challenge of quantum gravity. In: Callender, C. and Huggett, N. (eds.) (2001) *Physics Meets Philosophy at the Planck Scale*, Cambridge University Press, Cambridge, 33-89.

Chela-Flores, J. (2007) Fitness of the cosmos for the origin and evolution of life: From biochemical fine-tuning to the Anthropic Principle, in *Fitness of the Cosmos for Life: Biochemistry and Fine-Tuning*, J. D. Barrow, S. Conway Morris, S. J. Freeland and C. L. Harper, (eds.), Cambridge University Press, 151-166.

Chela-Flores, J. (2009). *A Second Genesis: Stepping stones towards the intelligibility of nature.* World Scientific Publishers, Singapore, 248 pp.

Dyson, F. (2004) The world on a string. Review of the book of Brian Green *The Fabric of the Cosmos: Space, Time and the Texture of Reality.* The New York Review of Books, 13 May.

Dyson, F. (2010) Lo scienziato come ribelle. Biblioteca delle Scienze, Milano. The chapter concerned refers to a book review appearing in (Dyson, 2004).

Feferman, S. (1998) In the Light of Logic. Oxford University Press, UK.

Feferman, S. (2004) The New York Review of Books, 15 July.

Flew, A. (ed.) (1979) *A dictionary of philosophy*, Pan Books, London, pp. 300-301.

Henderson, L. J. (1913) *The Fitness of the Environment An Enquiry into the Biological Significance of the Properties of Matter*, Peter Smith, Gloucester, Mass., 1970, p. 312.

Riess, A. G., Filippenko, A. V., Challis, P., Clocchiattia, A., Diercks, A., Garnavich,P.M., Gilliland, R. L., Hogan, C. J., Jha, S., Kirshner, R. P., Leibundgut, B., Phillips, M. M., Reiss, D., Schmidt, B. P., Schommer, R. A., Smith, R. C., Spyromilio, J., Stubbs,C., Suntzeff, N. B. and Tonry, J. (1998) Observational Evidence from Supernovae for an Accelerating Universe and a Cosmological Constant. *Astronomical J.* **116**, 1009–38.

Russell, R.J. (2001) Life in the universe: Philosophical and Theological Issues, in Chela-Flores, J., Owen, T. and Raulin, F., *The First Steps of Life in the Universe*. Proceedings of the Sixth Trieste Conference on Chemical Evolution. Trieste, Italy, 18-22 September. Kluwer Academic Publishers: Dordrecht, The Netherlands.

Snow, C. P. (1993) *The Two Cultures*, Cambridge University Press.

Wilson, E. O. (1998) *Consilience The Unity of knowledge*, Alfred A. Knopf, New York.

Witten, E. (2010) Beautiful Minds. A video series of contemporary science edited by G. Oddifreddi. Milan. Volume **20**.

Epilogue

Learning to read the Book of Life: An interdisciplinary process

The main message that we have attempted to transfer to our readers is that unlike most other branches of science, astrobiology is an interdisciplinary activity that crosses over the frontiers of science and the humanities. We have not intended to teach the readers how they can read the Book of Life, but rather how The Book of Life itself has began to be intelligible from the point of view of science since early in the 20th century. Although efforts to come to grips with the phenomenon of life have been part of human culture since Ancient Greece, in the Introduction we have shown that biology was not ready to lend us a hand, not even after the most significant and relevant scientific revolution that took place with the coming of Darwinism towards the middle of the 19th century. Darwin himself was of the opinion that time was not ripe for such an undertaking during his lifetime (cf., "Introduction", p. 7). The emergence of the science of astrobiology was a gradual process, in which a stunning range of scientific disciplines took part. Some aspects of this vast scientific discipline are still to be fully comprehended. We would like to take a stroll with the readers along the pathway that we have followed together.

In Chapters 1 and 2 astrophysics and cosmology, together with organic chemistry, showed us the way to the fundamental enquiry of how life has emerged in the cosmos. This was the main topic of what we have called the "First Part of the Book of Life" (cf., pp. 29-100). In these pages we have seen how Alexander Oparin, Stanley Miller, Joan Oro, Cyril Ponnamperuma, Sidney Fox brought to our attention the chemical evolution of the main ingredients of living microorganisms that later would evolve from bacteria to humans. But even though they may be considered pioneers of astrobiology, they were not working in isolation. We have attempted to show the reader that it was not just the card-carrying organic chemists the scientists that were creating the foundations of astrobiology. Geochronologists, such as Stephen Moorbath, together with geochemists, such as Manfred Schildowski and Gustaf Arrhenius, and micropaleontologists, such as J. William Schopf were making independent preliminary, but significant steps in the right direction. Already in Chapter 3 we discussed the early, as well as the current steps in planetary science, especially the "conquest of the Moon" that had caught for a brief period the imagination of the layman who was witnessing a race between two

J. Chela-Flores, *The Science of Astrobiology: A Personal View on Learning to Read the Book of Life*, Cellular Origin, Life in Extreme Habitats and Astrobiology 20, DOI 10.1007/978-94-007-1627-8, superpowers. Like Camelot it was an inspiring, but alas too brief a moment. We had to wait forty years—a biblical time—to return to the nearest body of the Solar System.

A search for biomarkers has been initiated with the hope that they could orient us in the search for our own origin. We have to search for samples not just in the very limited area on the visible side of the Moon probed by the Apollo missions, but also in the polar region and on the far side of the Moon. There was an intense traffic of meteorites in the heavy bombardment period in the early Earth-Moon system. We cannot exclude a rapid onset of life even in that early period (the Early Archean around 4 billion years before the present). This implies that eventually even the Moon should be a target for astrobiology, in spite of the high costs of exploration. We also need to probe the small bodies of the Solar System, with Rosetta and other missions. In addition we should bring back samples from Mars with the Mars Sample-Return mission and get further insights from oceanography beyond the terrestrial limits to which astrobiology has been confined. In this context, glaciology in Antarctica and northern Canada can lend us a hand in these analogs of the surface of the icy satellites of Jupiter, as we have discussed in Chapter3.

Beyond planetary science, in Chapters 5 to 6 we have completed a brief review of the "Second Part of the Book of Life" (cf., pp. 101-131). We have discussed how the other great revolution of last century—molecular biology—can orient us in putting biology solidly into astrobiology. We illustrate how the earth sciences atmospheric physics and climatology play a role in astrobiology. Similarly, we discuss the relevance of anthropology in Chapter 6.

A significant challenge while reviewing the whole spectrum of astrobiology is the "Third Part of the Book of Life". Not surprisingly it has taken the major part of the book (cf., pp. 133-214). A fleet of missions are currently either on Mars, or in their planning stages (Chapter 7). Spacecrafts and their probes have reached, or will reach the outer Solar System: Galileo, the Europa Jupiter System Mission and Cassini Huygens (cf., Chapters 8 and 9). Technologies such as the penetrators with their potential for transporting *in-situ* microlaboratories to the Solar System bodies can reach the Moon (not only the very limited area known to us since the Apollo era, but also the lunar poles and its far side). Subsequently, penetrators are in principle capable of probing the icy surfaces of the Jovian satellites (and beyond) in search of geophysical insights into the origin of the bodies of the Solar System and in search of reliable and unambiguous biomarkers. Since 1995 the discovery of Michel Mayor and colleagues of a large number of exoplanets in our galactic neighborhood adds hope that the search for biomarkers outside the Solar System will yield its fruits. The Kepler mission is widening surprisingly fast our repertoire of exoplanets beyond the early efforts of Jean Schneider and colleagues with CoROT (Chapter 10). This gigantic step forward in astronomy that is relevant for astrobiology has as a major consequence with which this part of the Book of Life ends. Indeed, for over half a century the search for extraterrestrial intelligence (SETI) has improved its technological capabilities well beyond all the expectations of its pioneer Frank Drake, when he first searched for intelligent signals with Project Ozma (Chapters 11 and 12).

In the concluding "Fourth Part of the Book of Life" (cf., pp. 215-280) we bring into focus additional aspects that lie beyond the frontiers of science, mainly philosophy and theology (Chapters 13 and 14 and 16). These considerations allow us to reflect on the larger issues that affect astrobiology. We have questioned our eventual ability to complete all the objectives of astrobiology. Is the distribution of life in the universe a project that can eventually run into conceptual difficulties? To answer these questions we bring in Chapter 15 other scientific disciplines into the science of astrobiology, including especially the physical sciences at the atomic, nuclear and subnuclear level. These scales are presently available for experiments at the Large Hadron Collider. We return to the some of the preliminary mathematical insights that were followed up at the end of the 1950s. This approach led the physicist Stanley Mandelstam to change the course of subnuclear physics away from the quantum mechanics and quantum field theory that had begun in the 1920s with Werner Heisenberg, Paul Dirac and Erwin Schrodinger and other physicists (Chapter 15, Sec. 15.7). These early efforts of Mandelstam led to a group of physicists Tullio Regge, Gabriel Veneziano, Miguel Angel Virasoro and others to a theory that eventually was given its final form by several scientists, prominent amongst which was Edward Witten. We complete our discussion of astrobiology in Chapter 16 with philosophical reflections on the pathways that our current research in astrobiology is heading. We present the reader with alternative interpretations of modern cosmology that do not necessarily close the door for us to complete the program of astrobiology, especially obtaining eventually a complete understanding of the distribution of life in a cosmos. We present a point of view, still not of general agreement, in which there may not be an intrinsic difficulty in cosmological models that would be an impediment for our understanding of the distribution of life in an evolving cosmos.

Having expressed the contents of the present volume in the last few pages, it may seem pretentious for a single author to have attempted a decade ago to cover the whole spectrum of this fascinating science in "The New Science of Astrobiology", the first edition of this book. A decade later we have made a modest attempt to repeat the previous effort hoping to eliminate the many shortcomings of our earlier book, even though the author is aware that the previous criticism addressed to a single author may be brought upon him. In spite of his single authorship, he has not proceeded in isolation. He has been fortunate to benefit from the experience of some of the scientists that were mentioned in the previous chapters. This opportunity was granted to him firstly as a graduate student attending the International Center for Theoretical Physics (ICTP) at a time when Bethe, Crick, Dirac and Heisenberg converged in Trieste for the inauguration of the present premises of the ICTP (cf., Figs. I.10, 1.3, 15.3 and 15.2). The Nobel Laureate Abdus Salam had founded the ICTP in 1964.

When the author completed his career at the Simon Bolivar University (Caracas) he returned to the ICTP. Salam introduced him to Cyril Ponnamperuma in February 1991. He was an American organic chemist born in Sri Lanka, who had done remarkable work in the origin of life. Salam was enthusiastic on astrobiology (cf., "Abdus Salam's vision of symmetry in nature", pp. 87-88) and suggested that Ponnamperuma should direct the conference that had previously been planned, but had not taken place, with the collaboration of the author. The regular meetings that followed went on for a decade (over 500 scientists attended them). This singular experience has shaped the work that is discussed in this book. In 1971 Ponnamperuma, who was born in Sri Lanka, had moved from NASA-Ames Research Centre to the University of Maryland at College Park. He established the Laboratory of Chemical Evolution (LCE) becoming its Director. This laboratory soon developed into a center of excellence for research on chemical and biological evolution. The work ranged from planetary science to biochemistry to geochemistry to physical chemistry. In this setting he gathered scientists from all over the world in his traditional *College Park Colloquium on Chemical Evolution.* According to Rafael Navarro Gonzalez "since 1992 this series of conferences were relocated in Italy and named *Conference on Chemical Evolution and the Origin of Life* (Navarro Gonzalez, 1998). The organization of the First Conference began in February 1991. The event itself was held from 26 till 30 October 1992 (Ponnamperuma and Chela-Flores, 1993). These events continued till the end of Ponnamperuma's life in 1994 (cf., Fig. E.1). The French astrobiologist, François Raulin, one of the visitors of the LCE Laboratory, agreed to co-direct the 4th Conference on Chemical Evolution and the Origin of Life and continued to do so till the year 2003. The American planetary scientist Tobias Owen became a co-director of these academic events for the 6th and 7th Conferences (Chela-Flores *et al.*, 2001; Seckbach *et al.*, 2004).

Fig. E.1 The First Trieste Conference, 1992. Sitting in the front row from right to left: Abdus Salam is in the fifth position and Cyril Ponnamperuma is in the fourth one

The Seventh Conference on Chemical Evolution and the Origin of Life focused on the 50th anniversary of the seminal experiment of Stanley Miller. Over 100 scientists joined Professor Miller in the recollections of one of the most exciting moments in the development of astrobiology (cf., Fig. E.2).

Besides Salam and Ponnamperuma the Trieste conferences had the participation of some of the pioneers in astrobiology described at the beginning of this Epilogue, including Drake, Sidney Fox, Miller and Oró. Others who were present were: Gustav Arrhenius, Vladik Avetisov, Harrick and Marghareta Baltscheffsky, Laurence Barron, Francesco Bertola, André Brack, Graham Cairns-Smith, Giorgio Careri, Mohindra Chadha, David Cline, Marcello Coradini, John Corliss, Cristiano Cosmovici, George Coyne, Paul Davies, Donald De Vincenzi, Steven J. Dick, Klaus Dose, Enzo Gallori, Giancarlo Genta, Georgi Gladyshev, Vitali Goldanskii, J. Mayo Greenberg, Kaoru Harada, Jean Heidmann, Gerda Horneck, Yoji Ishikawa, Jamal Islam, Michael Ivanov, Torrence Johnson, Otto Kandler, Lajos Keszthelyi, Richard D. Keynes, Kensei

Fig. E.2 Participants of the Seventh Trieste Conference

Kobayashi, Mikhail Kritsky, Igor Kulaev, Narendra Kumar, Doron Lancet, Antonio Lazcano, Guillermo Lemarchand, Alexandra MacDermott, Claudio Maccone, Koichiro Matsuno, Clifford Matthews, Michel Mayor, Christopher McKay, David McKay, Juan Perez Mercader, Michael Meyer, Stenlio Montebugnoli, Stephen Moorbath, Rafael

Fig. E.3 Some of the participants of the 1999 Iberoamerican School of Astrobiology in Caracas, R. B. Venezuela that was subsequently continued in Uruguay and Spain

Navarro-González, Alicia Negrón-Mendoza, Marc Ollivier, Tahiro Oshima, Cynthia Phillips, Daniel Prieur, Martino Rizzotti, Robert John Russell, Takeshi Saito, Manfred Schidlowski, J. William Schopf, Jean Schneider, Peter Schuster, Joseph Seckbach, Everett Shock, Jill Tarter, Margaret Turnbull, Peter D. Ward, Wang Wenqing, Frances Westall, Yu-Fen Zhao and many others.

The tradition of College Park colloquia was transferred in the form of a training event to Caracas at the Convention Center of the Instituto de Estudios Avanzados (Fundacion IDEA, November-December 1999). The same spirit that Cyril Ponnamperuma imprinted to the College Park Colloquia was preserved in the Caracas *School of Astrobiology* with the presence of two pioneers of the origin of life studies: the Spanish astrobiologist Juan Oro, and the American astronomer Frank Drake, who initiated the search for extraterrestrial intelligence (SETI) by means of radio astronomy (cf., Fig. E.3, above, and Chela-Flores *et al*., 2000).

The singular experience of the Chemical Evolution Series of conferences and the advanced school provided a unique environment among the wide spectrum of academic events on astrobiology that the author has attempted to transfer to the present book.

References

(For a complete list of the proceedings of the Trieste Conferences the readers should consult the Section "Books by the Author" on p. 336.

Chela-Flores, J., Owen, T. and Raulin, F. (2001) *The First Steps of Life in the Universe.* Kluwer Academic Publishers: Dordrecht, The Netherlands.□ http://www.wkap.nl/book.htm/1-4020-0077-4

Chela-Flores, J., Lemarchand, G.A. and Oro, J. (2000) *Astrobiology: Origins from the Big Bang to Civilisation.* Kluwer Academic Publishers: Dordrecht, The Netherlands. □http://www.wkap.nl/prod/b/0-7923-6587-9

Chela-Flores, J. (2004) The New Science of Astrobiology From Genesis of the Living Cell to Evolution of Intelligent Behavior in the Universe. Series : Cellular Origin, Life in Extreme Habitats and Astrobiology, Band 3 Kluwer Academic Publishers, Dordrecht, The Netherlands, 251 p., Softcover edition of the 2001 book, ISBN: 1-4020-2229-8□http://www.springeronline.com/sgw/cda/frontpage/0,11855,5-198-72-33595256-0,00.html

Navarro Gonzalez, R. (1998) In Memoriam Cyril Andrew Ponnamperuma, *Origins of Life and Evolution of the Biosphere* **28**, 105-108.

Ponnamperuma, C. and Chela-Flores, J. (1993) *Chemical Evolution: Origin of Life.* A. Deepak Publishing, Vol. **135**: Hampton, Virginia, USA. http://www.stcnet.com/adpub/135.html

Seckbach, J.; Chela-Flores, J.; Owen, T.; Raulin, F. (eds.) (2004) Life in the Universe □*From the Miller Experiment to the Search for Life on Other Worlds* Series: Cellular Origin, Life in Extreme Habitats and Astrobiology, Vol. **7**, 387 p.

Acronyms and abbreviations

AMASE: Arctic Mars Analog Svalbard Expedition.
APL: Applied Physics Laboratory (of John Hopkins University)
AU: Astronomical unit (the distance Sun-Earth).

b-Pic: Beta-Pictoris.
BIF: Banded iron formation.
BP: Before the present
BVC: Bockfjord Volcanic Complex.

CAM: surface cell-adhesion molecules.
cAMP: cyclic adenosine monophosphate.
CDA: Cosmic Dust Analyzer.
CERN: European Organization for Nuclear Research.
CMB: cosmic microwave background.
CME: Coronal mass ejections.
CMS: Compact Muon Solenoid.
CNSA: The China National Space Administration.
COBE: Cosmic Background Explorer
COROT: Convection Rotation and Planetary Transits

DNA: deoxyribonucleic acid.

EJSM: The Europa Jupiter System Mission.
EQ: encephlization quotient.
ESA: European Space Agency.
ev: electron volt

Ga: One billion years (cf., also Gyr).
GCMS: Gas Chromatograph Mass Spectrometer.
GFP: green fluorescent protein.
GR: General Relativity.
GRB: gamma-ray burst.
GRS: Odyssey's Gamma-Ray Spectrometer.
GUT: Grand Unified Theory.
Gy: gray.
Gyr BP: gigayear (10^9) years before the present.

HRMS: High Resolution Microwave Study.
HARPS: High Accuracy Radial velocity Planet Searcher at the ESO La Silla
HIPPARCOS: High Precision Parallax Collecting Satellite.
HGT: horizontal gene transfer.
H_0: Hubble constant.
HR: Hertzsprung-Russell,.
HST: Hubble Space Telescope.

IDP: Interplanetary dust particle.

ICTP: The Abdus Salam International Centre for Theoretical Physics, Trieste, Italy
IRAS: Infrared Astronomical Satellite.
ISAS: Japan Institute of Space and Astronautical Science.
ISRO: Indian Space Research Organization.
ISS: International Space Station.

JAXA: Japan Aerospace Exploration Agency.
JEO: the NASA-led Europa Orbiter.
JGO: the ESA-led Ganymede Orbiter.
JMO: Jupiter Magnetospheric Orbiter.
JPL: Jet Propulsion Laboratories.

kev: one thousand electron volt.

LHC: Large Hadron Collider.
LISA: Laser Interferometer Space Antenna.
LRS: Laser Raman Spectroscopy.
Ly: Light year (cf., Illustrated glossary)).

MAP: Microwave Anisotropy Probe.
MARIE: Mars Radiation Environment Experiment.
MARSIS: Mars Advanced Radar for Subsurface and Ionospheric

Sounding.
mas: milli-arcsec.
MGS: Mars Global Surveyor.
MIF: mass independent fractionation.
Mpc: Megaparsec.
MSSS: Malin Space Science Systems.
MSW: Mikheyev–Smirnov–Wolfenstein effect.
Myr BP: million years before the present.

NASA: National Aeronautics and Space Administration.
NGST: Next Generation Space Telescope.
NICMOS: Near Infrared Camera and Multi-Object Spectrometer of the HST.
NIMS: A Galileo Mission Near Infrared Mass Spectrometer.
NSF: National Science Foundation.

PAL: present atmospheric level
pc: parsec.
Phylogenetic tree, 223
ppm: parts per million.

QCD: quantum chromodynamics
QED: quantum electrodynamics

R: the scale factor of cosmology.
RNA: ribonucleic acid.
ROSCOSMOS: The Russian Federal Space Agency.

S-Matrix: Scattering Matrix (cf., Glossary).
SAM: substrate adhesion molecule. 223
SETI: acronym for search for extraterrestrial intelligence.
SIMS: Secondary Ion Mass Spectrometry
SNC: shergottites, nackhlites and chassignites meteorites.
SOHO: the Solar and Heliospheric Observatory.
SpW: Space weather.

SSI: A Galileo Mission Solid State Imaging experiment.
STEREO: Solar Terrestrial Relations Observatory.
SWEAP: Solar Wind Electron Alphas and Protons Investigation.

t_0: Age of the universe.
TEGA; Thermal and Evolved-Gas Analyzer.
TDV: transit duration variation.
THEMIS: Thermal Emission Imaging System.
TSSM: Titan Saturn System Mission.
TTV: transit time variation.

UV: ultraviolet.
UVS: A Galileo Mission Ultraviolet Spectrometer.

WMAP: Wilkinson Microwave Anisotropy Probe.

XAS: X-ray absorption spectroscopy.

yr BP: years before the present

Illustrated glossary

(*Words in italics in the definition of a given term are defined elsewhere in the glossary*)

Accretion. In the *solar nebula* it is the early stage of accumulation of mass into a protoplanet.

Age of the universe (t_0). Several methods for fixing the value of this fundamental cosmological parameter are in agreement with the value quoted as $t_0 = 13.7 \pm 0.5$ Gyr (Freedman and Madore, 2010)

Albedo. Percentage of incoming visible radiation reflected by the surface.

Alga. Any of a group of mainly aquatic organisms that contain chlorophyll and are able to carry out *photosynthesis.*

Amino acid. Any of the twenty organic compounds that are the building blocks of proteins, when they are synthesized at *ribosomes,* according to the rules dictated by the *genetic code.* Their precise chemical formula is not essential for following the arguments in this book (cf., Chapter 3 in the section on "The Genetic Code").

Analyticity in S-Matrix Theory. A series of assumptions made in S-Matrix theory that include *Crossing*, *Dispersion Relations* and *Causality.*

Angular momentum. A property of rotating objects expressed as **mvr**, where **m** denotes mass, **v** denotes velocity and **r** is the distance from the center of rotation.

Angular resolution or 'spatial resolution' describes the resolving power of any image-forming device such as an optical or radio telescope, a microscope, a camera, or an eye.

Anthropic Principle in physics **(the strong form)**
The laws of nature and the physical constants were established so that human beings would arise in the universe.

Anthropic Principle in physics **(the weak form)**
Change the laws (and constants of nature) and the universe that would emerge most likely would not be compatible with life.

Anthropocentrism. Doctrine that maintains that man is the center of everything, the ultimate end of nature.

Apollo Program. In 1962 President John F. Kennedy offered his support for a series of missions intended to land humans on the Moon. They ranged from an unsuccessful—tragic—Apollo 1 Mission in 1967 through Apollo 17 in 1972. Perhaps the most significant achievement was that the first sample-return mission from another part of the

Solar System took place during the Apollo 11 flight in 1969. The Moon rock samples have given astrobiologists, especially paleontologists deeper insights into the evolution of the Solar System.

Archaea. One of the three *domains* introduced in the *taxonomic* classification of all life on Earth by *Carl Woese*.

Archaebacteria. These are a group of single-celled organisms that are neither bacteria nor *eukaryotes*. They may be adapted to extreme conditions of temperature (up to just over 100 ° C), in which case they may be called thermophiles. They may also be adapted to extreme acidic conditions (acidophiles). The alternative expression of *extremophiles* is used sometimes for these organisms to distinguish different degrees of adaptability to such extreme ranges of conditions. All archaebacteria are said to form the domain *Archaea,* divided into kingdoms.

Archean. In geologic time, this is an era that spans from 4.5 to 2.5 Gyr BP. (The Hadean is the first of its suberas). We refer to this era together with the *Proterozoic* (2.5-0.57 Gyr BP) as the *Precambrian Eon.* (The eon in which multicellular eukaryotic life arose is called the *Phanerozoic and* it ranges from the end of the Precambrian till the present).

Aristotle (384-322 BC). He was a Greek philosopher and scientist. Together with Plato they are considered the most distinguished intellectuals of Ancient Greece. He had a wide range of his interests, which spread over virtually the whole of human culture: from physics to chemistry, biology, zoology, botany, psychology, political theory, ethics, logic and metaphysics, history and literary theory.

Arnothosite. A deep-seated rock found on Earth and retrieved with the Moon samples of the *Apollo missions*. It is formed from a solidified form of *magma.*

Astrometry. This subject concerns the observation of the position of celestial objects and its variation over time. Recently, astrometry has been applied to the measurable wobbling motion of a star, due to its rotating suite of planets, against a fixed background of stars. This phenomenon has been used to discover the existence of extrasolar planets.

Astronomical unit. The average distance between the Sun and the Earth, approximately 150 million kilometers, or equivalently 93 million miles. It is abbreviated as AU.

Banded iron-formation (BIF). This is a type of sedimentary rock with iron (15% or more). The structures consist of repeated thin layers of iron oxides, either magnetite (Fe_3O_4) or hematite (Fe_2O_3), alternating with bands of iron-poor *shale* and *chert.*

Basalt. Common *rock* that has solidified from exposed *magma,* composed mainly of silicon, oxygen, iron, aluminum and magnesium. It is abundant in the oceanic crust, and in general on the surface of the *terrestrial planets.*

Base. *Purines* or *pyramidines* in *DNA* or *RNA.*

Bioastronomy. A synonym of *exobiology,* but it emphasizes is more on research with the specific tools of the radio astronomers. In view of this duality the term '*Astrobiology'* is used to encompass both disciplines.

Biochemistry. The study of the chemistry of living organisms. (It overlaps to a certain extent with *molecular biology).*

Biogeocentrism. A term introduced in the text to reflect a tendency observed in some contemporary scientists and philosophers according to which life is only likely to have occurred on Earth.

Biogeochemistry. This is the science concerned with the chemical, physical, geological and biological phenomena that are relevant for the origin and evolution of the environment (biosphere, hydrosphere, atmosphere and lithosphere).

Biomarker. A characteristic biochemical substance or *mineral* that can be taken to be a reliable indicator of the biological origin of a given sample.

Biota. The totality of life on Earth.

Black body. A term for an idealized object that absorbs all electromagnetic radiation falling on it. Studying its laws led Max Planck (in 1900) to the concept of the quantum. Others contributed to this approach and developed a quantum mechanics: *Niels Bohr, Werner Heisenberg, Erwin Schrödinger* and *Paul Dirac*. Blackbodies absorb and re-emit radiation in a typical continuous spectrum. Because no light is reflected (or transmitted), the object's appearance is black at low temperatures (T). (cf., *Kirchhoff*).

Book of Life. An expression used in the text to set the science of astrobiology in its appropriate cultural context. Questions on the origin, evolution, distribution and destiny of life in the universe precede the emergence of this branch of science. These queries have been deep concern of humanity since the beginning of civilization. This remark is illustrated throughout this book by returning to the history of philosophy in the Western World (Russell, 1991). Only in very recent times has the Book of Life been a main objective of a scientific discipline (astrobiology): It coincides with the subject matter of the present book *The Science of Astrobiology.*

Brown dwarf. A very cool star that due to its low mass is incapable of sustaing stable hydrogen fusion in its core. This internal phenomenon differs from stars in the main sequence (cf., Fig. 1.7). Planets are known to orbit brown dwarfs, one example being in the constellation Centaurus, some 170 light-years from Earth. In terms of Jupiter masses (M_J) their range lies between $13M_J$ and $80M_J$ approximately.

C-type asteroid. Term that refers to dark, carbonaceous asteroid in a classification according to the spectra of reflected sunlight.

Cabibbo, Nicola (1935-2010).

An Italian physicist whose important contributions to physics include recognizing that in the electroweak interactions, there is of a new class of physical constants, whose first example is what is now known as the "Cabibbo angle". This angle has been measured experimentally and has played an important role in our understanding of the Standard Model, especially in what concerns the phenomenon of *CP violation*. Cabibbo headed the Pontifical Academy of Sciences.

Nicola Cabibbo attending a lecture at the ICTP, Trieste

Carbonaceous chondrite. A meteorite formed in the early solar system. Its constituents are *silicates,* bound water, carbon and organic compounds, including *amino acids.*

Carbonate. A mineral that releases carbon dioxide by heating: it is a compound containing the CO_3 group, for example calcium carbonate Ca CO_3.

Carbonic acid. H_2CO_3 is formed when carbon dioxide is dissolved in water.

Causality. An assumption in S-Matrix Theory stating that the singularities of the S-matrix can only occur in ways that prevent the future to influence the past.

Cell division. Separation of a cell into two daughter cells. In *eukaryotes* the *nucleus* divides as well *(mitosis).* This is followed by a division of the extranuclear contents inside the *lipid bilayer* of the plasma membrane.

Cell signaling. Communication between cells by extracellular chemical signals.

Cenancestor. (cf., *progenote).*

Cenozoic. The most recent era of the *Phanerozoic eon.*

Chandra X-ray Observatory. Cutting-edge space instrumentation that is changing our view of the universe through the X-ray window.

Chandrasekhar, Subrahmanyan (1910 – 1995) was an Indian American astrophysicist. He was a Nobel laureate in physics along with William Alfred Fowler for their work in the theoretical structure and evolution of stars.

Chandrasekhar limit. Sets a bound to the mass of bodies made from electron-degenerate matter, a dense form of matter which consists of nuclei immersed in a gas of electrons. The limit is the maximum nonrotating mass that can be supported against gravitational collapse by electron degeneracy pressure. It is named after Subrahmanyan Chandrasekhar, and is commonly given as being about 1.4 solar masses.

Chaos. Chaotic systems display a level of behavioral complexity that frequently cannot be deduced from knowledge of the behavior of their parts.

Chemical equilibrium is reached in a reversible chemical reaction when no net change in the amounts of reactants and products have occurred.

Chemotrophs. These are organisms using inorganic (chemoautotrophs) or organic substances (chemoheterotrophs), instead of light ('phototrophs') as energy sources.

Chemolithoautotrophs. These are organisms using inorganic substances derived form rocks as energy sources.

Chert. This is a siliceous rock that is geochemically relevant as it may contain the remains of siliceous organisms such as sponges or microorganisms of various kinds.

Chirality. The property of molecules that exist in two forms whose spatial configurations are mirror images of each other. An example repeatedly mentioned in the text is the protein *amino acids.*

Chloroplast. Any of the chlorophyll-containing organelles that are found in large numbers in photosynthesizing plant cells. (They are the sites of *photosynthesis.*) Chloroplasts are widely believed to be the relic of a once free-living *cyanobacterium.*

Chondrules. A millimeter-sized spheroidal particle present in some kinds of meteorites. Originally they were molten or partially molten droplets.

Chromosome. A cell structure that contains DNA and whose number and complexity depends on the degree of evolution of the cell itself. Typically its constituent DNA in eukaryotes folds around proteins called *histones,* forming a set of compact structures *'nucleosomes'*; each structure is separated from others by a length of 'linker' DNA. The degree of folding are referred to by means of the dimension of the nucleosomes (in Ångstroms Å).

Codon. A triplet of *bases* in the *nucleic acid DNA* and also *RNA* that takes part in the process of *protein* synthesis) that codes for a given *amino acid.*

Color index of a star. These parameters are formed by taking the differences between the magnitude of a star measured in one wavelength band and its magnitude in another.

Contingency. The notion that the world today is the result of chance events in the past.

Convergent evolution. Independent evolution of similar genetic or morphological features.

Coyne S.J., George (1933-). A member of the Society of Jesus since the age of 18. He joined the Vatican Observatory as an astronomer in 1969 and became an assistant professor at the LPL in 1970. Coyne became Director of the Vatican Observatory in 1978. He was adjunct professor in the University of Arizona Astronomy Department. Coyne's studied Seyfert galaxies, the polarization produced in cataclysmic variables, or interacting binary star systems that give off sudden bursts of intense energy. He has contributed significantly to the dialogue between science and faith.

CP violation. This term refers to a violation of the postulated combination of C symmetry and P symmetry. This symmetry states that the laws of physics should be the same if a particle were interchanged with its antiparticle (C symmetry, or charge conjugation symmetry), and left and right were exchanged (P symmetry, or parity symmetry). It was discovered in 1964. The Nobel Prize in Physics was awarded in 1980 to James Cronin and Val Fitch for the experimental discovery. From the point of view of our book CP violation is relevant, since it is relevant for our understanding of the dominance of matter over antimatter in the Cosmos. The Italian physicist *Cabibbo* made a significant contribution to the understanding of CP violation.

Crick, Francis Harry Compton (1916 –2004). He was one of the co-discoverers of the structure of the *DNA* molecule in 1953, together with James D. Watson. He, Watson and Maurice Wilkins received the Nobel Prize for Physiology or Medicine in 1962. Crick was a theoretical biologist whose most significant contribution after the DNA was the discovery of the genetic code.

Crossing. An assumption in S-Matrix Theory asserting that the amplitudes for antiparticle scattering are mathematically understood as the analytic continuation of particle scattering amplitudes.

Cross-section. This concept is used to express the likelihood of interaction between particles. When particles in a beam are thrown against a foil made of a certain substance, the *cross section* σ is a hypothetical area measure around the target particles of the substance (usually its atoms) that represents a surface. If a particle of the beam crosses this surface, there will be some kind of interaction. The term is derived from the purely classical picture of (a large number of) point-like projectiles directed to an area that includes a solid target. Assuming that an interaction will occur (with 100% probability) if the projectile hits the solid, and not at all (0% probability) if it misses, the total interaction probability for the single projectile will be the ratio of the area of the section of the solid (the *cross section* σ) to the total targeted area (cf., "Optical Theorem").

Cyanobacterium. Prokaryote capable of oxygenic *photosynthesis*. Some cyanobacteria appear in the fossil record in the *Archean* and are regarded to be ancestral to *chloroplasts.*

De Duve, Christian (1917-). A Belgian cytologist and biochemist who discovered cell organelles shared the Nobel Prize for Physiology or Medicine in 1974. Since 1962 he headed a research laboratory at the University of Louvain (until his retirement in 1985) and a laboratory at Rockefeller University (until 1988). His books on astrobiological topics have been very influential.

Diagenesis. Changes that take place in a sedimentary rock in conditions of both low temperature and low pressure after the deposition.

Diploblasts. Animals arising from an embryo consisting of two cell layers.

Dirac Equation. *Dirac* produced a wave equation for the electron that combined *Special Relativity* with *Quantum Mechanics*. This contribution, known as "the Dirac Equation" predicted new states of the electron: to every particle there is a corresponding antiparticle, differing only in charge. The positron is just such an antiparticle of the negatively charged electron, having the same mass as the latter but a positive charge.

Dirac, Paul (1902-1984).

An English physicist known for his theoretical work in Quantum Mechanics. In 1932 he became Professor of Mathematics in the Lucasian Chair at Cambridge University, which had been held by Sir Isaac Newton. A Member of the Royal Society since 1930, Professor Dirac shared the Nobel Prize for Physics with E. Schrödinger in 1933. *Dirac's (wave) equation* predicted antimatter. His further work includes his formulations of quantum field theory.

Paul Dirac lecturing at the "Contemporary Physics Symposium" in Trieste in 1968.

***Drake, Frank* (1929-)** In 1960 he conducted the first radio search for extraterrestrial intelligence with a radio telescope 26 m in diameter at the U.S. National Radio Astronomy Observatory at Green Bank, Virginia. The telescope was aimed at Sun-like stars that are about 11 light-years from the Earth.

Dispersion relations An assumption in S-Matrix Theory stating that the values of the S-matrix can be calculated by integrals over internal energy variables of the imaginary part of the same values.

Dissimilatory sulfate reduction. Sulfate-reducing bacteria use sulfate as an oxidizing agent, reducing it to sulfide. This type of metabolism is called dissimilatory, since sulfur is not assimilated into any organic compound.

DNA (Deoxyribose nucleic acid). A substance present in every cell, bearing its hereditary characteristics.

Dyson, Freeman (1923-). As a graduate student in 1947 he worked with Hans Bethe and Richard Feynman. He went on to contribute to the unification of the three versions of quantum electrodynamics invented by Feynman, Schwinger and Tomonaga. His wide-ranging interest go from nuclear reactors to solid state physics, astrophysics and astrobiology, having published "Origins of Life" in 1986. *Infinite in all directions* was published in1988 is a philosophical meditation based on Dyson's Gifford Lectures on Natural Theology given at the University of Aberdeen in Scotland. In 2000 he was awarded the Templeton Prize for progress in Religion.

Domain. In *taxonomy* it refers to the highest grouping of organisms, which include kingdoms and lower taxons, such as phyla or divisions, orders, families, genera and species.

Duality. A concept that when it is applied to string theories is a class of symmetries that link different string theories.

Dual resonance model. An *S-matrix theory* of the strong interaction based upon the observation that the amplitudes for the s-channel scatterings matched exactly with the amplitudes for the t-channel scatterings among mesons and also the Regge trajectory.

Einstein, Albert (1879- 1955).

A physicist who developed the theories of Special and General Relativity. He won the Nobel Prize for Physics in 1921 for his explanation of the photoelectric effect. His ideas laid the foundation of the current concepts of space and time, differing from our common experience. He introduced a constant into his equations of gravitation (the 'cosmological constant'), purely on theoretical grounds. In fact, according to recent work in astrophysics it may represent some form of gravitational repulsion, rather than attraction

Electroweak interactions. The interaction that is responsible for the decay of the neutron into a proton, an electron and a neutrino ("beta" decay) was called the weak force: It is five orders of magnitude smaller than the electromagnetic interaction. *Sheldon Glashow, Abdus Salam and Steven Weinberg* proposed a theory that unified both of these interactions, which we refer to as the electroweak interaction. It predicted the existence of the *Higgs boson.* With a field theory of the strong nuclear force that keeps the nucleus together (*quantum chromodynamics*), these theories constitute the Standard Model.

Element. A substance that is irreducible to simpler substances in the sense that all its atoms have the same number of protons and electrons, but may have a different number of neutrons (cf., also *'isotope')*.

Elementary particle. A particle that up to certain energy cannot be probed to reveal internal constituents.

Empiricism. This thesis maintains that all knowledge on matter of fact is based on experience.

Enantiomer. One of the two forms of mirror images of chiral compounds.

Enlightenment This was an intellectual movement of the 17th and 18th centuries in which ideas concerning God, reason, nature, and man went into a synthesis that had many supporters. Amongst the most distinguished thinkers of this period we have: Descartes, Diderot, Montesquieu, Pascal, Rousseau and Voltaire, This movement was influential on the development of art, philosophy, and politics. Reason was the main theme underlying most innovations of this period. The thinkers behind this movement searched a deeper understanding of the cosmos. Rationalists strived towards more freedom, knowledge and happiness.

Epoch (in Earth Sciences). A subdivision of a geological period.

Eukaryogenesis. The first appearance of the *eukaryotes*, usually believed to be partly due to the process of *symbiosis*, probably during the Late *Archean* 2.7 billion years before the present (Gyr BP), according to current views; but certainly during the *Proterozoic,* some 1.8 Gyr BP, eukaryotes were co-existing with *prokaryotes,* both bacteria and *Archaea.*

Eukaryotes. These are either single-celled, or multicellular organisms in which the genetic material is enclosed inside a double membrane, which is called nuclear envelope. Taxonomically the totality of such organisms is said to form the domain *Eucarya,* which contains kingdoms, such as Animalia.

Equilibrium constant of a reaction (K) is obtained at some given temperature and pressure, as a certain ratio of the amounts of products and reactants present at *chemical equilibrium*.

Evolution (biological). In the case of biology it is a theory that assumes various types of animals and plants to have their origin in other pre-existing types. In addition, evolutionary theory assumes that distinguishable differences between living organisms are due to modifications that occurred in previous generations

Evolution (cosmic). This is a theory that considers the universe itself to be under a continuous process of expansion. The first such theory that was called "Big Bang" by *Sir Fred Hoyle*, is based on the early 1922 theoretical model of the universe by the Russian mathematician *Alexander Friedmann*. It is also based on the 1929 observations of the American Astronomer *Edwin Hubble*. The Hubble Law states that large groups of stars or galaxies move away from each other and that the velocity of recession is proportional to their distance.

Evolution (Lamarckian). An early theory of evolution, enunciated by *Jean Baptiste Lamarck* that involves the inheritance of acquired characteristics.

Exobiology. An older term used for astrobiology, namely the study of extraterrestrial life based on astronomy, physics and chemistry, as well as the earth and life sciences (cf., *bioastronomy).*

Extremophiles. This term is explained under "*Archaebacteria*".

Feferman, Solomon (1928-). This American philosopher and mathematician has contributed significantly in mathematical logic. He is a Stanford University professor. Feferman has received many awards. In the context of our arguments of Chapter 16,

Feferman was the editor-in-chief of the *Collected Works* of Kurt Gödel.

Fermi Paradox is the apparent contradiction between our expectation of life being an intrinsic phenomenon of cosmic evolution and the lack of evidence from any of the hundreds of *exoplanets* discovered so far.

Feynman Richard (1918-1988). An American physicist who was responsible for many fundamental contributions to physical theory. Most widely recognized for his work in Quantum Electrodynamics. He shared the Nobel Prize for Physics in 1965 for this work with Julian Schwinger and Sin-itiro Tomonaga. His pictorial representations of particle interactions are known as *"Feynman diagrams"*. They have influenced the study of the high energy physics and the fundamental interactions, but also the Feynman diagrams have also been fundamental for the theoretical approach to condensed matter physics.

Feynman diagram. An intuitive graphical representation of a term in a series in which a physical quantity, like a scattering cross section, is calculated by representing it as an infinite series of ever decreasing terms. In this diagram time runs from left to right of

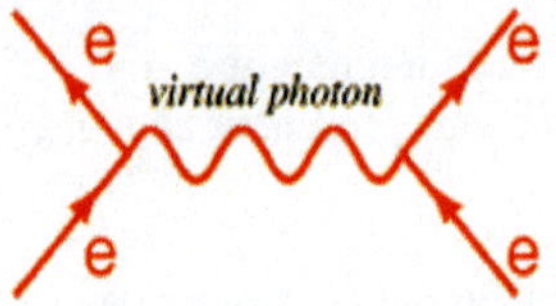

the diagram. The interaction of two charged particles (e^+e^- in a series of processes of increasing complexity. In the simplest case illustrated only one virtual photon is involved; in a second-order process, there are two and so forth. The processes correspond to all the possible ways in which the particles can interact by the exchange of virtual photons, and each of them can be represented graphically by means of these diagrams, which in addition allows calculating the variable involved. Each subatomic process becomes computationally more difficult than the previous one, and there are an infinite number of processes. The QED theory states that for each level of complexity, a factor of $(1/137)^2$ decreases the contribution of the process, and thus, after a few levels the contribution is negligible. This factor α is called the fine-structure constant and is a measure of the strength of the electromagnetic interaction. .

Flow cytometry allows the counting and microscopic examination of particles, for example cells and its internal components. The particle to be analyzed is suspended in a stream of fluid to be inspected by an electronic detection apparatus.

Fluorescence. A physical property of some substances, which consists of being able to absorb light at a given frequency and re-emitting it at a longer wavelength. *DNA* and *RNA* are able to bind certain dyes and are thereby subject to detection by means of fluorescent microscopy.

Galileo Mission. The Galileo spacecraft and probe traveled as one for almost six years. In July 1995, the probe was released to begin a solo flight into Jupiter. Five months later, the probe sliced into Jupiter's atmosphere at 106 miles per hour. The implication of this mission is the main topic of Chapter 8.

Gas chromatography. A chemical technique for separating gas mixtures, in which a gas goes through a column containing an absorbent phase that separate the gas mixture into its components.

Gene expression. The process that uses *RNA* by means of which a gene translates the information it codifies, according to the rules of the *genetic code,* into *proteins.*

Genetic code. This is a small dictionary that relates the four-letter language of *nucleic acids* to the twenty-letter language of the *proteins*. RNA [composed of four nucleotides: adenine (A), guanine (G), cytosine (C), and uracil (U)] is involved in assembling proteins. Three adjacent nucleotides (a "*codon*") code for a specific *amino acid.*

Genetic drift. Evolutionary change in small populations produced by random effects, not by *natural selection.*

Genome. The set of all genes contained in a single set of *chromosomes* of one species.

Geocentric. An old hypothesis that maintained that the Earth was at the center of the universe. It maintained that the Sun was in orbit around the Earth.

Geochronology deals with application and interpretation of isotopic dating methods in geology.

Glashow, Sheldon (1932-). American physicist who, with *Steven Weinberg* and *Abdus Salam*, received the Nobel Prize for Physics in 1979 for their efforts in formulating the electroweak theory, which explains the unity of electromagnetism and the weak force.

Glycerol. A colorless, sweet-tasting viscous liquid, widely distributed as part of molecules of all living organisms. (Its atomic formula is that of an *alcohol.)*

Greene, Brian (1963-), He is a physicist, Professor of Physics and Mathematics at Columbia University. Green has made multiple contributions to string theory. As the author of successful book in science communication that is focused on string theory, *The Elegant Universe* (published in 1999) he became widly known to the general public.

Hadean. The earliest subera of the *Archean.*

Hadrons. Elementary particles that are subject to the strong nuclear force. In this sense they differ form *leptons*.

***Hegel, Georg Wilhelm Friedrich* (1770 - 1831).** He was a German philosopher whose main contribution was to develop a dialectical scheme stressing the progress of history and the progress of ideas, firstly from thesis, secondly to antithesis and finally to a synthesis. Hegel was the leading philosopher in Germany of the post Kantian period between the 18th and 19th centuries, together with the two idealists Johann Gottlieb Fichte (1762-1814) and Friedrich Schelling (1775-1854). One key to Hegel's philosophy lies in the central role played by the history of thought.

Heidmann, Jean (1923-2000).

French-born astronomer Jean Heidmann, an enthusiastic supporter of the search for extraterrestrial intelligence (SETI). He began his career at Cornell University, USA, conducting research on cosmic rays. He then moved to the Paris Observatory, Meudon, France, to pursue his interests in radio astronomy. He advocated the design and ultimate construction of a lunar SETI radio telescope fully shielded from terrestrial interference.

Jean Heidmann during the 5th Trieste Conference.

Heisenberg, Werner (1901–1976).

A German physicist who made fundamental contributions to quantum mechanics, especially for the uncertainty principle of quantum theory. Heisenberg was awarded the 1932 Nobel Prize in Physics for his contribution to quantum mechanics.

W. Heisenberg, first on the right in the front row, the 1968 Contemporary Physics Symposium.

Hematite. A *mineral* of ferric oxide.

Hertzprung-Russell diagram. This useful plot of luminosity of stars versus surface temperature (or equivalent physical parameters) was introduced independently by Ejnar Hertzprung and Henry Russell in the early part of the 20th century. At various stages of *stellar evolution,* stars will occupy different parts on this diagram.

Heteroatom is any atom that is not carbon or hydrogen (for example, nitrogen, oxygen, sulfur, phosphorus, boron, chlorine and others).

Heterochromatin. A densely packed form of *chromatin* in eukaryotic chromosomes.

Higgs boson. The Higgs boson is a massive scalar elementary particle predicted to exist by the *Standard Model* in particle physics. In terms of the Higgs boson we explain the difference between the massless photon, which mediates electromagnetism, and the massive W and Z bosons, which mediate the weak force.

Histone. A protein belonging to one of the four classes H2A, H2B, H3 and H4. A couple of representatives of each class, together with a stretch of double helix DNA, form a nucleosome.

Hominid (Hominidae). The name given to the taxon that comprises humans and the great apes.

Hominoid (Hominoidea). The name given to the taxon that comprises humans and all apes (great apes and gibbons).

Hoyle, Sir Fred (1915-2001).

British astronomer who was a proponent of the steady-state theory of the universe: the universe is expanding and that matter is being continuously created to keep the mean density of matter in space constant. His comment on the synthesis of carbon atoms in stars led to the current interest in the *Anthropic Principles*. He introduced the term "big bang".

Sir Fred Hoyle lecturing at the ICTP in 1985

Hubble, Edwin Powell (1889-1953). American astronomer who is considered the founder of extragalactic astronomy and who provided the first evidence of the expansion of the universe. In 1929 Hubble demonstrated that there was a linear correlation between the apparent distance to galaxies and their velocity of recession. The constant of proportionality H_0 bears his name. With a large sample of relatively near supernovae H_0 is expressed in units of kilometers per second per million *parsec* (megaparsec, Mpc) with the value 74.2 ± 3.6 kms^{-1} Mpc^{-1}.

Hydrocarbon. A class of organic compounds in which only hydrogen and carbon take part.

Hydrothermal vent. These structures occur at the crest of oceanic ridges producing ascension of a very hot suspension of small particles of solid *sulfide* in water (cf., *plate tectonics, sea-floor spreading)*. When they occur at mid-ocean ridges they become ecosystems where the production of biomass is considerable. Its inhabitants are *chemolithoautotrophic* microorganisms exploiting the difference in chemistry between hot vent fluids and the surrounding seawater to gain metabolic energy.

Interphase. The period following the completion of *cell division* when its *nucleus* is not dividing.

Interstellar dust particle (IDP). The ensemble of solid particles borne by the gas that occupies the space between the stars.

Intron. A typically eukaryotic sequence in a gene that does not code for amino acids of a protein.

Ion. An atom or molecule that is electrically charged due to the loss or gain of one or more electrons.

Isotope. One of two or more atoms of the same chemical element that have the same number of protons in their *nucleus*, but differ in their number of neutrons.

Isotopic fractionation. This concept means the enrichment of one *isotope* relative to another in a chemical or physical process. Isotopes of a given element may have slightly different *equilibrium constants* for a particular chemical reaction, so that a fractionation of the isotopes results from that reaction (cf., *Mass-independent fractionation*).

Jovian planet. Collective name for Jupiter, Saturn, Uranus and Neptune. These are all giant planets; hence very different form the *terrestrial planets.*

Kant, Immanuel (1724 - 1804). A German philosopher who worked in the theory of knowledge, ethics, and aesthetics. He influenced all subsequent philosophy. Kant was the leading philosopher of the *Enlightenment* and one of the greatest philosophers of all time. In him were subsumed new trends that had begun with *Rationalism* (stressing reason) and *Empiricism* (stressing experience).

Kepler mission. Launched on March 6, 2009, from Cape Canaveral Air Force Station in Florida, it is a mission designed to observe more than 150,000 stars in the search for exoplanets.

Keynes, Sir Richard Darwin (1919-2010). A distinguished electrophysiologist, he was born into two illustrious families. His uncle was the economist John Maynard Keynes. His mother, Margaret, was the daughter of Darwin's son, the astronomer and mathematician Sir George Darwin. Collaborating with Alan Hodgkin, Keynes discovered that sodium ions flow into a nerve cell and potassium ions flow out when the cell is stimulated. This research forms part of the theory of the nerve impulse for which in 1963 Hodgkin and Andrew Huxley were awarded the Nobel Prize physiology. His interest in the basis of the theory of evolution mentioned in our introductory chapter (Emergence of Astrobiology) was triggered by several research trips to South America, where he came under the influence of the biologist Carlos Chagas figlio, the son of the physician who discovered the Chagas disease, that Charles Darwin may have suffered from, as a consequence of his Beagle voyage, while visiting South America.

Kirchhoff, Gustav (1824–1887). A German physicist who contributed to many fields in physics but we remember him in our work especially for his studies in the 1860s of the emission of blackbody radiation by heated objects. His contributions to spectroscopy are relevant.

Kuiper belt. A belt of some billion (10^9) or more comets beyond the orbit of Neptune. It is the source of short-period comets. It extends possibly as far as 1000 *astronomical units*. Some dwarf planets lie in this part of the Solar System beyond Pluto: Sedna, Eris, Haumea, Makemake.

Lagrangian points are the five positions in an orbital configuration where a small object affected only by gravity can be stationary relative to two larger objects, such as an artificial satellite with respect to the Earth and Moon. They label positions where the combined gravitational force of the two large masses provides exactly the centripetal force required to rotate with them. They allow an object to be in a "fixed" position in space, rather than an orbit in which its relative position changes continuously.

Lava. Molten *rock* material that is expelled from volcanoes. It consolidates once it reaches the surface of the Earth or the sea floor.

Lee, Tsung-Dao (1926-). He is a Chinese-born American physicist who, who together with *Chen Ning Yang*, discovered *parity violation*. They shared the Nobel Prize for Physics in 1957.

Lepton. A class of elementary particles including electrons and neutrinos. Unlike the *quarks* and *hadrons* in general, leptons are not subject to the strong nuclear force.

Light-year. The distance light travels in a year, some ten trillion miles (10^{13} Km).

Linea. Elongate markings observed on the surface of Europa by the *Voyager mission,* which are a few kilometers wide and of lower albedo than the surrounding terrain.

Lipid. A wide group of organic compounds having in common their solubility in organic solvents, such as *alcohol.* They are important in biology, as they are constituents of the cell *membrane* and have a multitude of other important roles.

Lipid bilayer. A characteristic fabric of all biological *membranes*. It consists of two monolayers of *lipids.*

Logical positivism. A philosophical current that maintains that scientific knowledge is the only kind of factual knowledge; all traditional doctrines are to be rejected as meaningless. These ideas were mainly developed in the 1920s in Vienna.

Lyell, Charles (1797-1875). He was Scottish geologist concerned with "uniformitarianism", namely demonstrating that that in geologic time physical, chemical, and biological processes are the main causes of surficial terrestrial characteristics. He developed his main arguments in the book *Principles of Geology* (1830-1833) that was a major influence in Charles Darwin's thinking.

M ***theory***. An extension of *string theory* by *Edward Witten* requiring the assumption of 11 dimensions. Because the dimensionality exceeds the dimensionality of superstring theories in 10 dimensions, it is believed that the 11-dimensional theory generalizes all five string theories. Its existence was used to explain a number of previously observed dualities (cf., Dualities). It had been shown previously that the various accepted versions of superstring theories were related by dualities. These relationships imply that each of the superstring theories is a different aspect of a single underlying "M theory".

Magma. Liquid or molten *rock* material, called *lava* when it reaches the Earth's surface.

Mass-independent fractionation (MIF). This term refers to any process that tends to separate isotopes, where the amount of separation is not proportional to the difference in the masses of the isotopes. Most isotopic fractionations are caused by the effects of the mass of an isotope on atomic phenomena, including the corresponding atomic or molecular velocities.

Mass spectrometer. Instrument used for the separation of a beam of gaseous *ions* into components with different values of mass divided by charge. The ion beams are detected with an electrometer.

Maria. It is the plural of mare. Latin for sea. It refers to the dark patches on the surface of the Moon; the darkness arises from *basaltic lava* flows.

Maxwell, James Clerk (1831–1879).

He was a Scottish physicist whose most remarkable contribution to science was the electromagnetic theory that unified electricity, magnetism and optics into a set of equations. In fact, in 1864 Maxwell published *"A Dynamical Theory of the Electromagnetic Field"*. In this book he suggested his unification. With the publication of the 1873 *Treatise on Electricity and Magnetism* the interactions of positive and negative charges were shown to be regulated by a single force.

Membrane. A tissue consisting of *lipids, protein* and sugars (polysaccharides) that cover biological cells and some of their organelles. The cell nucleus is covered by an envelope made of two membranes.

Mesozoic. An era of the *Phanerozoic eon,* which was characterized by reptiles together with brachiopods, gastropods and corals. The best-known period of this era is the Jurassic period (208-146 Myr BP) with rich flora and warm climate when reptiles, mainly dinosaurs, were dominant on land.

Messenger RNA (mRNA). An *RNA* molecule that specifies the *amino acid sequence* of a protein.

Metaphase. The second stage of *cell division,* during which the envelope around the *nucleus* breaks down in some eukaryotes such as animals.

Mitochondrion. A eukaryotic *organelle* that is the main site of energy production in most cells.

Mitosis. Division of the *nucleus* of the *eukaryotes*. During this process the DNA is condensed into visible *chromosomes.*

Molecular biology. The study of the structure and function of the macromolecules associated with living organisms.

MSW effect. The **Mikheyev–Smirnov–Wolfenstein or "matter" effect** is a process that can modify neutrino oscillations in matter. Lincoln Wolfenstein in 1978 and Stanislav Mikheyev and Alexei Smirnov in 1986 led to an explanation of this effect.

Nambu, Yoichiro (1935-).

Yoichiro Nambu is one of the pioneers in the formulation of spontaneous symmetry breaking and in particular, chiral symmetry breaking in relativistic particle physics. His contributions to the quark model in the sixties and, later, his geometrical formulation of the dual resonance models as the dynamics of a *relativistic string* are of fundamental importance. The scope and intensity of current research in string theory are witness to the profundity of Nambu's contributions to particle physics. He shared the Nobel Prize in Physics 2008 for his discovery of the mechanism of spontaneous broken symmetry in subatomic physics.

Yoichiro Nambu during his award of the 1986 Dirac Medal

Natural selection. Darwin and Wallace suggested this term for the reproductive success. It refers to the selection that eliminates most organisms that over reproduce, and hence allows adaptation to changing environmental conditions.

Natural theology. The body of knowledge about religion that can be obtained by human *reason* alone, without appealing to *revelation.*

Neutron star. A relic of a *supernova* explosion. It consists of subnuclear particles called neutrons in various states of condensed matter.

Nitrile. A class of chemical compound that contains the group CN, which on hydrolysis yields ammonia and a carboxylic acid.

Nucleic acid. These are organic acids whose molecular structure consists of five-carbon sugars, a phosphate and one of the following five *bases:* adenine, guanine, uracil, thymine and cytosine (sometimes modified).

Nucleosome. A bead-like subunit of *chromatin* in the *chromosomes* of *eukaryotes.* It consists of a length of *DNA* double-helix wrapped around a core of eight *histones.*

Nucleosynthesis. The generation of the natural *elements* starting from hydrogen. Three possibilities are: at an early stage of the big bang, after temperatures had cooled down sufficiently to allow the electromagnetic force to produce atoms; at the center of stars; and during the *supernova* stage of *stellar evolution.*

Nucleus of an atom (plural: nuclei). Massive central body composed of protons neutrons.

Nucleus of a cell (plural: nuclei). The major *organelle* of *eukaryotes* containing the *chromosomes.*

Oort cloud. A group of comets surrounding the solar system more than a thousand times more populated than the *Kuiper belt.* It is estimated to extend to interstellar distances, perhaps 50,000 AU or more.

Optical Theorem. This is a general law of the scattering of waves. It is usually written in the form that the total scattering cross section is proportional to the imaginary part of the complex function that represents the "scattering amplitude" f(0), that is the amplitude of the wave scattered to the center of a distant screen. From the simple case of an incident plane wave the Optical Theorem states:

$$\sigma_{tot} = \frac{4\pi}{k}\,\mathrm{Im}\, f(0)$$

where k is the momentum of the plane wave. The Optical Theorem relates the forward scattering amplitude to the total cross section of the scatterer.

Organelle. Structures inside the biological cell that has a given function. Examples are *nuclei, mitochondria* and *chloroplasts.*

Orion Nebula. The largest complex region of interstellar matter known in our galaxy. It is in the Constellation of Orion, some 1,300 *light years* away form the Earth. It is found in the "sword" of Orion surrounding a multiple star system of four hot stars (Theta Orionis, otherwise known as the 'Trapezium'), not older than 100,000 years.

Ozma, Project. The first search for extraterrestrial intelligence by *Frank Drake* at the National Radio Astronomy Observatory in Green Bank, West Virginia, following the earlier suggestion of physicists Giuseppe Cocconi and Philip Morrison in a paper published in Nature in 1959. Drake focused only on two nearby stars.

Paleozoic. This is the earliest era of the *Phanerozoic eon.*

Parity. The symmetry between physical phenomena occurring in right-handed and left-handed coordinate systems

Parity violation. Usually it refers to the subnuclear world. In quantum mechanics it means an asymmetry of the wave function of a system of particles. A parity transformation replaces such a system with a type of mirror image. This is not symmetry of nature as discovered by *Chen Ning Yang* and *Tsung Dao Lee* in 1956. They were awarded soon afterwards the Nobel Prize in Physics.

Parsec. This is a measure of astronomical distance equal to 3.26 light-years. Compare with astronomical unit.

Pasteur, Louis (1822—1895). He was a French chemist and microbiologist who demonstrated that microbes could be the source of disease, attempted to cure them with vaccines and did some remarkable work in stereochemistry. He is also remembered by the process known as pasteurization.

Phanerozoic. It is the most recent eon of geologic time extending from the *Paleozoic* (570-230 Myr BP) to the *Mesozoic* (230-62 Myr BP) and the *Cenozoic* (62 Myr BP to the present).

Photosphere. The apparent solar surface, where the gas of the atmosphere becomes opaque.

Photosynthesis. An ancient metabolic process, first used by *prokaryotes,* which today produces *hydrocarbons* and liberates oxygen with the source of carbon dioxide and water. In the deep ocean and underground an alternative metabolic process is used (chemosynthesis).

Phylogenetic tree. This is a graphic representation of the evolutionary history of a group of organisms. This may be extended from organisms to species or any higher ordering, such as kingdoms. In *molecular biology* this concept has been extended to genes as well. The main lesson we may draw form such trees is the path followed by evolution.

Phylogeny. The evolutionary history of an organism, or the group to which it belongs.

Pioneers.

This was a series of spacecrafts launched since the 1960s, in particular from the point of view of astrobiology we should recall that Pioneer 10 flew by Jupiter in December 1973, the first space probe to do so. This mission was responsible for the discovery of the Jovian magnetic tail. Pioneer 11 flew within 20,900 km of Saturn in September 1979.

Planck, Max (1858-1947).

A German physicist considered as the founder of quantum mechanics. Planck attempted to derive *"Wien's law"* on the basis of the second law of thermodynamics from first principles: (a) the second law of thermodynamics was a statistical law (as had been assumed by Ludwig Boltzmann) and (b) the oscillators forming the blackbody could only absorb energy in discrete amounts, in "quanta of energy". Planck was awarded the Nobel Prize in Physics in 1918.

Planck Space Observatory. This is an instrument that was designed to observe the anisotropies of the *cosmic microwave background* over the entire sky. Planck was the third Medium-Sized Mission of ESA's Horizon 2000 Scientific Programme. The project is named in honor of the German scientist *Max Planck* (1858–1947). Planck was launched in 2009, reaching the L_2 *Lagrangian point*. It completed a second all-sky survey the following year.

Plane-polarized light. Light in which the typical electromagnetic vibrations are rectilinear, parallel to a plane ("plane of polarization") and transverse to the direction of travel.

Plate tectonics. This is the standard theory, with convincing experimental support that maintains that the Earth is composed of thick plates floating on submerged viscous material. Each plate moves slowly: its leading edge sinks slowly at regions called subduction zones, while the rear of the plate produces an observed phenomenon called *sea-floor spreading,* where phenomena relevant to the origin of life may occur such as *hydrothermal vents.*

Plato (428/427-348/347 BC). A Greek philosopher, who helped significantly to the philosophy of Western Civilization. Starting with the achievements of Socrates, Plato made significant contributions to logic, epistemology, and metaphysics; but his main thrust is in ethics. Essentially Plato is a rationalist, with complete confidence on reason.

Plinian eruption. A volcanic eruption that produces repeated explosions.

Plume. Ascending partially molten mantle material.

Progenote. A term introduced by the American evolutionist Carl Woese to denote the earliest common ancestor of all living organisms (also called *'cenancestor')*.

Prokaryotes. These unicellular organisms lack a nuclear envelope around their genetic material. Normally they are smaller than nucleated cells *(eukaryotes)*. Well-known examples are bacteria. All prokaryotes are encompassed in two *domains*: Bacteria and *Archaea.*

Protein. These are organic compounds that are an essential for all living organisms. Its elements are: hydrogen, carbon, oxygen, nitrogen and sulfur. It is made up of a series of *amino acids*. (A medium-size protein may contain 600 amino acids.)

Proterozoic. In geologic time it is an era that ranges form the *Archean* era (4.5-2.5 Gyr BP) to the *Phanerozoic Eon* (570 Myr BP to the present).

Pulsar. These are highly magnetized, rotating neutron stars that emit a beam of electromagnetic radiation.

Purine. One of two categories of nitrogen-containing ring compounds found in *DNA* or *RNA*. (Examples are guanine and adenine.); cf., also *pyrimidine.*

Pyrimidine. One of two categories of nitrogen-containing ring compounds found in *DNA* or *RNA*. (Examples are cytosine, thymine and uracil.)

Pyrolysis. Chemical decomposition occurring as a result of high temperature.

Quantum chemistry. A fundamental approach of quantum mechanics applied to chemistry that is based on a mathematical physical theory dealing with the mechanics of atomic systems.

Quantum chromodynamics (QCD). A part of what has been called the standard theory deals with the strong nuclear force constructed in analogy with *quantum electrodynamics* (QED). Instead of the emission and subsequent absorption of photons, the strong force acts upon quarks that are bound together in the protons and neutrons of the atomic nucleus. By analogy with QED, in QCD quarks interact through the strong force by a strong charge called color. Since the electron has instead an electric charge it is not subject to the strong interaction.

Quantum electrodynamics **(QED).** The theory of the interactions of charged subatomic particles with the electromagnetic field. The main contributors were *Richard Feynman, Julian Schwinger* and *Sin-itro Tomonaga.* QED is often called a perturbation theory because of the smallness of the characteristic constant (cf., Feynman diagrams) and the resultant decreasing size of higher order contributions.

Quantum field theory. A theory by *Paul Dirac* that unifies *quantum mechanics* and the *Special Theory of Relativity*.

Quantum mechanics is the physical science that studies the behavior of matter and light on the atomic, nuclear and subnuclear scale. Its main objective is the description of the properties of molecules, atoms and their constituents—electrons, protons, neutrons, and other elementary particles, for example *leptons* and *hadrons*. These properties include the interactions of the elementary particles with one another and with electromagnetic, weak, nuclear interactions. Eventually the integration of the gravitational forces is one of its objectives, for example within the scope of the *M-theory*.

Quark is a class of 6 sub-nuclear particles, which make up the proton and neutron, as well as other elementary particles subject to the strong nuclear force, the so-called *hadrons*.

Rationalism. The philosophical concept is concerned with the possibility that knowledge of nature can be obtained by reason alone.

Realism. This philosophical concept is related to the belief that reality is independent of our theories, perceptions or beliefs. Realists maintain that our present beliefs are only an approximation of reality Realism is contrasted with idealism and anti-realism. Other doctrines that do not support this view are various forms of idealism

Reason. A faculty contrasted with experience, passion or *faith*.

Red bed. It is a type of sedimentary rock whose color is due to its ferric oxide *minerals*.

Red giant. This is a late stage in *stellar evolution* in which the star increases in size and undergoes a change in surface temperature, responsible for its red color.

Redox. This is shorthand for oxidation-reduction reactions, which describe all chemical reactions in which atoms have their oxidation number (oxidation state) changed. This can be either a simple process, such as the oxidation of carbon to yield carbon dioxide (CO_2) or the reduction of carbon by hydrogen to yield methane (CH_4).

Reductionism The ideas that physical bodies are collections of atoms or that thoughts are combinations of sense impressions are forms of reductionism. Two general forms of reductionism have been held by philosophers: (1) Logical positivism has maintained that expressions referring to existing things or to states of affairs are definable in terms of directly observable objects, or sense-data, and, hence, that any statement of fact is equivalent to some set of empirically verifiable statements. (2) Proponents of the unity of science have held the position that the theoretical entities of particular sciences, such as biology or psychology, are definable in terms of those of some more basic science, such as physics. The logical positivist version of reductionism also implies the unity of science insofar as the definability of the theoretical in terms of the observable would constitute the common basis of all scientific laws.

Refractory. Material that vaporizes only at high temperatures.

Regge poles and trajectories. Tulio Regge studied potential scattering in the Schrödinger equation discovering that the scattering amplitude can be thought of as an analytic function of the angular momentum. Especially relevant was his remark that the position of the poles determines power-law growth rates of the amplitude in the region of large values of the cosine of the scattering angle. Regge's concept of a trajectory had profound implications in later developments of theoretical physics by attempting to rationalize the large number of subnuclear particles in terms of simpler concepts.

Regge, Tulio (1931-). He reinterpreted the analytic growth rate of the scattering amplitude as a function of the cosine of the scattering angle as the power law for the falloff of scattering amplitudes at high energy. Along with the double *dispersion relation* of *Stanley Mandelstam*, Regge Theory allowed the identification of sufficient analytic constraints on *scattering amplitudes* of bound states to formulate a theory in which there are infinitely many particle types, none of which are fundamental.

Reversible chemical reaction. If products, as soon as they are formed, react to produce the original reactants we recognize this special type of reaction. At chemical equilibrium, the two opposing reactions go on at equal rates and consequently there is no net change in the amounts of substances involved.

Ribosome. One of many small cellular bodies in which protein synthesis is carried out. It consists of *RNA* and proteins.

RNA (Ribonucleic acid). Substance present in all organisms in three main forms: *messenger RNA,* ribosomal RNA (cf., *ribosome),* and transfer RNA (used in the last stage of *translation d*uring protein synthesis). Their general function is related with the *translation* of the genetic message from *DNA* to *proteins.*

Rock. These are natural aggregates of one or more *minerals,* and sometimes they may include some non-crystalline substances.

Russell, Bertrand (1872-1970). Russell was an English logician and philosopher, whose seminal work in mathematical logic was published in the early 20th century. Russell collaborated with Alfred North Whitehead on "Principia Mathematica" (1910-1913). He received the Nobel Prize for Literature in 1950. His interests ranged from philosophy and mathematics to religion. His seminal books on the history of philosophy, as well as his book on science and religion, have influenced the present work.

Rydberg's formula. In atomic physics this formula describes the wavelengths of spectral lines of many chemical elements. It was invented by the Swedish physicist Johannes Rydberg in1888.

S-matrix principles. Include firstly, the invariance of the S-Matrix in the *Theory of Special Relativity*. Secondly, *unitarity*, thirdly, analyticity, and finally all singularities of the S-matrix correspond to production thresholds of physical particles.

Sagan, Carl (1934-1996). An American astronomer who searched for intelligent life in the cosmos. Sagan began researching the origins of life in the 1950s and went on to play a leading role in every major U.S. spacecraft expedition to the planets of last century.

Salam Abdus (1926-1996). He made many contributions to the creation of the theory that unifies the electromagnetic and weak nuclear interactions, sharing the 1979 Nobel Prize in Physics with *Glashow* and *Weinberg*. He was the Founding Director of the International Centre for Theoretical Physics thinking that scientific thought is the common heritage of mankind. His permanent concern was to reduce the gap between the emerging South and the industrialized North. From the point of view of the main topic of this book, he encouraged *Ponnamperuma* and the present author to initiate a series of conferences on astrobiology that spread over a period of 11 years. Salam focused on the question of the origin of life feeling "*particularly proud of my last paper on the role of chirality on the origin of life*" (cf., Chapter 4).

Sartre, Jean-Paul (1905-1980). A French philosopher, who became a leader of existentialism. He published his main philosophical work in 1943 (Etre e le Néant). As a novelist he also expressed concern with the nature of human existence.

Scattering Amplitude. In physics the concept of amplitude includes the maximum displacement, or distance moved by a point on a wave measured from its equilibrium position. In nuclear collisions, where complicated situations have to be considered, the *Optical Theorem* gives us an insight (for simple scattering configurations) of the significance of a complex function (the "scattering amplitude") that embodies the essential information of the magnitude of the collision.

Schopf, J. William (1942-).

An American paleobiologist who has pointed out that in the abundant microflora of the 900-million-year-old Bitter Springs Formation of central Australia, some eukaryotic algae have cells in various stages of division into tetrahedral sporelike forms. His most influential work has been the identification and extensive discussion of the Lower Archean stromatolitic microfossils from Australia.

J. W. Schopf at the Sixth Trieste conference in 2000

Schwarz, John H. (1941-).

His basic contributions to superstring theory included his discovery (with Michael B. Green) that certain anomalies are absent for a class of ten-dimensional superstring theories. This provided a strong indication that superstring theory with a specific type of symmetry ("gauge") may provide a consistent unified quantum theory of the fundamental forces including gravity. It led to an explosion of interest in string theory

John Schwarz (on the right) receiving the Dirac Medal at the ICTP from Abdus Salam (on the left)

Schwinger, Julian (1918-1994). An American physicist who formulated *Quantum Electrodynamics* with *Feynman* and *Tomonaga*. He suggested to his graduate student *Sheldon Glashow* that an electroweak unified theory should be possible.

Sequence. A linear order of the monomers in a biopolymer. In general knowing the sequence is helpful in knowing the function of the molecule.

Series (in the Earth Sciences). A time stratigraphic unit corresponding to an *epoch* of geologic time.

Shale. Mainly clay that has hardened into *rock.*

Silicate is a *mineral,* common in terrestrial *rocks,* containing silicon and oxygen.

Smirnov, Alexei Yuryevich (1951-). He is a Russian physicist at the Abdus Salam ICTP High Energy Physics Section. Smirnov is one of the discoverers of the MSW effect. He is a co-recipient of the 2005 Bruno Pontecorvo Prize.

SNC meteorites. These meteorites are included in a class of *basaltic* meteorites, which are of Martian origin.

SOHO is a project of international cooperation between ESA and NASA to study the Sun, from its deep core to the outer corona, and the solar wind.

Solar nebula. Disk of gas and dust around the early Sun that gave rise to the planets and small bodies of the solar system.

Special Theory of Relativity. The realization by Einstein that Newton's mechanics fails when the objects described move too fast. It was formulated in 1905.

Spectral type of a star. This concept refers to a classification of stars according to the appearance of its spectrum.

Spectroscopy. Observational technique that breaks light from an object into discrete wavelength bins for subsequent physical analysis.

Standard model. Combines into a single theory the electroweak theory, describing interactions via the electromagnetic and weak forces, and quantum chromodynamics, the theory of the strong nuclear force.

Stellar evolution. Stars constantly change since their condensation from the interstellar medium. Eventually they exhaust their nuclear fuel and the *supernova* stage may occur. At this stage *elements* synthesized in its interior are expelled, enriching the interstellar medium, out of which new generations of stars will be born.

Steranes. These are *biomarkers* (i.e., chemical compounds that may be used in dating fossils). Steranes have been found in oils and sediments with ages predating animal fossils (cf., Chapter 5).

STEREO is a NASA mission that employed two nearly identical space-based observatories - one ahead of Earth in its orbit, the other trailing behind - to provide stereoscopic measurements to study the Sun and the nature of coronal mass ejections.

Stromatolite. A geological feature consisting of a stratified rock formation that is essentially the fossil remains of bacterial mats. The bacteria that gave rise to these formations were mainly *cyanobacteria.* Similar mat-building communities can develop analogous structures in the world today. The microfossils occur in cherts and shales.

Sulfide. Inorganic compound of sulfur with more *elements* that tend to lose electrons and form positive *ions.*

Sulfur-reducing bacteria get their energy by reducing elemental sulfur to hydrogen sulfide. They couple this reaction with the oxidation of organic compounds.

Super-Earth is an extrasolar planet with a mass between that of Earth and smaller than a gas giant such as Jupiter. For example, the first super-Earth around a main sequence star was discovered in 2005 orbiting the nearby star Gliese 876.

Supernova. A late stage in stellar evolution. An exploding star that has exhausted its nuclear fuel. It enriches the *interstellar dust.* Its remnant is a *neutron star.*

Supersymmetry. This symmetry would be observed between subatomic particles with half-integer values of intrinsic angular momentum, or spin (called fermions) and particles with integer values of spin (called bosons). With supersymmetry, fermions can be transformed into bosons without changing the structure of the underlying theory of the particles and their interactions. Supersymmetry relates transformations in an internal property of particles (spin) to transformations in space-time.

Symbiosis. The living together of two (or more) species with presumed mutual benefit for the partners.

Tarter, Jill Cornell (1944-). Jill Cornell Tarter is an American astronomer and the current director of the SETI Institute. Together with Margaret Turnbull she has created the HabCat in 2002, a principal component of Project Phoenix.

Taxonomy concerns the study of the theory, practice and rules of classification of living or extinct organisms into groups, according to a given set of relationships.

Teilhard de Chardin, Pierre (1881-1955). He was a French philosopher and also a paleontologist. His work appeals to an overlap of science and *natural theology*. Most of Teilhard's scientific output was concerned with mammalian paleontology. Teilhard's approach included theology, science and philosophy. Teilhard regarded basic trends in physics as being ordered toward the production of progressively more complex entities of atoms, molecules, cells and organisms, until the human body evolved, with a nervous system to permit rational reflection, self-awareness, and moral responsibility.

Terrestrial planets. One of the inner rocky planets (Mercury, Venus, Earth, Mars). Due to the similar characteristics of the Moon it is convenient to include it in this group. Some terrestrial-like planets are known around other stars (cf., Super-Earths).

Tidal heating. In the case of the vicinity of a satellite to a giant planet heating may be produced due to repeated stressing arising from orbital motion in the planetary field. This source of heating may be enhanced due to the radioactive dacay of nuclei, a phenomenon known as radiogenic heating.

Tomonaga, Sin-itiro (1906-1979). A Japanese physicist, with Richard Feynman and Julian Schwinger won of the Nobel Prize for Physics in 1965 for developing the basic principles of quantum electrodynamics demonstrating the compatibility of the Special Theory of Relativity with the Quantum Mechanics of the electromagnetic field.

Ultramafic rocks. These are igneous rocks containing more than 90% of dark minerals (olivine, pyroxene, micas). The term 'mafic' refers to rocks containing a high proportion of ferromagnesian minerals.

Unitarity. A principle of S-matrix Theory, which is essentially equivalent to assuming the conservation of probability.

Veneziano, Gabriel (1942-). An Italian physicist who constructed a scattering amplitude describing infinitely many particle types. This work was widely recognized as a key precursor of string theory.

Viking.

US spacecrafts that were able to soft-land two vehicles on opposite hemispheres of Mars: Viking 1 landed in Chryse Planitia on July 20, 1976; Viking 2 landed in Utopia Planitia seven weeks later. The mission attempted to detect signs of life. They consisted of an orbiter and a lander.

Very Long Baseline Interferometry is used in radio astronomy, allowing observations of an object that are made simultaneously by many telescopes to be combined, The net effect is to emulate a telescope with a size equal to the maximum separation between the telescopes.

Virasoro Algebra (1970). This term refers to an algebra developed by the Argentinian-born physicist Miguel Angel Virasoro, which is used in string theory. Virasoro wrote down some operators generating his algebra while studying dual resonance models. A significant extension of the Virasoro algebra was rediscovered in physics shortly after by *J. H. Weis* (Brower and Thorn, 1971, footnote on page 167).

Virasoro, Miguel Angel (1940-)

An Argentine physicist who was responsible for creation of the Virasoro algebra. He was a director of the International Centre for Theoretical Physics (ICTP) from 1995 to 2002. He taught physical-mathematical models for economy at Università di Roma "La Sapienza".

[Miguel Angel Virasoro, as Director of the Abdus Salam International Centre for Theoretical Physics at the Centre .

Voyager 1 and 2.

Two identical space probes were launched in 1970s, so that Voyager 1 would be kept in the plane of the Solar System to make close approaches to Uranus and Neptune by means of utilizing gravity assists during its fly-by of Saturn in 1981 and of Uranus in 1986. Up to the present time the Voyagers have been the only missions that attempted to provide a complete pictorial recording of the Solar System.

Weak interaction (cf., the electroweak interaction).

Weinberg, Steven (1933-). An American physicist who in 1979 shared the Nobel Prize for Physics with *Sheldon Lee Glashow* and *Abdus Salam* for work in formulating the theory, that explains the unity of electromagnetism with the weak nuclear force. His books on science have been influential.

***Weis, Joseph H.* (1942-1978)**

He was an American scientist who made many significant contributions to S-Matrix theory during his short lifetime.

Joseph Weis, first on the right during a break of the 43rd Course of the Enrico Fermi International School of Physics, July 1967

***Wien, Wilhelm* (1864-1928).** A German physicist who discovered a law that concerns the radiation emitted by the perfectly efficient *blackbody*. This law is known as the "Displacement Law" and for this work he received the Nobel Prize for Physics in 1911.

Wigner, Eugene (1902-1995).

A Hungarian-born American physicist, He shared the 1963 Nobel Prize for Physics for his remarkable contributions to nuclear physics, which included the formulation of the law of conservation of parity.

Eugene Wigner (front row, first on the left) attending a lecture at the "Contemporary Physics Symposium" in Trieste in 1968

Wilson*, *Edward Osborne (1929-). He is an American biologist. By applying the evolutionary principles he attempted to explain the behavior of social insects, in order to understand the social behavior of animals, including humans. These were the bases of Wilson socio-biology. In his 1998 book "Consilience: The Unity of Knowledge", Wilson discusses methods that might be able to unite the sciences with the humanities.

Wilson, Robert Woodrow (1936-). American radio astronomer who shared, with *Penzias*, the 1978 Nobel Prize for Physics for a discovery that supported the big-bang model of the universe

Witten, Edward (1951-). An American physicist, professor at the Institute for Advanced Study. His field of expertise is superstring theory, *supersymmetric* quantum field theories and other areas of mathematical physics. He was awarded the Fields Medal. A major contribution to theoretical physics was his formulation of *M-Theory.*

***WMAP* (*Wilkinson Microwave Anisotropy Probe*).**

By making accurate measurements of fluctuations in the *CMB*, WMAP was able to measure basic parameters of the Big Bang model, including the density and composition of the universe, including the density of *baryonic* and non-baryonic matter.

Woese, Carl (1928-). He is an American microbiologist who defined the Archaea as a new *domain* in the *phylogenetic* tree. His three-domain system is based upon genetic relationships. According to his work we began calling the three new branches of life as Bacteria, *Archaea*, and Eucarya. His taxonomy is central for modern biology.

Wolfenstein, Lincoln (1923-).

An American physicist who provided an insight in understanding the *MSW effect* (electrons in terrestrial and Solar matter could affect neutrino propagation). In 1992, he was awarded the Sakurai Prize for his contributions to CP violation and the properties of neutrinos.

L. Wolfenstein (first on the right) the 1968 Ettore Majorana Summer School, its director A. Zichichi is on the left

Wu, Chien-Shiung (1912-1997). She was a Chinese-born American physicist who provided at NBS the proof of parity violation in the weak (electroweak) interactions by demonstrating that beta particles from cobalt 60 atoms would be emitted asymmetrically. She was awarded the Wolf Prize in Physics (1978).

Yang, Chen Ning (1922-). A Chinese-born American physicist who together with Tsung-Dao Lee demonstrated by theoretical means that *parity* is violated when particles decay through the electroweak interactions. These two scientists won the Nobel Prize for Physics for 1957.

References

Brower, R. C. and Thorn, C. B. (1971) Eliminating spurious states from the dual resonance model. *Nucl. Phys.* **B31** 163-182.

Freedman, W. L. and Madore, B. F. (2010) The Hubble Constant. *Annu. Rev. Astro. Asttrophys*. **48**, 673-710.

Russell, B. (1991) *History of Western Philosophy and its Connection with Political and Social Circumstances from the Earliest Times to the Present Day,* Routledge, London, p.13.

General Index

(a) Index of Illustrations

Introduction

Fig. I.1 The heliocentric universe. Courtesy of Professor Francesco Bertola 5

Fig. I.2 Sir Richard Darwin Keynes. Courtesy The ICTP Photo Archives 7

Fig. I.3 Alexander Oparin and Cyril Ponnamperuma. Courtesy of Mrs. Valli Ponnamperuma 9

Fig. I.4 Stanley Miller. Courtesy Massimo Silvano, The ICTP Photo Archives 10

Fig. I.5 Sidney Fox. Courtesy of Massimo Sterle, The ICTP Photo Archives 11

Fig. I.6 Cyril Ponnamperuma. Courtesy of the ICTP Photo Archives 12

Fig. I.7 John Oro. Courtesy of ICTP Photo Archives 14

Fig. I.8 The Earth as seen by astronauts. Courtesy of NASA 15

Fig. I.9 The Earth as part of the Solar System. Courtesy of NASA 16

Fig. I.10 Sir Francis Crick and Abdus Salam. Courtesy of the ICTP Photo Archives 19

Fig. I. 11 Galileo's encounter with Jupiter's satellite Io. Courtesy of NASA 21

Fig. I.12 The Huygens probe detached itself from Cassini late in 2004. Courtesy of NASA 21

Fig. I.13 A mission for Europa. Left: courtesy of ESA. Right: courtesy of ESA and NASA 23

Part1

The Kepler telescope. Credit: NASA/Kepler mission/Wendy Stenzel 29

The components of the Kepler telescope. Credit NASA 30

Chapter 1

Fig. 1.1 The large spiral galaxy NGC 4414. Credit: NASA/HST 33

Fig. 1.2 Planck image of the cosmos. Credit ESA, High Frequency Instrument and Low Frequency Instrument consortia, July 2010 35

Fig. 1.3 Hans Albrecht. Credit ICTP Photo Archive0 38

Fig. 1.4 The Ring Nebula. Courtesy of NASA, HST 39

Fig. 1.5 The Crab Nebula. Courtesy of NASA/ ESA 40

Fig. 1.6 A supernova within the galaxy M100. Credit: X-ray: NASA/CXC/SAO/D.Patnaude *et al.*, Optical: ESO/VLT, Infrared: NASA/JPL/Caltech 41

Fig. 1.7 The Hertzprung-Russell diagram 44

Chapter 2

Fig. 2.1 The Rosetta mission's target. Credit ESA 52
Fig. 2.2 The Rosetta spacecraft. Credit ESA 53
Fig. 2.3 The Jovian system. □□Credit: the SSI system aboard NASA's Galileo 54

Chapter 3

Fig. 3.1 The Moon. Courtesy of NASA 58
Fig. 3.2 The Moon in orbit around the Earth. Courtesy of Galileo, NASA 59
Fig. 3.3 The Lunar Reconnaissance Orbiter. Credit NASA 61
Fig. 3.4 The ESA proposed Moon lander. Credit ESA 61
Fig. 3.5 The ISRO lunar orbiter Chandrayaan-1. Courtesy ISRO 62
Fig. 3.6 The asteroid Ida and its satellite. Courtesy of NASA 63
Fig. 3.7 Impact of the comet Shoemaker-Levy 9. Courtesy of NASA/Hubble Space Telescope 64
Fig. 3.8 The Murchison meteorite. Credit NASA 68
Fig. 3.9 Lake Joyce. Credit NASA 73

Chapter 4

Fig. 4.1 Chien-Shiung Wu. From the author's Photo Archives 83
Fig. 4.2 Tsung-Dao Lee and Chen NingYang. From the author's Photo Archives 84
Fig. 4.3 Abdus Salam. Courtesy of the The ICTP Photo Archives 87
Fig. 4.4 The SOHO telescope. Credit: ESA/NASA 94
Fig. 4.5 The STEREO probes. Credit NASA 94
Fig. 4.6 Ulysses spacecraft. Credit JPL and David Hardy 95
Fig. 4.7 One of the Io volcanos. Credit NASA 95
Fig. 4.8 The Solar Probe. Credit NASA, Johns Hopkins University APL 96
Fig. 4.9 The Solar Dynamics Observatory (SDO) 97

Part 2

A seventeenth century representation of the Solar System. Courtesy of Francesco Bertola 101

Chapter 5

Fig. 5.1 Stratigraphic classification 109
Fig. 5.2 Epochs of the Cenozoic 110

Chapter 6

Fig. 6.1 The three domains of life. Adapted from Chela-Flores *et al.,* 2010 120

Fig. 6.2 Eras and suberas of the Phanerozoic Eon 123

Part 3

The Nanedi Vallis canyon, Mars. Credit NASA/JPL/Malin Space 133

Chapter 7

Fig. 7.1 Vallis Marineris, Mars. Courtesy of NASA 137

Fig. 7.2 The Sojourner rover on the surface of Mars. Courtesy NASA 139

Fig. 7.3 Mars Global Surveyor over the extinct volcano Olympus Mons. Courtesy of NASA 140

Fig. 7.4 The Mars Express Orbiter. Courtesy NASA. 142

Fig. 7.5 The Martian northern polar cap. Courtesy of NASA 143

Fig. 7.6 Image from the Mars Global Surveyor of the Newton Crater. Courtesy of NASA 145

Fig. 7.7 2001 Mars Odyssey. Courtesy NASA 146

Fig. 7.8 The Allan Hills meteorite ALH84001.0. Courtesy NASA 148

Chapter 8

Fig. 8.1 A Galileo image of the morphology of the Europan ice. Courtesy of NASA 153

Fig. 8.2 The Minos Linea region of Jupiter's moon Europa. Courtesy of NASA 154

Fig. 8.3. A full hemisphere of Europa. Courtesy of NASA 155

Fig. 8.4 The subsurface structure of the Jovian moon Europa. Courtesy of NASA 156

Fig. 8.5 Io, moon of Jupiter in an early Voyager image. Courtesy of NASA 158

Fig. 8.6 A hydrobot at the bottom of Europa's ocean. Courtesy of Jet Propulsion Laboratory 161

Fig. 8.7 ENDURANCE for the exploration of Europa's ocean. Courtesy of NASA 162

Chapter 9

Fig. 9.1 A Voyager 2 image of Titan. Courtesy of NASA 173

Fig. 9.2 An artist's conception of the landing of the Huygens probe 174

Fig. 9.3 Triton the largest moon of the planet Neptune. Courtesy of NASA 176

Fig. 9.4 Enceladus. Courtesy of NASA 177

Fig. 9.5 A feature that may be a lake. Courtesy of NASA/JPL/Space Science Institute 178

Fig. 9.6 The first colour view of Titan's surface. Courtesy NASA 180

Fig. 9.7 The plumes in the South Polar Terrain of Enceladus. Courtesy NASA 182

Fig. 9.8 A mongolfiere and would be a challenge to space exploration. Courtesy of NASA 184

Chapter 10

Fig. 10.1 Jupiter, that serves as a reference for the size of extrasolar planets. Courtesy of NASA 190
Fig. 10.2. The planetary system around the dwarf star Kepler 11. Credit: NASA/Tim Pyle 193
Fig. 10.3 An exoplanet as seen form its exomoon. Image credit: NASA, ESA and G. Bacon (STSci) 195
Fig. 10.4 The Orion Nebula. Courtesy of NASA/Hubble Space Telescope 196
Fig. 10.5 The planet, Fomalhaut b, orbiting its parent star. Credit: NASA, ESA 197

Chapter 11

Fig. 11.1 Frank Drake. Courtesy of the The ICTP Photo Archives 203
Fig. 11.2 Jill Tarter, the Trieste Conference, 1998. Courtesy of The ICTP Photo Archives 206
Fig 11.3 The 32 m antenna in Medicina. With kind permission of INAF-IRA 208
Fig. 11.4 Saha E on the lunar farside. Credit: NASA/GSFC/Arizona State University 209

Part 4

Cupola at the West Atrium of St. Mark's Basilica. By kind permission of the Procuraturia 215

Chapter 12

Fig. 12.1 The cracks of Europa's surface. Credit Artwork by Barbara Aulicino/American Scientist 227
Fig12.2 A penetrator. Image courtesy of Mullard Space Science Laboratory, University College London 229
Fig. 12.3 A penetrator in an Europan orbit. Image courtesy of Surrey Satellite Technology Limited, UK 230
Fig. 12.4 Image of Ganymede, taken by the Galileo spacecraft in 1996. Credit NASA 231

Chapter 13

Fig. 13.1 George Coyne at the Sixth Trieste Conference. Courtesy of The ICTP Photo Archives 240

Chapter 14

Fig. 14.1 Steven Weinberg. Courtesy of The ICTP Photo Archives 253

Chapter 15

Fig. 15.1 The 2002 Dirac Medal. Courtesy of Massimo Silvano, The ICTP Photo Archives 258
Fig. 15.2 Werner Heisenberg. Courtesy of The ICTP Photo Archives 261
Fig. 15.3 Julian Schwinger and Paul Dirac. Courtesy of The ICTP Photo Archives 262
Fig. 15.4 Sheldon Glashow. Courtesy of Marino Sterle and The ICTP Photo Archives 263
Fig. 15.5 Stanley Mandelstam receiving the Dirac Medal. Courtesy of The ICTP Photo Archives 265
Fig. 15.6 Tullio Regge. Courtesy of Massimo Silvano, and the ICTP Photo Archives 267
Fig. 15.7 Gabriele Veneziano. Courtesy of The ICTP Photo Archives 268

Chapter 16

Fig. 16.1 Edward Witten lecturing at the ICTP. Courtesy of The ICTP Photo Archives 278

Epilogue

Fig. E.1 First Trieste Conference, 1992. Courtesy of The ICTP Photo Archives 284
Fig. E.2 Final Trieste Conference, 2003. Courtesy of Massimo Silvano, The ICTP Photo Archives 285
Fig. E.3 The IASA School. Courtesy of Public Information Office, IDEA 285

Illustrated glossary and short biographies

Nicola Cabibbo. Courtesy of The ICTP Photo Archives 292
Paul Adrien Maurice Dirac. Courtesy of The ICTP Photo Archives 295
Albert Einstein. Courtesy Editorial Equinoccio, (Chela-Flores *et al.*, 1981) 296
The Galileo Mission. Courtesy of NASA 299
Jean Heidmann at the 5th Trieste Conference. Courtesy of Marino Sterle and The ICTP Photo Archives 301
Werner Heisenberg. Courtesy of the author Photo Archives 301
Sir Fred Hoyle Hoyle. Courtesy of The ICTP Photo Archives 302
James Clerk Maxwell. (Chela-Flores and Ladera, 1979) 305
Yoichiro Nambu. Courtesy of The ICTP Photo Archives 306
The Pioneer Missions. Courtesy of NASA 309
Max Planck. (Chela-Flores, 1977) 310
J. William Schopf. Courtesy of Massimo Silvano and The ICTP Photo Archives 312
John H. Schwarz and Abdus Salam. Courtesy of The ICTP Photo Archives 313
The Viking Mission. Courtesy of NASA 316
Miguel Angel Virasoro. Courtesy of The ICTP Photo Archives 316
Voyagers 1 and 2. Courtesy of NASA 316
Joseph Weis. From the autor's Photo Archive 317
Eugene Wigner. Courtesy of The ICTP Photo Archives 317

Wilkinson Microwave Anisotropy Probe. Courtesy of NASA 317
Wolfenstein, Lincoln. From the author's Photo Archives 317

References

Chela-Flores, J. (1977) *Significación del trabajo de Max Planck*. Instituto Venezolano de Investigaciones Cientificas (Centro de Estudios Avanzados), Caracas.

Chela-Flores, J. and Ladera, C. L. (1979) *Maxwell Einstein Ensayos Biograficos de Dos Grandes Fisicos*. Ediciones de C. A. La Electricidad de Caracas y C. A. Luz Electrica de Venezuela, Caracas.

Chela-Flores, J., Leite Lopes, J., Bunge, M., Garcia-Sucre, M., Kalnay, A. J., Herrera, L. A. and Aragone, C. (1981) *Einstein*. Editorial Equinoccio, Caracas.

Chela-Flores, J. *et al.* (2010) Evolution of plant-animal interactions. In: *All flesh is grass*. J. Seckbach and Z. Dubinsky (eds.). COLE Series, Springer, The Netherlands.

(b) Index of Tables

Chapter 1

Table 1.1 Some of the present and future space missions 36

Table 1.2 Chemical composition of the Sun. 40

Chapter 2

Table 2.1 Abundance of the elements in the solar photosphere and in a carbonaceous (C) chondrite 49

Table 2.2 A few precursor biomolecules in interstellar dust particles 50

Table 2.3 Dust in the Halley Comet: The organic elements are 33% of the total mass 50

Table 2.4 Dust in the Halley Comet: The inorganic elements are 67% of the total mass 51

Chapter 3

Table 3.1 Statistics on the later Apollo Program that allowed the first sample-return missions 58

Table 3.2 Physical parameters of the Moon 60

Table 3.3 Lunar biogenic element abundance mean value, parts per million 60

Table 3.4 Statistics of the largest C-type asteroids 63

Table 3.5 Estimated volatile abundance in comet nuclei 65

Table 3.6 Some important meteorites including the SNC meteorites 66

Table 3.7 Composition of the Martian atmosphere. 67

Table 3.8 Comparison of minerals in shergottites and Martian soil 67

Table 3.9 Amino acids from an extraterrestrial source and from chemical evolution experiments. 69

Table 3.10 Precursors of the biomolecules in the Murchison meteorite 70

Table 3.11 Statistics of the Dry Valleys lakes in Antarctica 72

Table 3.12 Microorganisms living in the Dry Valleys lakes, Antarctica. 73

Table 3.13 A few examples of eukaryotes present in Antarctica 74

Chapter 4

Table 4.1 The 20 amino acids and their three-letter abbreviations 80

Table 4.2 Reactions relevant to chemical evolution 86

Table 4.3 The standard Genetic Code following the standard three-letter notation. 90

Chapter 6

Table 6.1 Archaeological classification with a reference to the European cultures. 124

Chapter 7

Table 7.1 A selection of some of the early missions sent to Mars 135

Table 7.2 Physical parameters of Mars 136

Table 7.3 Elemental composition of Mars soil 138

Table 7.4 Factors favoring rapid evolution of the Martian environment 141

Table 7.5 Minerals in Martian samples 143

Table 7.6 The question of life on Mars 144

Table 7.7 Some of the missions for Mars in the second decade of the 21st century 149

Chapter 8

Table 8.1 Europa, some information and statistics compared with the Moon 152

Table 8.2 Some outstanding features of Europa 154

Table 8.3 Comparison between densities of the Jovian satellites and carbonaceous chondrites 157

Table 8.4 Chemical composition of carbonaceous chondrites 158

Table 8.5 The question of life on Europa. 159

Chapter 9

Table 9.1 Characteristics of Titan 172

Table 9.2 Minor components of Titan's atmosphere 174

Table 9.3 Some experiments for the Cassini Huygens mission 175

Table 9.4 Comparison of the main physical parameters of Triton and Titan 176

Table 9.5 Major components of Titan's atmosphere 179

Chapter 10

Table 10.1 Stars where exoplanets were first detected at the end of the 20th century 192

Table 10.2. Baseline distances needed for ExoMoon direct imaging 196

Table 10.3 Some properties of Fomalhaut b 197

Table 10.4 The Planetary system of Gliese 581 198

Chapter 11

Table 11.1 The Drake Equation 203

Table 11.2 A simplified form of the Drake Equation 204

Table 11.3 A simplified form for the equation for the f_i parameter. 204

Table 11.4 Some SETI Projects 207

Chapter 12

Table 12.1 Some physiological responses at the lowest level of the phylogenetic tree 221

Table 12.2 The emergence of cerebral ganglions at the lowest level of the phylogenetic tree 223

(c) Alphabetical Index

A
Abdus Salam International Centre for Theoretical Physics, The (ICTP), 283
Accretion, 289
Acetylene, C_2H_2,
Age of the universe, 289
Albedo, 289
Allan Hill meteorite, 148, 242
Allen Telescope Array (ATA), 213
Amino acid, 289
Analyticity in S-Matrix Theory, 289
Angular momentum, 266, 289
Antarctica, 67, 70-76, 115
Anthropic Principle, 289
Anthropocentrism, 289
Anthropoid, 127
Apollo Program, 13, 15, 57-59, 290
Archaea (previously called Archaebacteria), 104, 290
Archean, 121, 290
Aretxaga-Burgos, Roberto, xxi, 241
Aristarchus of Samos, 4, 31-32, 202
Aristotelian cosmology, 4
Aristotelian philosophy, 8
Ash Wednesday Supper, The, 6
Arnothosite, 57, 290
Arrhenius, Gustaf, 281
Asteroid, 63-64
Astrobiology, frontiers, 235
Astrometry, 290
Astronomical unit, 290
Astronomical unit (AU), 291
Astrotheology, 238
AU (cf., Astronomical unit)
Aurignatian, 123, 128
Australopithecines, 125
Astralopithecus africanus, 126
Astrobleme, 126
Augustine of Hippo, Saint, 249

B
Bacteria (a domain), 290
Banded-iron formation, 20, 107, 291
Bartolomé de las Casas, 240
Basalt, 291
Base, 291
Behavior genetics, 276
Bernal, J. D., 13
Bertola, Francesco, 4
Bethe, Hans, 37-38
Bhattacherjee, Aranya, xxi, 72
Big bang, 34, 37, 257-258, 264
Biochemistry, 291
Biogeocentrism, 250, 275, 291
Biogeochemistry, 291
Biomarker (cf., biosignature), 291
Biomineralization, 76
Biosignature, 291
Biota, 291
Black body, 291
Black hole, 41
Black hole (massive), 260
Bohr, Niel, 260, 291
Book of Genesis, 216, 249
Book of Life, 3, 17, 216, 257, 269, 279, 281
Borup Fiord Pass, 76
Brain evolution, 129, 220-221
Brown dwarf, 291
Bruno
Giordano, 5-6, 32, 202
Oxford dialogues, 6, 32

C
C-type asteroid, 291
Cabibbo, Nicola, 292
Calvin, Melvin, 13
CAM: surface cell-adhesion molecules, 223
cAMP: cyclic adenosine monophosphate, 223
Cambrian, 109-110
Camus, Albert, 273
Carbonaceous chondrite, 157-158, 292
Cassini-Huygens Mission, 20, 95, 173
Cataclysmic events, 126
Causality, 292
Cenancestor (cf., progenote), 219, 292
Cenozoic, 110, 122-123, 292
Chandra X-ray Observatory, 41
Chandrasekhar limit, 41, 292
Chance and Necessity, 114, 240, 249
Chemical equilibrium, 293
Chemical evolution, 122
Chemolithoautotrophs, 293
Chemothroph, 105, 292
Chert, 293
China National Space Administration (CNSA), 62

Chirality, 85-86, 293
Chloroplast, 112-113, 293
Chomsky, Noam, 128
Chondrite, (cf., Carbonaceous chondrite)
Chondrules, 293
Christian faith, 254
Chromosome, 110, 293
City of God, The, 249
Cnidarians, 220
Cocconi, Giuseppe, 202
Codon, 89, 293
Coelenterate, 109
Cognitive ability, 128
Comas Sola, Josep, 171
Comets
    Hale-Bopp, 65
    Hartley, 64
    Shoemaker-Levy, 64
Comparative planetology, 37
Communication, 127-128
Consilience, 274
Constraints on chance, 114
Contingency, 218-220, 293
Convergence,
    biology, 218-220, 251
    evolutionary, 114,
    SETI, 228-229
Convergent evolution, 293
Convection Rotation and Planetary Transits mission, (cf., COROT mission)
Copernicus, Nicholas, 4, 31, 201
CoROT mission, 20, 192-193, 197, 224
Cosmic
    ecumenism, 244
    microwave background (CMB), 34, 259
    Background Explorer (COBE), 34
Cosmological constant, 296
Cosmology, 32
Cosmotheology, 244
Cosmovici, Cristiano, xxi
Coyne SJ., George, xxi, 239-240, 293
CP violation, 84-85, 294
Crab nebula, 40
Crick, Sir Francis, 3, 18-19, 261, 294
Crossing, 294
Cross section, 294
Crucible of Creation, The, 251
Cryobot, 160
Cusa, Nicholas of (Cusanus), 4
Cyanidiophyceae, 112
Cyanidium caldarium, 115
Cyanobacterium (pl. cyanobacteria), 112, 294
Cytometry, 225

D

Dark energy, 260
Dark matter, 259, 264
Darwin
    Charles Robert, 3, 6, 18, 45, 104, 106, 114, 238, 273
    Darwin-Keynes, Sir Richard, 6
    Erasmus, 6
    finches, 76
Dawkins, Richard, 129
De Duve, Christian, 120, 295, 306
De Vladar, Harold P., xxi, 224
Delbrück, Max, 18
Delphinid, 127
Descent from a common ancestor (cf., "cenancestor" and "progenote"),
Descent of Man, The, 7
Design argument, 248
Destiny of life in the universe, 215
DeVincenzi, Donald, 36
Diagenesis, 295
Digges, Thomas, 31
Diploblasts, 109, 220, 295
Dilational areas, 228
Dirac
    Equation, 262295
    Paul, 262, 295
Dispersion relations, 295
Dissimilatory sulfate reduction, 24, 295
Distribution of life in the universe, 133
Divine Action, 273
DNA (deoxyribonucleic acid), 113, 296
Dolphin, 127
Domain, 104, 296
Down House, 8
Drake
    Equation, 202, 218, 228
    Frank, 203, 218, 220, 282, 295, 307
Dry Valley Lakes, 72, 166
Duality, 269, 296
Dual resonance model, 296
Dudeja, Suman, xxi, 72
Dyson, Freeman, 264, 276, 279, 295

E

Earth sciences, 19
Ediacaran fauna, 109
Einstein, Albert, 33, 243, 259, 273, 296
Electroweak theory, 263, 296
Ellesmere Island, 76
Empiricism, 128, 272, 296, 302

Enantiomer, 297
Enceladus, 20, 115, 175-177, 182
Enlightenment, 273, 297
Eocene, 110
Eris, 303
ESA, The European Space Agency,
Eucarya (a domain), 104
Eukaryogenesis, 108, 121, 128, 297
Eukaryote, 103, 107, 110, 120, 297
European Space Agency (ESA), 282
Europa,
    atmosphere, 113
    Drake Equation, 228
    habitability, 15, 115, 159
Europa Jupiter System Mission" (EJSM), 165
Europa(terrestrial analogs), 72, 75, 76
European Space Agency (ESA), 34
Evolutionary convergence (cf., Convergence)
Evolution,
    biological, 297
    convergence, 293
    cosmic, 297
    of intelligent behavior, 129, 222
    of life, 144
    theistic, 118
Evolution (Lamarckian), 297
Existentialism, 273
Exobiology, 297
Exomoon, 114
Exoplanet (cf., extra-solar planet).
Extrasolar planet, 191, 205, 241
Extreme environment, 115
Extremophile, 93, 105, 113, 115, 298
Exo-Mars Mission, 36, 149
Exomoonology, 194-198

F

Feferman, Solomon, 276, 298
Feinberg, Gerald, 159
Feynman diagram, 298
Feynman, Richard, 261, 298
Fermi Paradox, 211, 298
Fitness of the Environment, The, 275
Five-Kingdom classification, 111
Flow cytometry, 298
Fluorescence, 299
Flute, 128
Formalhaut, 196-197
Fox, Sidney Walter, 10, 11-12, 30, 82, 281
Friedmann, Alexander, 34
Friedmann model, 34, 258
Fulle, Marco, xxi

G

Galaxy NGC 4414, 33
Galileo
    Galilei, 6
    mission (1995-2003), 15, 35, 154, 298
    Near-Infrared Mapping Spectrometer (cf., NIMS)
Gamma-ray bursts, 41
Ganymede, 22, 164, 228, 231
Gas Chromatograph Mass Spectrometer, 138
Gas chromatography, 299
General Relativity, 279
Genesis cupola, 216
Genesis, First book of the Torah, 216
Genetic code, 90, 299
Geocentrism, 210, 243, 299
Geochemistry, 23, 123
Geochronology, 123, 299
Glashow, Sheldon, 84, 263, 299
Gliese 581, 198
Glycerol. 300
Gödel's Incompleteness Theorem, 276
Gowen, Robert, xxi, 164, 228
Grand Unified Theory, 269
Gray (Gy), 93
Great Oxidation Event, 106
Green, Brian, 276, 299
Greenberg, Richard, xxi, 157, 164, 226
Green fluorescent protein (GFP), 225
Grypania spiralis, 108
Guth, Allan, 258

H

Habitability, 159, 184
Habitable zone, 190, 194
Hadean. 300
Haldane, John Burdon Sanderson, 12, 237
Hadrons, 263, 300
Hegel, Georg Wilhelm Friedrich, 272, 300
Hegel's Absolute Idealism, 274
Heidmann, Jean, 208, 300
Heisenberg, Werner, 261, 265, 273, 300
Henderson, Lawrence J., 275
Hertzprung-Russell diagram, 300
Heavy bombardment period, 103, 105
Heliocentric hypothesis, 5, 31
Hematite, 300
Henderson, Lawrence J., 275
Hertzprung, Ejnar, 43
Hertzsprung-Russell diagram, 44
Heterochromatin, 114, 301

Higgs boson, 263, 301
Higgs, Peter, 264
High Resolution Microwave Study, HRMS, 206
Hinode (Solar-B), 97
Hipparchus, 4
Histone, 301
Holley, Robert W., 90
Holocene, 110, 126
Hominid, 45, 122, 301
Hominoid, 45, 122, 301
Homo (genus)
- *habilis*, 124-125
- genus, 44, 122, 126
- *neanderthalis,* 44, 126
- *sapiens*, 44, 110, 126

Hooker, Sir Joseph Dalton, 7
Horizontal gene transfer (HGT), 113
Hoyle, Sir Fred, 34, 301
HR diagram (cf., Hertzsprung-Russell diagram), 43-44
Hubble
- Edwin, 32, 277, 301
- Law, 32
- Space Telescope (HST), 33, 42, 259

Humanities, 249
Haumea, 303
Hydrobot, 160, 162
Hydrocarbon, 301
Hydrothermal vent, 24, 104, 301
Hyperthermophiles, 137
Huygens, probe, 174

I

IDPs (cf., Interplanetary dust particles)
Inflation, 257-258
Instituto de Estudios Avanzados (Fundacion IDEA), 285
Intelligence in the universe, 112
Intelligent behavior, 202
Intelligibility, 253, 271, 274
International Space Station (ISS), 93
Interstellar dust, 302
Interplanetary dust particles (IDPs), 48
Io, 54, 158
Ion, 302
Isham, Christopher, 88, 271
Isotope, 302
Isotopic fractionation, 22, 26, 302
ISSOL: The International Society for the Study of the Origin of Life/The International Astrobiology Society, 10, 12-13
Isua, 24

J

Japan Aerospace Exploration Agency (JAXA), 60, 166
John Paul II, 236, 238
Jovian
- planets,
- accretion, 55
- origin of, 53
- System, 54, 302

Jupiter, 151-168

K

Kant, Immanuel, 259, 272, 302
Kennedy, John Fitzgerald, 57
Kepler
- -11, 193, 198
- Johannes, 5
- Mission, 29-30, 189, 192-193, 224, 282, 302
- Space Telescope, 193

Keynes, Richard Darwin, 7, 302
Khorana, Har Gobind, 90
Kirchhoff, Gustav, 303
Kuiper Belt, 43, 303
Kuiper, Gerard, 172
Kumar, Narendra, xxi

L

Lagrangian points, 35, 303
Lake
- Fryxell, 72
- Hoare, 72
- Joyce, 73
- of liquid hydrocarbon, 178
- Vanda, 72
- Vostok, 75, 159

Language, 128
The Language Instinct, 128
LAPLACE mission to Europa, 164
Large Hadron Collider (LHC), 264, 283
Large NIMS area, 229
Laser
- Laser-induced breakdown spectroscopy (LIBS), 225
- Laser Interferometer Space Antenna (LISA), 260, 279
- Laser-Raman Spectroscopy, 26

Lee, Tsung-Dao, 84, 303
Levy, Bernard-Henri, 17
Lepton, 303
Light-year, 303

Linea, 154-155, 303
Linde, Andre, 258
Lipid bilayer, 303
Logical positivism, 236, 274, 303
Lyell, Sir Charles, 106, 205, 250, 304

M
M theory, 268, 279, 304
Magdalenian culture, 123
Magma, 304
Makemake, 303
Mandelstam, Stanley, 265-266, 283
Mandelstam Representation, 266
Mare Smythii (Moon), 208-209
Maria, 208-209, 309
Mariner, 135
Mars,
  Express, 141, 149
  Global Surveyor, 36, 139
  Odyssey, 146
  physical parameters, 136
  search for life, 15, 113, 115
  Sample Return Mission, 149
  Science Laboratory, 36, 149
Martini, Cardinal Carlo Maria, xxi
Mass independent fractionation (MIF),
Marius, Simon, 151
Mass-dependent fractionation (MDF), 24
Mass-independent fractionation (MIF), 24, 107-108, 304
Mass spectrometer, 304
Mass Spectrometry, 26, 127
Maxwell, James, 304
Mayor, Michel, 189
Mayr, Ernst, 18
Mayz Vallenilla, Ernesto, xxi
McMurdo Sound, 76
Medawar, Peter, 120
Membrane, 304
Messenger RNA, 89
Messerotti, Mauro, xxi, 93
Metazoan, 109
Methane as a biomarker, 147, 185
Methanogen, 25, 147
Methanotrop, 163
Methylotroph, 108
Mesozoic, 113, 122, 304
Messenger RNA (mRNA), 305
Micropaleontology, 281
Microwave Anisotropy Probe (MAP), 34
Middle Ages, 3, 254
MIF, cf., mass-independent fractionation.
Miller, Stanley, 10-11, 13, 30, 104, 284
Miocene, 110
Miret-Magdalena, E., 244
Mitochondrion, pl. Mitochondria, 112, 305
Molecular
  biology, 305
  clock, 111
  cloud, 42
Monod Jacques, 114, 240, 249
Monotheistic view of Deity, 238, 249
Moon, 57-63
Moon, SETI, 208
Moorbath, Stephen, 20, 281
Morrison, Philip, 202
Mousterian industry, 124, 128
mRNA (cf., messenger RNA)
MSW effect, 305
Mullard Space Science Laboratory, 229
Multicellularity, 121-122, 209, 223-224
Murchison meteorite, 68
Music, 128

N
Nambu, Yoichiro, 267, 305
Nanedi Vallis canyon (Mars), 134
NASA (cf., National Aeronautics and Space Administration)
National Aeronautics and Space Administration, NASA, 13, 34
Natural fission reactors, 109
Natural selection, 113, 305
Natural theology, 306
Navarro-Gonzalez, Rafael, 13
Neogene, 110
Neuron (cf., origin of), 128, 217, 220-221
Neutrinos, 41
Neutron star, 306
New Horizons Mission, 167
Newton Crater, (on Mars) 145
Newton, Sir Isaac, 5, 259-260, 276
NIMS (The Galileo Near-Infrared Mapping Spectrometer), 167
NIMS area, 229, 231
Nirenberg, Marshall W., 90
Nitrile, 306
Noble gas elements, 92
Nuclear
  fusion, 37-38
  physics, 20
Nucleomorph, 111
Nucleic acid, 306
Nucleosynthesis, 37, 306
Nucleus of an atom (plural: nuclei), 306
Nucleus of a cell (plural: nuclei), 306

O
Oligocene, 110
Oliver, Bernard, 206
On the Infinite Universe and Worlds, 6
Oort cloud, 306
Oparin, Alexander Ivanovich, 8-10, 30, 237, 248
Oparin Medal, 13, 14
Optical Theorem, 306
Organelle, 306
Origin of the neuron, 217
Origin of metazoans, 218, 225
Origin of Species, The, 6, 18, 106
Orion Nebula, 306
Oró, Marques Joan (John, Juan), 13-14, 30
Owen, Tobias, xxi, xxiii, 284
Ozma, Project, 307
Ozone (trioxygen, O_3), 108

P
PAL, cf., Present atmospheric level, 108-110
Paleocene, 110
Paleogene, 110
Paleolithic, 123
Paleoproterozoic, 109
Paleozoic, 110, 113, 122-123, 307
Parity, 307
Parity violation, 307
Parsec (pc), 307
Particle accelerator, 264, 307
Pasteur, Louis, 247, 307
Pathfinder, 135, 139
Pauling, Linus, 104
Pegasi 51, 189, 206
Penetrators, 105,163, 167, 229, 282
Penzias, Arno, 34
Phanerozoic, 110, 122-123, 307
Phenomenon of Man, The, 119, 252
Philosophy,
    (and) astrobiology, 17, 247
    definition, 17
    (and) science, 235
Phobos-Grunt mission, 36, 149
Phoenix Mission, 142, 149
Phoenix Project, 206
Photosphere, 307
Photosynthesis, 307
Phylogenetic tree, 104, 307
Phylogeny, 308
Plicher, Carl, 152
Pioneers 10 and 11, 308
Pinker, Steven, 128-129
Planck
    Max, 43, 260-261, 273, 276, 308
    Space Observatory, 34, 308
    radiation, 43
    telescope, 35
Plane-polarized light, 308
Planet like Earth, 198
Planetary nebula, 66
Planetary protection, 35
Plate tectonics, 308
Plato, 259, 308
Pleistocene, 110, 128
Plinian eruption, 309
Pliocene, 110
Plume, 228, 309
Pluto, 303
Polarization, 308
Polkinghorne, John, 21
Ponnamperuma, Cyril Andrew, xxi, xxiii, 8-10, 12-13, 30, 69, 283-284
Popper, Karl, 249
Positivism, 236
Potential scattering, 266
Precambrian, 109
Present atmospheric level (PAL), 108-110
Principles of Geology, 205
Progenote or cenancestor, 309
Protozoa, 112
Protein, 309
Proterozoic Eon, 109-110, 309
Prokaryogenesis, 122
Prokaryote, 103, 110, 116, 309
Proterozoic, 109
Pugliese, Nevio, xxi
Pulsar, 309
Purine, 70, 309
Pyrimidine, 70, 309
Pyrolysis, 309
Purcell, Edward Mills, 202

Q
Quantum
    chemistry. 309
    chromodynamics, 263-264, 309
    electrodynamics, 263, 309
    field theory, 265-266, 310
    mechanics, 260, 262, 276, 310
Quark, 268, 310
Quaternary, 110
Queloz, Didier, 189

R
R (cf., radius of the universe)
Race, Margaret, 36

Radio astronomy, 16
Radius of the universe, 33, 258
Rago, Hector, 33
Rationalism, 128, 272, 310
Raulin, Francois, xxi, xxiii, 174, 284
Realism, 310
Reality, 271-272, 278
Reason, 310
Red bed, 310
Red giant, 38, 310
Redox, 227, 310
Reductionism, 274-275, 279, 310
Reductionist dream, 279
Refractory, 311
Regge poles, 266, 311
Regge, Tullio, 265-267, 283, 311
Regge trajectories, 311
Religion and Science, 250
Reversible chemical reaction, 311
Ribosomal RNA (rRNA), 104, 112
Ribosome, 311
Ring Nebula (M57), 39
RNA (ribonucleic acid), 104, 112
Robertson, H.P., 34
Rock, 311
Rosetta mission, 36
Rummel, John, 36
Russell
 Bertrand, 5, 31, 236, 311
 Henry Norris, 43
 Robert J., 276
Russian Federal Space Agency (Roskosmos), 166
Rydberg formula, 260, 311

S
S-matrix, 266, 312
Sagan, Carl, 189, 312
Saha crater (Moon), 208
Salam, Abdus, xxi, xxiii, 19, 87, 263-264, 312
SAM: substrate adhesion molecule, 223
Sample return missions, 57-59, 149
Santillán, Moisés, xxi
Sartre, Jean Paul, 312
Saturn, 171-183
Scattering amplitude, 266, 312
Schidlowski, M., 19, 22, 281
Schopf, J. William, 8, 19, 312
Schrödinger, Erwin, 18, 261, 266, 273
Schrodinger Equation, 261
Schwarz, John H., 313
Schwinger, Julian, 262, 313
Scott, Sir Robert, 72
Search for extraterrestrial intelligence (cf., SETI)
Seckbach, Joseph, xxi, 115
Sedna, 303
Sepúlveda, Juan Ginés de, 240
Sequence, 313
Series (in the Earth Sciences), 313
SETI, 202, 252
SETI Institute, 206
SETI Italy, 208
SETI on the Moon, 208
Shale, 108, 313
Shapiro, Robert, 159
Sickle band (on Europa's surface), 229
Silicate, 313
Smirnov, Alexei Yuryevich, 313
SNC meteorites, 67, 145, 313
Snow, C.P (Lord),
Sociobiology, 276
Socrates, 308
SOHO, the Solar and Heliospheric Observatory, 93-94, 313
Solá, Josep Comas, 171
Solar Dynamics Observatory, 97
Solar Missions, 93
Solar Orbiter, 98
Solar Probe Plus, 96
Solid State imaging experiment (SSI), 167
Sojourner Rover, 139
Solar Terrestrial Relations Observatory (STEREO), 94, 314
Space weather, 95, 167
Special Theory of Relativity, 262, 313
Spectral type of a star, 314
Spectroscopy, 225, 314
Spectroscopy X-ray wavelengths, 224
Square Kilometer Array (SKA), 213
Sreenivasan, K.R., 258
Standard model, 264, 314
Stellar evolution, 43, 314
Steranes, 314
STEREO, cf., Solar Terrestrial Relations Observatory.
String theory, 266-269
Stromatolite, 314
Sulfide, 314
Sulfur
 fractionation, 163
 patches (on Europa), 167
 -reducing bacteria, 164, 314
Super-Earths, 189, 194, 198, 224, 314
Supernova (pl. supernovae), 39, 126, 260, 314
Supersymmetry, 314

Swallow, 251
Swift, 251
Symbiosis, 111-112, 315

T
Tardigrade, 115-116
Tarter, Jill, 206, 220, 315
Taxonomy, 119, 315
Teilhard de Chardin, Pierre, 119, 252, 315
Terrestrial analog to Europa, 72, 75, 76
Terrestrial Planets, 315
Tewari, Vinod, xxi
Thales of Miletus, 259
Theory of Common Descent, 105-106,
Tidal heating, 156, 315
Timiryazev, K. A., 8
Tinto River, Spain, 242
Titan, 15, 113, 115, 171
  atmosphere, 177
  ocean, 181
  surface, 179
Titan Saturn System Mission (TSSM), 183
Tomonaga, Sinitiro, 263, 315
Tree of life (cf., phylogenetic tree), 104
Triploblastic phyla, 110
Tuniz, Claudio, xxi
Two Cultures, 274

U
UK Penetrator Consortium, 163
Ultramafic rocks, 315
Ultraviolet Spectrometer (UVS), 167
Ulysses (solar mission), 95
Unitarity, 315
Unity of science, 274
Universality of biology, 259
Upper Paleolithic, 128
Urey, Harold, 11

V
Van Allen, James, 155
Veneziano, Gabriel, 265, 267, 283, 315
Very Large Array (VLA), 212
Very Long Baseline Interferometry, 316
Vienna Circle, 236
Viking missions, 13-14, 35-36, 135, 137, 148-149, 316
Villegas, Raimundo, 223
Virasoro, Miguel Angel, 264, 283, 316
Vladilo, Giovanni, xxi
Volatile, 316
Voyagers 1 and 2, 13, 16, 35, 152, 155, 172, 316

W
Wallace, Alfred Russell, 7, 18, 247
Walker, Arthur Geoffrey, 34
Watson, James, 3, 261
Wave mechanics, 261
Weak interaction, 263, 316
Weinberg, Steven, 253, 263, 273, 316
Weis, Joseph H., 317
Wells, Herbert George, 45
Western civilization, 249
Western culture, 308
Western thought, 254, 274
What is Life? 18, 261
Wheeler, John Archibald, 265
White dwarf, 39
Wien, Wilhelm, 317
Wigner, Eugene, 83, 317
Willson
  Edward O., 275, 317
  Robert, 34, 317
Wilkinson Microwave Anisotropy Probe (WMAP), 317
Whipple, Fred, 65
Witten, Edward, 268, 278, 283, 317
Woese, Carl, 119-120, 318
Wolpert, Lewis, 19
Wolfenstein, Lincoln, 84, 305, 318
Wu, Chien-Shiung, 83, 318
Würm glaciation, 124

X
X-ray absorption spectroscopy (XAS), 225

Y
Yang, Chen Ning, 84, 318

Z
Zichichi, A., 318
Zircon, 103
Zuckerkandl, Emil, 104
Zwicky, Fritz, 259

About the author

Julian Chela-Flores was born in 1942 in Caracas, República Bolivariana de Venezuela. He studied in the University of London, England, where he obtained a Ph.D. in Quantum Mechanics (1969). He was a researcher at the Venezuelan Institute for Scientific Research (IVIC) from 1971 till 1978 and Professor at Simón Bolívar University (USB, Caracas) until his retirement in 1990. During his USB tenure he was Dean of Research from 1979 till 1985. He is a Fellow of The Latin American Academy of Sciences (Caracas), The Academy of Sciences of the Developing World (Trieste), the Academy of Creative Endeavors (Moscow) and a Corresponding Member of the Venezuelan Academy of Physics, Mathematics and Natural Sciences (Caracas).

His current positions are Staff Associate of the Abdus Salam International Center for Theoretical Physics (ICTP), Trieste, Research Associate, Dublin Institute for Advanced Studies (DIAS) and Profesor-Titular, Institute of Advanced Studies (IDEA), Caracas. His area of expertise is astrobiology, in which he is the author of numerous papers including some articles published in the frontier between astrobiology and the humanities (philosophy and theology).

He organized a series of conferences in Trieste on Chemical Evolution and the Origin of Life (now called Conferences on Astrobiology) from 1992 till 1994 with Cyril Ponnamperuma; from 1995 till 1998 with Francois Raulin and, finally from 2001 till 2003 with Raulin and Tobias Owen.

Chela-Flores is also the author of *The New Science of Astrobiology From Genesis of the Living Cell to Evolution of Intelligent Behavior in the Universe* (the first edition of the present book published in 2001), and *A Second Genesis: Stepping-stones towards the intelligibility of nature* that was published in 2009.

Books by author

Ponnamperuma, C. and Chela-Flores, J. (eds.), (1993) *Chemical Evolution: Origin of Life,* A. Deepak Publishing, Hampton, Virginia, USA.

Chela-Flores, J., Chadha, M., Negron-Mendoza, A. and Oshima, T. (eds.), (1995) *Chemical Evolution: Self-Organization of the Macromolecules of Life,* A. Deepak Publishing, Hampton, Virginia, USA.

Ponnamperuma, C. and Chela-Flores, J. (eds.), (1995) *Chemical Evolution: The Structure and Model of the First Cell,* Kluwer Academic Publishers, Dordrecht.

Chela-Flores, J. and Raulin, F. (eds.), (1996) *Chemical Evolution: Physics of the Origin and Evolution of Life,* Kluwer Academic Publishers, Dordrecht.

Chela-Flores, J. and Raulin, F. (eds.), (1998) *Chemical Evolution: Exobiology. Matter, Energy, and Information in the Origin and Evolution of Life in the Universe,* Kluwer Academic Publishers, Dordrecht.

Chela-Flores, J., Lemarchand, G. A. and Oro, J. (eds.), (2000) *Astrobiology From the Big Bang to Civilization,* Kluwer Academic Publishers, Dordrecht.

Chela-Flores, J., Owen, Tobias and Raulin, F. (eds.), (2001) *The First Steps of Life in the Universe.* Kluwer Academic Publishers, Dordrecht, The Netherlands.

Chela-Flores, J. (2001) *The New Science of Astrobiology From Genesis of the Living Cell to Evolution of Intelligent Behavior in the Universe.* Kluwer Academic Publishers, Dordrecht, The Netherlands (279 p.) First Edition.

Seckbach, J., Chela-Flores, J., Owen, T., Raulin, F. (eds.), (2004) *Life in the Universe From the Miller Experiment to the Search for Life on Other Worlds* Series: Cellular Origin, Life in Extreme Habitats and Astrobiology, Vol. **7**, Springer, Dordrecht, The Netherlands 387 pp.

Chela-Flores, J. (2009) *A Second Genesis: Stepping-stones towards the intelligibility of nature.* World Scientific Publishers, Singapore, 248 pp.